T0401141

NEUROMETHODS

Series Editor
Wolfgang Walz
University of Saskatchewan
Saskatoon, SK, Canada

For further volumes:
http://www.springer.com/series/7657

Mood and Anxiety Related Phenotypes in Mice

Characterization Using Behavioral Tests, Volume II

Edited by

Todd D. Gould

University of Maryland School of Medicine, Department of Psychiatry, Baltimore, USA

 Humana Press

Editor
Todd D. Gould
University of Maryland
School of Medicine
Department of Psychiatry
Baltimore, Maryland, USA
gouldlab@me.com

ISSN 0893-2336 e-ISSN 1940-6045
ISBN 978-1-61779-312-7 e-ISBN 978-1-61779-313-4
DOI 10.1007/978-1-61779-313-4
Springer New York Dordrecht Heidelberg London

Library of Congress Control Number: 2011935540

Printed on acid-free paper

Humana Press is part of Springer Science+Business Media (www.springer.com)

Preface to the Series

Under the guidance of its founders Alan Boulton and Glen Baker, the Neuromethods series by Humana Press has been very successful since the first volume appeared in 1985. In about 17 years, 37 volumes have been published. In 2006, Springer Science + Business Media made a renewed commitment to this series. The new program will focus on methods that are either unique to the nervous system and excitable cells or which need special consideration to be applied to the neurosciences. The program will strike a balance between recent and exciting developments like those concerning new animal models of disease, imaging, in vivo methods, and more established techniques. These include immunocytochemistry and electrophysiological technologies. New trainees in neurosciences still need a sound footing in these older methods in order to apply a critical approach to their results. The careful application of methods is probably the most important step in the process of scientific inquiry. In the past, new methodologies led the way in developing new disciplines in the biological and medical sciences. For example, Physiology emerged out of Anatomy in the nineteenth century by harnessing new methods based on the newly discovered phenomenon of electricity. Nowadays, the relationships between disciplines and methods are more complex. Methods are now widely shared between disciplines and research areas. New developments in electronic publishing also make it possible for scientists to download chapters or protocols selectively within a very short time of encountering them. This new approach has been taken into account in the design of individual volumes and chapters in this series.

Wolfgang Walz

Preface

Preclinical research related to mood and anxiety disorders relies extensively upon mouse behavioral tests and models. The use of these approaches continues to increase as a greater number of underlying susceptibility genes are discovered, new targets for medications are identified, and clinical studies reveal novel neurobiological risk factors.

The rationale for this second volume is straightforward: to include tests that were not covered in the first volume. This book is meant, therefore, as a complement to Volume I. Together, the two volumes offer a comprehensive resource for the behavioral approaches that are valuable for the characterization of mood and anxiety disorder-related behaviors in mice, and that are utilized in the development of medications that are effective in their treatment. The contributing authors are world-renowned scientists with broad experiences in the development and application of mouse behavioral tasks. Each chapter provides a brief background and review of the test or model as well as a complete and up-to-date protocol narrative.

The chapters are primarily intended as a resource for scientists actively pursuing or interested in establishing behavioral protocols in their laboratories and who have prior experience with studying rodent behavior. However, it will also serve as a reference for those scientists and practitioners who seek to better understand the behavioral methods used in preclinical mood and anxiety research. Though the chapters provide an important resource for reference material and detailed methods, there are a number of subtleties in mouse husbandry, handling, and testing procedures that cannot be acquired solely from following a book. It is therefore strongly encouraged that those with limited experience with rodent behavior seek collaboration with an experienced behavioral neuroscientist to lend a hand in addressing experimental and analysis issues that will, without a doubt, arise.

Baltimore, USA   *Todd D. Gould*

Contents

Contributors

SANDRA BEESKÉ • *Sanofi-Aventis, Chilly-Mazarin, Paris, France*

YOAV BENJAMINI • *Department of Statistics and Operation Research, Tel Aviv University, Tel-Aviv, Israel*

MICHEL BOURIN • *Neurobiologie de l'anxiété et de la dépression, Faculté de Médecine, University of Nantes, Nantes, France*

RICHARD E. BROWN • *Psychology Department and Neuroscience Institute, Dalhousie University, Halifax, NS, Canada*

JONATHAN CACHAT • *Department of Pharmacology and Neuroscience Program, Tulane University Medical School, New Orleans, LA, USA*

DILLON CARLOS • *Department of Pharmacology and Neuroscience Program, Tulane University Medical School, New Orleans, LA, USA*

FRANCES A. CHAMPAGNE • *Department of Psychology, Columbia University, New York, NY, USA*

HAGIT COHEN • *Israel Ministry of Health Mental Health Center, Anxiety and Stress Research Unit, Faculty of Health Sciences, Ben-Gurion University of the Negev, Beer-Sheva, Israel*

JAMES P. CURLEY • *Department of Psychology, Columbia University, New York, NY, USA*

CLAIRE L. DENT • *Behavioural Genetics Group, MRC Centre for Neuropsychiatric Genetics and Genomics, Neuroscience and Mental Health Research Institute, Cardiff University, Cardiff, UK*

ELISABETH DOW • *Department of Neuroscience, Connecticut College, New London, CT, USA*

EHUD FONIO • *Department of Neurobiology, Weizmann Institute, Rehovot, Israel*

BECCA FRANKS • *Department of Psychology, Columbia University, New York, NY, USA*

CHERYL A. FRYE • *Department of Psychology, The University at Albany, SUNY, Albany, NY, USA; Department of Biological Sciences, The University at Albany, SUNY, Albany, NY, USA; The Center for Life Sciences Research, The University at Albany, SUNY, Albany, NY, USA; Department of Life Sciences, The University at Albany, SUNY, Albany, NY, USA*

GEORGETTE M. GAFFORD • *Department of Psychiatry and Behavioral Sciences, Center for Behavioral Neuroscience, Yerkes National Primate Research Center, Emory University, Atlanta, GA, USA*

SIDDHARTH GAIKWAD • *Department of Pharmacology and Neuroscience Program, Tulane University Medical School, New Orleans, LA, USA*

MARK A. GEYER • *Department of Psychiatry, University of California San Diego, La Jolla, CA, USA*

ILAN GOLANI • *Department of Zoology, Tel Aviv University, Tel-Aviv, Israel*

SHANNON L. GOURLEY • *Department of Pediatrics, Emory University, Atlanta, GA, USA*

GUY GRIEBEL • *Sanofi-Aventis, Chilly-Mazarin, Paris, France*

RHIAN K. GUNN • *Psychology Department and Neuroscience Institute, Dalhousie University, Halifax, NS, Canada*

MARTINE HASCOËT • *Neurobiologie de l'anxiété et de la dépression, Faculté de Médecine, University of Nantes, Nantes, France*

PETER HART • *Department of Pharmacology and Neuroscience Program, Tulane University Medical School, New Orleans, LA, USA*

RENÉ HEN • *Departments of Psychiatry and Neuroscience, New York State Psychiatric Institute, Columbia University, New York, NY, USA*

MOLLY HOOK • *Department of Pharmacology and Neuroscience Program, Tulane University Medical School, New Orleans, LA, USA*

ANTHONY R. ISLES • *Behavioural Genetics Group, MRC Centre for Neuropsychiatric Genetics and Genomics, Neuroscience and Mental Health Research Institute, Cardiff University, Cardiff, UK; Department of Psychological Medicine and Neurology, Cardiff University School of Medicine, Cardiff, UK*

ALLAN V. KALUEFF • *Department of Pharmacology and Neuroscience Program, Tulane University Medical School, New Orleans, LA, USA*

MARTIEN J. KAS • *Department of Neuroscience and Pharmacology, Rudolf Magnus Institute of Neuroscience, University Medical Center Utrecht, Utrecht, The Netherlands*

TAKEFUMI KIKUSUI • *Companion Animal Research, Azabu University, Sakamihara, Kanagawa, Japan*

ORSOLYA J. KUTI • *Department of Brain and Cognitive Sciences and Picower Institute for Learning and Memory, Massachusetts Institute of Technology, Cambridge, MA, USA; UC Davis School of Veterinary Medicine, Davis, CA, USA*

EVAN KYZAR • *Department of Pharmacology and Neuroscience Program, Tulane University Medical School, New Orleans, LA, USA*

LINDSAY M. LUEPTOW • *Department of Behavioral Medicine and Psychiatry, West Virginia University Health Sciences Center, Morgantown, WV, USA*

OZ MALKESMAN • *Laboratory of Molecular Pathophysiology, Intramural Research Program, National Institute of Mental Health, National Institutes of Health, Bethesda, MD, USA*

ATHINA MARKOU • *Department of Psychiatry, School of Medicine, University of California San Diego, La Jolla, CA, USA*

ALISON L. MARTIN • *Psychology Department and Neuroscience Institute, Dalhousie University, Halifax, NS, Canada*

MICHAEL A. MATAR • *Israel Ministry of Health Mental Health Center, Anxiety and Stress Research Unit, Faculty of Health Sciences, Ben-Gurion University of the Negev, Beer-Sheva, Israel*

ALAN NEWMAN • *Department of Pharmacology and Neuroscience Program, Tulane University Medical School, New Orleans, LA, USA*

JAMES M. O'DONNELL • *Department of Behavioral Medicine and Psychiatry, West Virginia University Health Sciences Center, Morgantown, WV, USA*

DAMON T. PAGE • *Department of Brain and Cognitive Sciences and Picower Institute for Learning and Memory, Massachusetts Institute of Technology, Cambridge, MA, USA; Department of Neuroscience, The Scripps Research Institute, Jupiter, FL, USA*

JIN HO PARK • *Department of Psychology, University of Massachusetts, Boston, MA, USA*

VALERIE PIET • *Department of Cell Biology and Anatomy, Louisiana State University Health Sciences Center, New Orleans, LA, USA*

KERRY J. RESSLER • *Department of Psychiatry and Behavioral Sciences, Center for Behavioral Neuroscience, Yerkes National Primate Research Center, Emory University, Atlanta, GA, USA*

KATHRYN RHYMES • *Department of Pharmacology and Neuroscience Program, Tulane University Medical School, New Orleans, LA, USA*

MICHELLE ROCHE • *Department of Physiology, School of Medicine, National University of Ireland, Galway, Ireland*

BENJAMIN ADAM SAMUELS • *Departments of Psychiatry and Neuroscience, New York State Psychiatric Institute, Columbia University, New York, NY, USA*

LARRY D. SANFORD • *Department of Pathology and Anatomy, Sleep Research Laboratory, Eastern Virginia Medical School, Norfolk, VA, USA*

ADAM STEWART • *Department of Pharmacology and Neuroscience Program, Tulane University Medical School, New Orleans, LA, USA*

OLIVER STIEDL • *Center for Neurogenomics and Cognitive Research, VU University Amsterdam, Amsterdam, The Netherlands*

ASTRID K. STOKER • *Department of Psychiatry, School of Medicine, University of California San Diego, La Jolla, CA, USA; Department of Psychopharmacology, Utrecht Institute for Pharmaceutical Sciences, Utrecht University, Utrecht, The Netherlands*

JANE R. TAYLOR • *Division of Molecular Psychiatry, Departments of Psychiatry and Psychology, Interdepartmental Neuroscience Program, Yale University, New Haven, CT, USA*

ELI UTTERBACK • *Department of Pharmacology and Neuroscience Program, Tulane University Medical School, New Orleans, LA, USA*

ALICIA A. WALF • *Department of Life Sciences, The University at Albany, SUNY, Albany, NY, USA*

LAURIE L. WELLMAN • *Department of Pathology and Anatomy, Sleep Research Laboratory, Eastern Virginia Medical School, Norfolk, VA, USA*

JEFFREY M. WITKIN • *Psychiatric Drug Discovery, Neuroscience Discovery Research, Lilly Research Laboratories, Eli Lilly and Company, Indianapolis, IN, USA*

NADINE WU • *Department of Pharmacology and Neuroscience Program, Tulane University Medical School, New Orleans, LA, USA*

LINGHUI YANG • *Department of Pathology and Anatomy, Sleep Research Laboratory, Eastern Virginia Medical School, Norfolk, VA, USA*

JARED W. YOUNG • *Department of Psychiatry, University of California San Diego, La Jolla, CA, USA*

JOSEPH ZOHAR • *The Chaim Sheba Medical Center, Sackler Medical School, Tel-Aviv University, Tel Hashomer, Israel*

Chapter 1

Longitudinal Assessment of Deliberate Mouse Behavior in the Home Cage and Attached Environments: Relevance to Anxiety and Mood Disorders

Martien J. Kas, Ilan Golani, Yoav Benjamini, Ehud Fonio, and Oliver Stiedl

Abstract

Understanding behavioral regulation can further progress by developing new approaches that allow refinement of behavioral phenotypes. The current availability of several thousand different mutant mice and of human candidate genes for emotional (affective) disorders challenges behavioral neuroscientists to extend their views and methodologies to dissect complex behaviors into behavioral phenotypes and subsequently to define gene–behavioral phenotype relationships. Here, we put forward multiday automated behavioral and physiological observations in carefully designed environments to assess evolutionary conserved behavioral strategies in mice. This offers the opportunity to design experimental setups that allow the animals themselves to regulate their own behavior, using representations of continuous kinematic variables, studying the dynamics of behavior (change across time or change across activity); i.e., growth or decay processes of behavior and concomitant physiological adjustments such as heart rate. The measures characterizing these processes should have discriminative power (across strains or treatments) and be replicable (across laboratories). Furthermore, cross species genetic studies for these neurobehavioral and physiological traits may provide a novel way toward identifying neurobiological mechanisms underlying core features of complex psychiatric disorders.

Key words: Genetics, Physiology, Replicability, Exploration, Psychiatry, Conditioned fear, Heart rate, Sequences of repeated motion, Dynamics of behavior, Behavioral growth, Arousal

1. Introduction

Behavioral testing in animals is a crucial feature of phenotyping in neuroscience research. In particular, animal models of human chronic diseases or behavioral symptoms, such as anxiety and mood disorders, should be tested in situations where long-lasting stable

Todd D. Gould (ed.), *Mood and Anxiety Related Phenotypes in Mice: Characterization Using Behavioral Tests, Volume II*, Neuromethods, vol. 63, DOI 10.1007/978-1-61779-313-4_1, © Springer Science+Business Media, LLC 2011

features ("chronic") of such behavior are manifested. It implies measuring the behavior for days rather than minutes. Testing animals in situations in which the animals respond to novel or transient stimuli is appropriate for studying "states," not "traits." Behavioral testing is usually based on measuring behavioral responses to environmental events that are induced by the experimenter. The frequently used open field test, e.g., is commonly employed to study general activity and fear-related behaviors in mice and rats (1, 2). In this test, movements of the animal are monitored up to 1 h after the animal has been placed in a novel open arena from which it cannot escape. Additional tests, such as the elevated plus maze (3) and the light–dark box (4), allow external validity of the observed open-field behaviors. However, these tests are short-lasting and depend on individual locomotor activity levels and novelty responsiveness of the animal and require human interference. This hampers their use for determining gene–behavioral phenotype relationships and stresses the need for new analytical procedures addressing the complex behaviors. Although some ideas for overcoming these problems have been put forward, such as improving currently available tests, using test batteries and increasing test information density (5, 6), behavioral complexity and gene–environment interactions require new methodologies in this field of research.

1.1. Interacting Physiological Processes

Behavior is triggered by internal and external motivational signals (such as hunger and food availability, respectively) and is guided by the ability of the animal to execute proper behavioral responses. For instance, a hungry mouse that searches for new food resources relies on an efficient exploration strategy in which finding the food resource in time and taking the risk of being exposed to predators need to be balanced. Furthermore, in the wild, mice face the risk of spending more energy to obtain food than this food gives them in return on any given day. Because exploration for food is influenced by different integrated physiological processes (e.g., energy balance, motor action, and fear), as well as by environmental factors (e.g., variations in ambient temperature, in food availability, and in photoperiod), the design of behavioral laboratory methods that dissociate the various behavioral components is a challenge and should focus on ethologically relevant behavior and appropriate environmental conditions for the species selected for the behavioral studies (7).

Conventional laboratory tests, such as the open-field test, touch upon different aspects of exploratory behavior, including locomotor activity and fear-related processes. However, during the relative short testing episode generally employed, it is impossible to discriminate between gene function in novelty-induced and baseline behaviors. For example, mice that lack the dopamine transporter gene have locomotor activity levels under baseline conditions that are comparable to those in wild-type animals, but they exhibit a 12-fold increase of locomotion following placement in a novel environment. Although dissociation of novelty-induced and baseline locomotor behavior in this mutant was observed during a relative short-lasting testing procedure, the time of day that these tests are performed can highly influence the outcome of the observations (8–10). Furthermore, the response to novelty induces a dynamic transient response whose measurement provides variable results depending critically on the time interval measured and the size of the time window sampled. Thus, characterization of gene functions in exploratory behavior requires behavioral paradigms that allow dissection of this complex behavior into different components in view of circadian-induced variations of these components.

1.2. Interference and Order Effects

Executing multiple behavioral tests usually involves experimenter interference, such as handling or transport of animals (11), and cues from the experimenter that influences the behavioral performance of an animal (12). For example, measuring pain responses in mice revealed that experimenter effects account for more trait variability than genotype (13). In addition, problems of replicability across laboratories such as those reported in (14) could reflect the effects of forced testing. Mice exposed to a battery of various behavioral tests expressed significant lower levels of locomotor activity in the open field than mice that were naïve to behavioral testing (15). These order effects could even be amplified in animals with selective mutations in genes that are involved in physiological processes, such as coping strategies to changing environments. Circumvention of these interfering procedural aspects is required to reduce nonspecific environmental influences on the gene–behavioral phenotype relationship. Especially, the adverse impact of the experimenter as uncontrollable variable or even confounder of experiments should be taken into consideration.

1.3. Dissection of Behavioral Phenotypes

Studies in the field of biological rhythms have revealed that behavioral observations during several consecutive days or weeks in the home cage of an animal allow reliable assessment of stable behavioral circadian rhythms that are highly sensitive to environmental signals, such as light and human interference (16, 17). Because behavioral observations during several days can also dissociate novelty-induced and baseline behaviors at different phases of the light–dark cycle, behavioral monitoring in the home cage will significantly contribute to the refinement of behavioral phenotypes. In addition, by carefully designing a home cage environment, with or without additionally attached compartments or arenas that addresses different behavioral characteristics of interest, complex behaviors can be further dissected into behavioral phenotypes with minimum human interference. In this chapter, we would like to view recent developments in the fields of behavioral neuroscience that uses the home cage and attached compartments as a basis for the assessment of behavioral exploration strategies in mice. Integration of longitudinal automated behavioral measures with physiological measures allows further refinement of these neurobehavioral traits. Furthermore, we provide an example on how interspecies trait genetics using home cage behavioral assessment in mice offers a basis for identifying novel neurobiological mechanisms underlying anxiety and mood disorders.

2. New Method Developments

In this chapter we introduce setups used in our own work, which attempt to separate state from trait anxiety by using side-by-side the home cage environment for long-term observation that might be more appropriate for longitudinal studies of trait, and environments attached to the home cage, which are most appropriate for the study of how mice manage deliberately novel input, but can also serve for long-term studies. In addition, we complement the type of information provided by the common assays and models with a large set of novel mouse-centered kinematic variables which imply active management of perceptual input. We suggest three requirements that should guide us in improving our choice of behavioral measures: measure kinematic variables that appear to be actively managed by the animal; demonstrate the discriminative power of these measures between strains and preparations; and demonstrate the replicability of these measures across laboratories (18–20). In what follows we briefly demonstrate what we do to fulfill these three requirements.

2.1. Segmenting Behavior into Animal-Centered Sequences of Repeated Approach-and-Avoid Motions

One way to obtain a view on the functional organization of exploratory behavior is to examine it in situations involving behavioral growth. To study and quantify this growth, we connect the mouse's home cage through a doorway to a large circular arena for an extended period of time, and allow the mouse to explore the arena at a self-regulated rate (Dimensionality Emergence assay, or DIEM assay; see (21)). In this setup the familiarity of the mouse with the environment increases gradually, allowing a correspondingly gradual, stretched out growth of behavior. This process exposes the elementary building blocks of behavior as they are progressively added to the animal's repertoire. The moment-to-moment developmental dynamics of exploratory behavior discloses its presumed function: a systematic active management of perceptual input acquired during the exploration of a novel environment, and active management of the arousal associated with the acquisition of that novel input (20, 21).

Having access to a technology that allows us to track and record a time series of locations occupied by a mouse during free exploration, and having developed analytical methods for quantifying continuous kinematic variables (https://www.tau.ac.il/~ilan99/see/help/), we segment the path, based on its intrinsic statistical and geometrical properties, into processes involving approach and avoidance: repetitive Peep and Hide motions from the home cage into the arena, repetitive Cross and Retreat motions performed across the doorway, repetitive Borderline round trips consisting of outbound–inbound movement along the wall, and repetitive incursions from the wall toward the center of the arena and back to the wall. All these are examples of what we term "sequences of repeated motion." The motions are performed in relation to specific reference values from which the motion commences and to which it returns: the inside of the home cage for Peep and Hide, the doorway for Cross and Retreat, the inside of the home cage plus the "garden," an area at the proximity of the doorway, for Borderline Roundtrips, and the Wall ring in the proximity of the arena wall for incursions. We further identify a growth of behavior that is manifested through a buildup in the extent of each of these motion types separately and an increase in complexity through the recruitment of additional sequences of repeated motion that are superimposed on previously emerged sequences of repeated motion. We finally quantify this process by computing the rates of growth in extent in each of the sequences of repeated motion, and by estimating the complexity of the sequence of sequences.

2.2. Management of Perceptual Input as Indicated by Buildup in Extent in Sequences of Borderline Roundtrips

A session of free exploration commences with peeping, where the mouse crosses the doorway into the arena, always leaving part of its body behind the doorway, and retreats back. The Peep and Hide sequence is followed by a Cross and Retreat sequence, Circle in Place, and Entry Head On, before commencing with the Borderline Roundtrip Motion sequence, which, in the BALB/c mice, commences strictly near and along the wall until the exhaustion of the borderline dimension (Fig. 2). The reference area near the doorway that we term garden is defined algorithmically by plotting a density of cumulative dwell time across the entire arena. This plot highlights a two-dimensional Gaussian located by the doorway, whose boundary defines the "garden" (21).

As illustrated in Fig. 1, Borderline movement builds up in maximal angular distance from home almost monotonically from one roundtrip to the next. This increase in borderline roundtrip amplitude is joined next by the option not to return all of the way into the home cage, as expressed by the emergence and subsequent proliferation of Cage skips and Home-related shuttles (blue dots in Figs. 1 and 2). The simple Borderline Roundtrips turn in this way into complex ones including one to several home-related shuttles. The buildup in the Borderline roundtrips in the other direction,

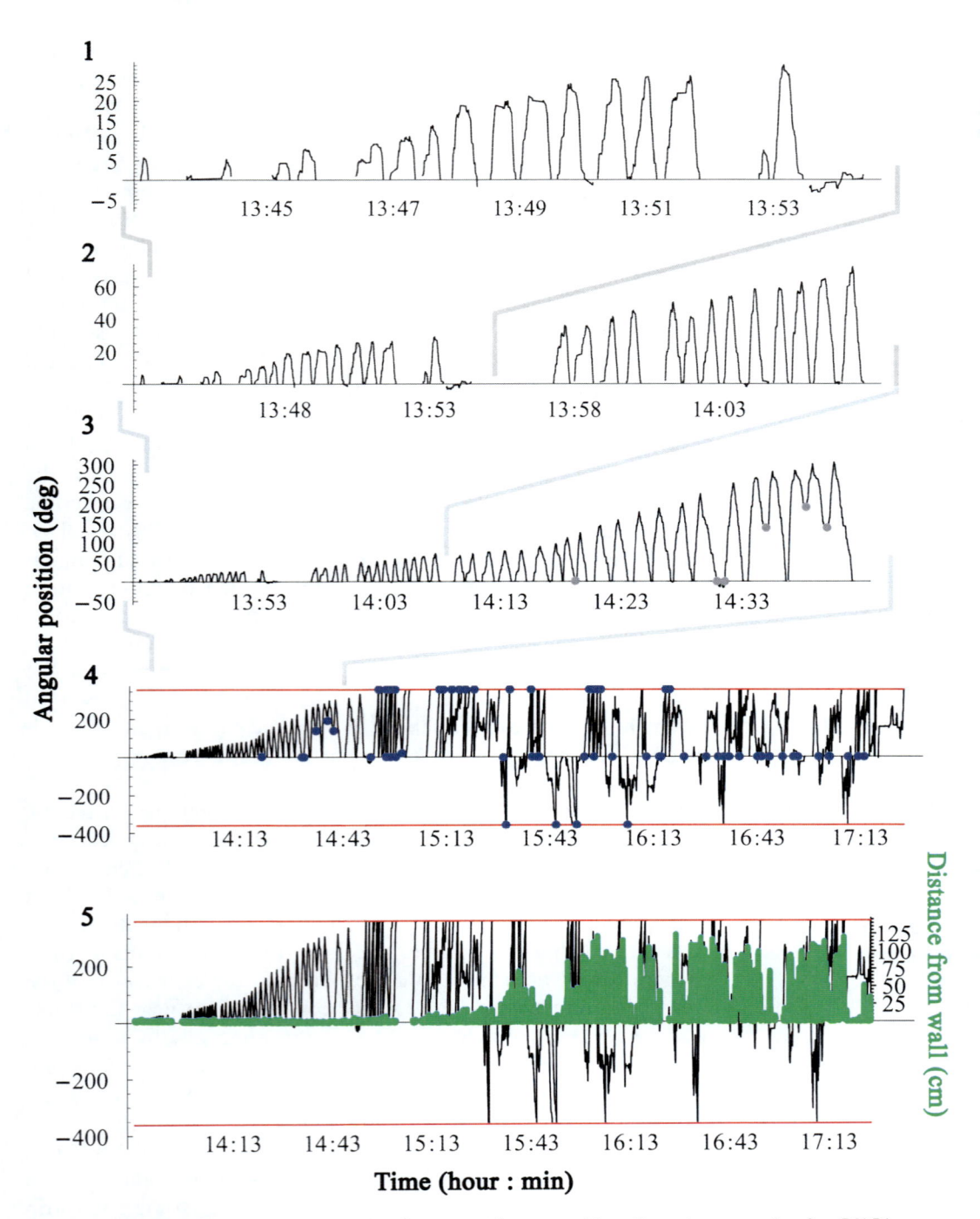

Fig. 1. The buildup of amplitude and complexity of movement in one-, and two-dimensions space in a free BALB/c mouse. (1–4): the buildup of angular positions reached across roundtrips. Note change of time scale from (1) through (4). *Black line* – borderline movements. *Blue* data points – cage skips. Positive values designate *right* and negative values *left* borderline directions. *Red lines*, angular positions of doorway at 360°. Graph lines between *X*-axis and *red line* represent full circles. All graphs start with the same initial roundtrip, progressively incorporating later roundtrips. (5): emergence and buildup of radial movement away from wall (in *green*), superimposed on the plot of angular positions (in *black*). Significant radial movements (incursions) are added only after 1.5 h.

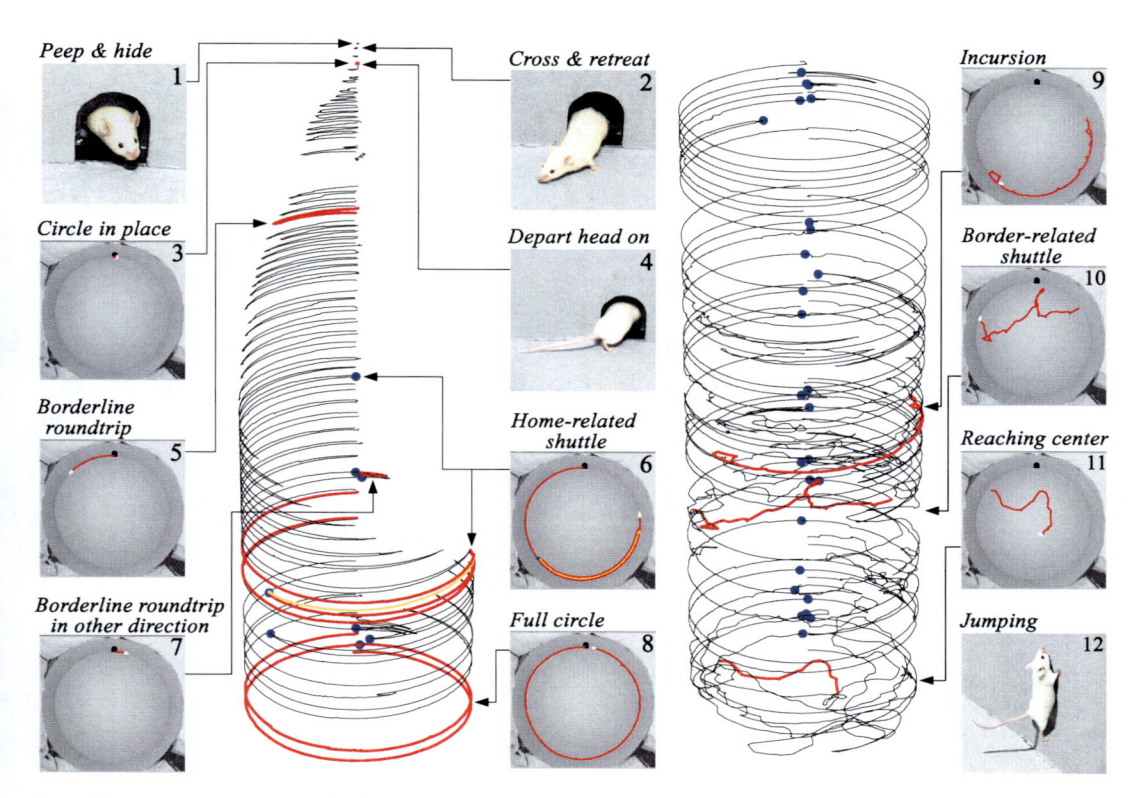

Fig. 2. The moment-to-moment developmental sequence of landmarks of free exploration of a selected BALB/c mouse-session performed across a 3-h period. The spiral proceeding from top to bottom, first in the left and then in the right column, presents the time series of two-dimension locations on the path traced by the mouse. The enumerated figure-inserts show the 12 landmarks described in the text, traced in *red* within the arena, and on the spiral. *Blue* dots indicate instances in which the mouse approached the cage doorway and did not enter the cage (cage-skips), or stopped short of returning all of the way home during a return (home-related shuttle). Absence of a blue dot implies departure into home cage. *Yellow* path stands for the return portion within a home-related shuttle.

which follows the extended sequence of one-sided roundtrips (Fig. 2, top part of left spiral), is steep in comparison to the corresponding buildup in the main direction (Fig. 1). One to several full circles in one or both directions herald the end of the one-dimensional stage and the emergence of the two-dimension movement stage (consisting of radial movement, plotted in green in Fig. 1).

2.3. Mouse Exploratory Behavior Is Composed of a Sequence of Sequences of Repeated Motion

Sequences of repeated motion are progressively added on top of each other, generating increasingly richer and more complex behavior, ultimately consisting of 13 types of sequences of repeated motion exposed so far (Figs. 2 and 3). The first occurrence of a new type of motion is a developmental landmark. It heralds the repeated performance of that type of motion in the immediate period that follows, and often across the rest of the session. The sequence of landmarks, the buildup in extent within sequences,

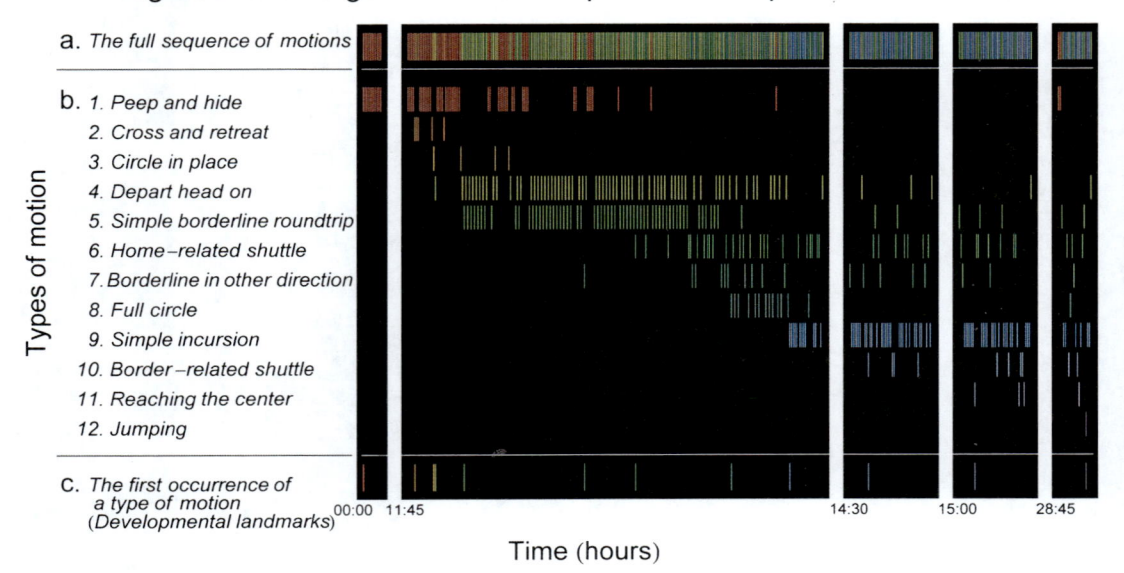

Progressive emergence of new sequences of repeated motion

a. *The full sequence of motions*

b. 1. *Peep and hide*
   2. *Cross and retreat*
   3. *Circle in place*
   4. *Depart head on*
   5. *Simple borderline roundtrip*
   6. *Home–related shuttle*
   7. *Borderline in other direction*
   8. *Full circle*
   9. *Simple incursion*
   10. *Border–related shuttle*
   11. *Reaching the center*
   12. *Jumping*

Types of motion

c. *The first occurrence of
   a type of motion
   (Developmental landmarks)*

00:00 11:45   14:30   15:00   28:45

Time (hours)

Fig. 3. Successive intervals of free exploratory behavior of a selected BALB/c mouse performed in a novel arena. The sequence of sequences is represented in the *bottom horizontal line* by the first performance of each of the motion types (developmental landmark) belonging to a sequence.

and the increase in dimensionality and freedom of movement across the session are all illustrated in Fig. 2. The growth of behavior proceeds across stages from staying-in place behavior (Fig. 2, developmental landmarks 1–4), to movement along one (Fig. 2, landmarks 5–8) then along two (landmarks 9–11), and then along three dimensions (landmark 12). It is accompanied by a buildup in amplitude and in complexity of the motions within and across the sequences of repeated motion. The regularity of the growth and the stability of the order within the BALB/c and C57BL/6 strains suggest *active management* of the measured kinematic variables (21).

The top horizontal line in Fig. 3 presents the original, "raw" sequence of motions of a selected BALB/c mouse. This sequence is algorithmically screened, yielding multiple sequences, each presented within an especially dedicated horizontal line in Fig. 3 (this particular BALB/c mouse performed only 12 sequences of repeated motion).

2.4. Quantifying the Buildup in Extent in Sequences of Repeated Motion

The quantification of the buildup in extent can be achieved in various ways. Figure 4a details one such way to quantify the buildup in the maximal arc reached during a Borderline Roundtrip, where the time to reach some threshold and the rate of growth at that threshold are calculated from the smoothed buildup curve.

2.5. Discriminative Power

The comparison between two different strains of the quantified buildup in Borderline Roundtrips demonstrated in Fig. 4 reveals that the differences in the times to reach the threshold and in the growth rates in the two strains are large and highly statistically significant (the two groups are almost entirely separated). Variations on the parameters estimated from this sequence of repeated motion, as well as the comparisons of measured buildup in the other sequences discussed, can all be assessed as to their discriminative

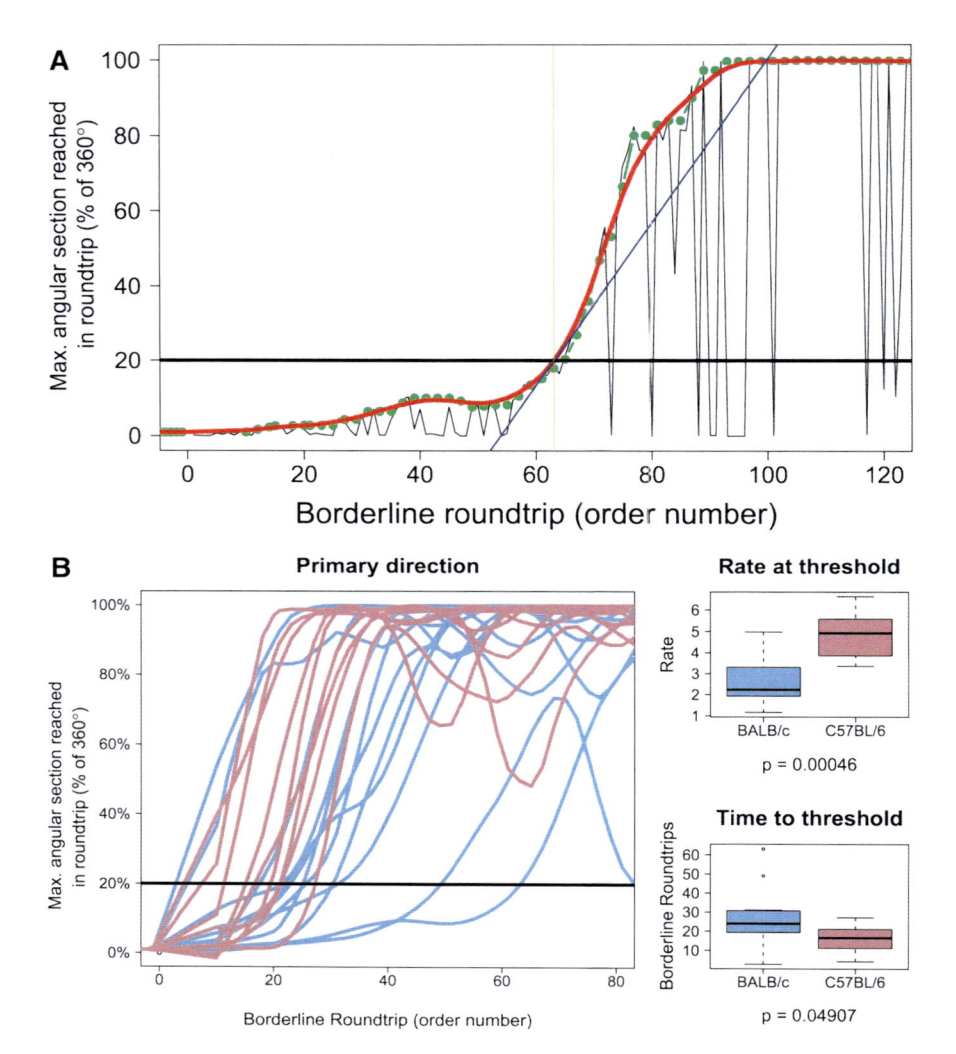

Fig. 4. (**a**) Quantifying the buildup in maximal arc during successive borderline roundtrips in the main direction of a mouse's exploratory session. (**b**) A quantitative comparison of the rate of growth of the maximal angular amplitude reached during successive borderline roundtrips in two strains. On the *left*, the smoothed percentile functions for all mice (*pink* for C57BL/6, *blue* for BALB/c) and the 20% threshold used (*horizontal line*). *Top right*: box plots comparing the growth rates of the mice in the two groups (rates are measured as additional percent of circle covered per roundtrip). *Bottom right*: box plots comparing the time to reach the threshold of the mice in the two groups (time is measured in terms of roundtrips performed). *P*-values indicate significant differences in magnitude between the two strains using Wilcoxon test.

power: comparing between strains (as was done here), treatments, and preparations. Measures with demonstrated high discriminative power should be chosen at this stage.

2.6. Replicability Across Laboratories

Having chosen measures of actively managed behavior enjoying discriminative power, the concern about replicability of results across laboratories should enter the decision which measures to adopt (14, 18, 19). In particular Kafkafi et al. (18) lay out a statistical approach for deciding whether a measure is replicable across laboratories, by utilizing the variation across laboratories in the construction of the yardstick for discrimination (mixed model analysis). Rather than entering the statistical details, we demonstrate the idea behind it with an example taken from a forced exploration experiment assessing inter-strain differences that was conducted in three laboratories (22).

For the number of incursions per session, the across laboratories analysis revealed that the strain difference was not significant ($p > 0.08$), which deemed this measure of little value. However, the aggregate of all incursions was identified to be a mixture consisting of three relatively distinct incursion types. Ignoring the "Near Wall" incursion type, for which the strain differences were found to be highly nonreplicable across laboratories, left us with two distinct incursion types, intermediate incursions and arena-crossing ones, and the inter-strain comparison of their numbers per session was found to be highly replicable across laboratories.

2.7. Implications for Phenotyping

This type of analysis thus illustrates our approach to the design of improved measures for quantified phenotyping of behavior. An aspect of behavior is worth phenotyping if its quantification supports the hypothesis that it is actively managed by the animal, indicating functionality in the animal's own *Umwelt* (operational environment) (23). A specific quantification is useful if it has discriminative power and if it is replicable across laboratories. Using these criteria jointly may require a search through many candidate measures, but it can guide the design of better ways to describe and quantify behavior.

2.8. Implications for the Study of Animal Models of Anxiety

The analysis presented by us provides new types of information regarding the common open field animal model of anxiety. It also casts some light on the reported failure to separate state from trait anxiety (24). At least two hypotheses that concern motivational and cognitive mechanisms stem from our descriptive model:

1. The periodic return to the home cage is suggestive of a perceptual input cutoff mechanism (25), whereby after being exposed to a given amount of novel environmental input the mouse appears to rush back home so as to cut off the novel input. Hence, the incremental growth between two successive round-trips reflects the amount of input the mouse can take in before

having to cut it off by returning home. It follows that a mouse having a low capacity for novel input, or, for that matter, lower information processing capacity, would cover smaller stretches of new terrain than a mouse with a higher capacity. It is therefore expected that low input capacity mice (e.g., mice that are highly aroused) would have gentler growth curve slopes than mice with high input capacity (e.g., mice that are experiencing low arousal).

2. Faced with the challenge of mapping a novel environment, periodic return to the home cage may reflect the need to parse environmental input into manageable chunks collected during a roundtrip. An animal with a lower information processing capacity would correspondingly be expected to parse the novel input into smaller chunks (such as BALB/c mice) than an animal with a higher information processing capacity (such as, perhaps, C57BL/6 mice).

Taken together, these two hypotheses expose two presumed functions indicated by our analysis of free mouse exploration: (1) active management of the arousal associated with the acquisition of the novel input, and (2) active management of dimension-specific perceptual input acquired during the exploration of a novel environment. Regardless of the validity of these two hypotheses, our results beg for experimental manipulations that would modify the magnitude of the incremental input managed by the animal.

3. Integrative Behavioral and Physiological Assessment of Emotional Behavior

The assessment of the emotional state of rodents and many animals is hampered by the fact that it cannot be measured directly. Instead the emotional state has to be inferred indirectly from, e.g., facial expressions (e.g., as in humans, wolf and dogs [canidae]) or other species-specific behavioral expressions based on postures (e.g., freezing in rats and mice). However, fear and anxiety are accompanied by a wide range of physiological adjustments mediated by the central and peripheral autonomic nervous system (26). Cardiovascular function is of major importance because heart rate and blood pressure are generally elevated during emotionally challenging conditions representing key symptoms of many affective disorders (e.g., posttraumatic stress and general anxiety disorder). Affective disorders such as depression show comorbidity with cardiovascular disease for various reasons (27). Particularly, reduced heart rate variability that is attributed to increased sympathetic and decreased parasympathetic tone is considered a risk factor in the clinical setting.

3.1. Physiological Adjustments by Aversive Emotional Challenge

A wide range of adjustments is elicited by an adverse emotional challenge ranging from fast startle response modulation, pupil reflex (dilation), sweating (e.g., in humans) and piloerection through heart rate, blood pressure, and body temperature increase, as well as endocrine responses such as corticosterone increase in rodents (28). All these adaptive response adjustments serve as additional independent physiological measures to support the interpretation of emotional states such as fear and anxiety (29). However, these responses show different response dynamics after the onset of an aversive stimulus. While pupil reflexes are very fast (in the ms range), other autonomic responses are considerably slower (e.g., body temperature and corticosterone elevations) with a lag of rise of more than 2 min after stimulus onset. In addition, the return of these responses to baseline conditions is generally slow, which may pose a problem for experimental conditions of repetitive stimulation at high frequency. Finally, the activation of many physiological parameters is not specific to aversive conditions. Appetitive conditions and physical activity also contribute to substantial elevations of many autonomic parameters (30). Therefore, the interpretation of observed physiological changes still has to be made with great care and should be guided by information on the responses range of, e.g., heart rate values observed throughout the circadian cycle when animals have been active. A rise of heart rate of mice by 70 beats per minute [bpm] from baseline values that is attributed to increased attention (31, 32) is unlikely to serve as index of a stressor. Relatively small autonomic changes such as a mild heart rate increase may be statistically significant but physiologically irrelevant unless persistent for long times. In general, the problem of nonstationarity and interdependence of heartbeat intervals in their temporal sequence confounds the use of linear statistics for the quantification of measures such as heart rate and blood pressure (see (33)).

3.2. Specificity of Physiological Responses

Unfortunately, the understanding of the consequences of behavioral actions such as grooming, running, or rearing on autonomic adjustments is very limited and detailed information on the temporal relations is lacking. Another major complication is the fact that traditional behavior experiments with rodents require handling. Handling is considered at least a mild stressor in these animals and will alter the baseline values of many, if not all, physiological measures although habituation to handling procedures leads to habituation, i.e., a reduction of the responses (34). Therefore, it is necessary to eliminate all unspecific interventions to be able to determine the proper responses. For that purpose, home cage-based behavioral assessment has been combined with physiological assessment using advanced telemetric methods to avoid any unspecific interference for optimal signal-to-noise ratio of, e.g., heart rate measures. The downside of this approach is that suppression of ongoing behavior, a useful index of conditioned and unconditioned fear in specific environments, cannot be quantified in the home cage under habituated conditions, since basal activity is generally low. This leaves little room for the quantification of active suppression as commonly observed in the conditioning context. The positive side of this is that the impact of physical activity on the autonomic measures can be largely neglected in the home cage.

3.3. Fear-Induced Heart Rate Adjustments in Mice

A telemetric approach to determine heart rate and blood pressure dynamics has been used in mice during expression of fear conditioned to an auditory cue (reviewed in (35)). These experiments, partly combined with genetic and pharmacological interventions, indicated fear- and extinction-specific heart rate responses. Their dynamical features show initially fast acceleration of heart rate (half-time: ~3 s) after the onset of the aversively conditioned tone. When starting out from stress-free conditions, baseline heart rate is in the range of less than 600 bpm and heart rate variability is high (36, 37). Tone presentation drives heart rate to maximum levels close to 800 bpm with substantially reduced heart rate variability. Heart rate is maintained at that level for some time before it slowly recovers before the offset of the 180-s tone (38). The heart rate response contrasts with blood pressure adjustments that show slower rise (half-time of mean arterial blood pressure: ~50 s) to peak values of 130 mmHg and prolonged recovery (38). The maximum heart rate and blood pressure values measured during fear retention tests are not observed during circadian measurements under undisturbed conditions in the home cage in the absence of arousal and fear. Furthermore, mouse strains such as DBA/2 show low conditioned freezing levels and only mild conditioned tachycardia (39) indicating similar consequences of deficient fear learning on various readouts. Finally, repeated nonreinforced exposure to the tone causes a decline in the response magnitude indicative of extinction in C57BL/6J mice (29, 35).

While home cage measurements are well suited for phasic stimulation of animals by, e.g., auditory cues, the issue of handling complicates the assessment of heart rate responses during retention of conditioned contextual fear. A study with handling and novelty exposure in C57BL/6J mice indicated that all mice tested showed initially almost maximum heart rate (close to 800 bpm) irrespective of shock experience during training. Heart rate is generally inversely related to heart rate variability with minimal heart rate variability at maximum heart rate range. Starting out from maximum values, heart rate dropped faster in mice either not shocked or subjected to an immediate shock during training (which does not induce fear learning) despite substantially higher physical activity (exploration) than mice that received a late shock during training thereby being aversively conditioned to the conditioning context (40). The difference in heart rate emerged after 5 min and persisted for another 20 min. This indicates that all commonly used anxiety tests of short duration (≤5 min) that require handling and novelty exposure are essentially unsuited for the assessment of autonomic functions such as heart rate. Therefore, new experimental approaches exploiting home cage-based deliberate behavior are crucial to determine the effect of anxiety and physical activity on basic physiological functions in mice and rats.

Based on these behavioral and autonomic constrains, a new experimental approach has been developed following concepts of deliberate novelty exploration starting out at home cage conditions (21). Following the approach of deliberate open field exploration described before, we investigated the effects of novelty exploration on heart rate in C57BL/6J mice. Preliminary experiments (Stiedl and Golani; unpublished results) indicated that heart rate of mice increases to maximum physiological levels during the first approach of the open field (Fig. 5) confirming the interpretation of high arousal or anxiety-like behavior as concluded from the behavioral performance.

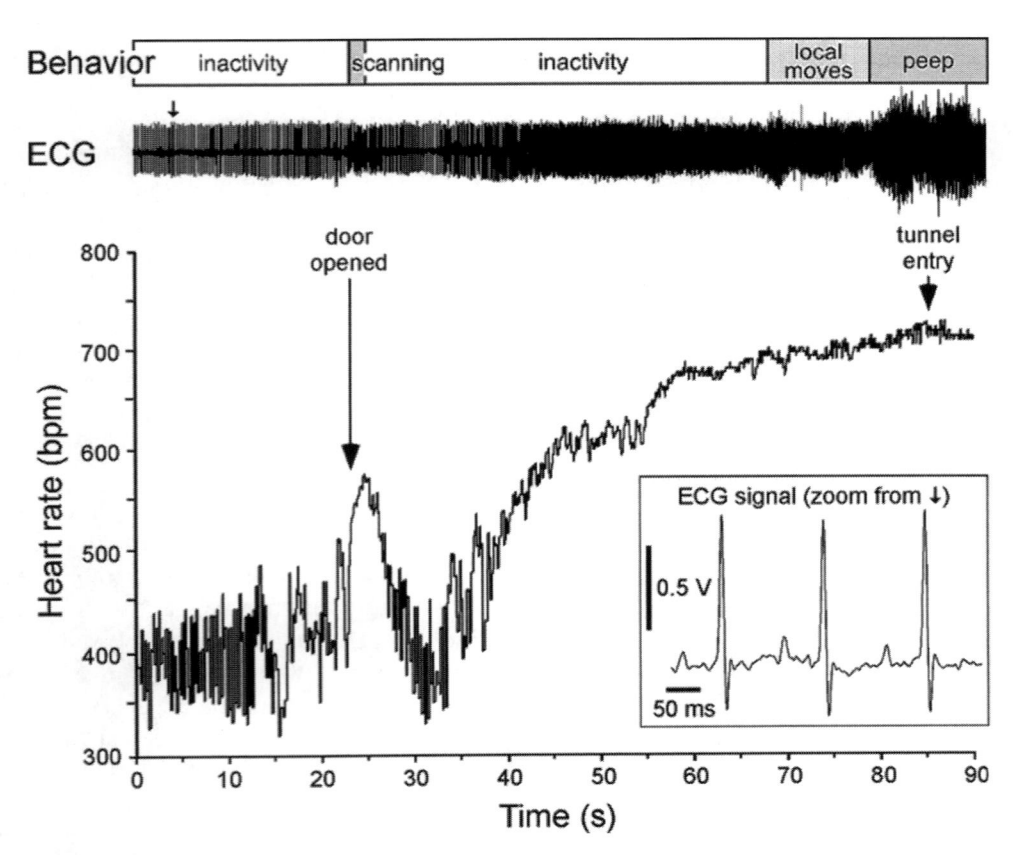

Fig. 5. Behavioral response, concomitant ECG pattern, and derived instantaneous heart rate of a male C57BL/6J mouse in its home cage before entering for the first peep and hide motion into a freely accessible open field. Interestingly, heart rate increased before the mouse moved toward the tunnel suggesting anticipatory arousal before the first physical activity (motion) by entering the tunnel and peeping into the open field. The zoom in the inset depicts a high resolution ECG signal from the location of the *arrow* above the ECG trace at the *top*. Local moves and peeping of the mouse lead to amplitude changes of the ECG signal due to changes of signal strength detected by the ECG receiver.

For the investigation of conditioned fear, another approach is required. Instead of connecting the home cage to a large open field, access is provided to a shock compartment via an automatically controlled door (41), similarly as used in passive avoidance experiments (42). Based on this design, deliberate exploration of the conditioning environment and contextual fear conditioning is investigated by exploiting two natural opposing incentives, fear vs. curiosity (novelty-seeking). Initial studies indicate that fear leads to avoidance of the shock compartment with stretch-attend postures and peeking into the shock compartment. Gradually, mice become more boldly, first entering the shock compartment partially only, before the first full body entry. Finally, mice explore the shock compartment at longer bouts and shorter intervals, a behavior that is indicative of fear extinction. It is important to note that this behavior is almost completely confined to the activity phase of the mice, i.e., the dark phase of the 12-h light-dark cycle. This experimental approach is ideal to quantify the progression of the behavioral responses from the first peep to full exploration and its concomitant heart rate responses in mice. Thereby, the progression of behavioral responses and concomitant heart rate changes can be monitored to better understand physiological states expressed during different phases of the test and determine potential pathological states such as delayed or impaired extinction as index of posttraumatic stress disorder-like behavior. Furthermore, this new approach serves as refined method to investigate the action of anxiolytic compounds with a deliberate choice of mice without experimenter influence on a long time scale of several days. Under these conditions, it is possible to investigate latent inhibition, the retardation of subsequent learning based on prior nonreinforced exposure, which indicates a schizophrenia-like phenotype when impaired.

3.4. Translational Value of Heart Rate Dynamics

An important feature of the heart rate dynamics is the highly dynamical beat-by-beat fluctuation that is essentially determined by the autonomic control originating from brain function. This has been shown by advanced nonlinear measures such as the *detrended fluctuation analysis* (33, 43) indicating persistent long-range correlation of heart beat intervals under baseline stress-free conditions. The long-range correlation is determined by tonic parasympathetic function that slows down heart rate similarly as observed in rats and humans. Blockade of this tonic parasympathetic function by atropine elevates heart rate similarly as functional denervation, as investigated in human heart transplant recipients (44). Both states compromise the control of the beat-by-beat fluctuation and lead to short-term correlation of the beat-by-beat fluctuation. Upon aversive stimulation, heart rate increases due to fast parasympathetic withdrawal and delayed sympathetic activation, thereby leading to reduced long-range correlation of heartbeats in their temporal sequence. This state is also induced upon handling when combined with novelty exposure, indicating that the proper characterization of autonomic function in mice and rat models of physiological and pathological emotional states will have to be performed under home cage conditions in future experiments. The combination of these experimental conditions with nonlinear assessment of heart rate dynamics will be crucial for the qualitative assessment of physiological and pathological dynamics as observed in disease states (33, 43).

4. Cross Species Genetics of Neurobehavioral Traits; Relevance to Anxiety and Mood Disorders

Mood disorders have a major impact on the quality of life of many people, with a prevalence of 10–20% worldwide (45). Finding the mechanisms underlying these heterogeneous psychiatric disorders and obtaining valid animal models is essential for the development of selective pharmacological treatments (46). Interspecies genetic analysis of mood disorder endophenotypes is an important approach to the discovery of novel insights in causality and to identify translational preclinical models (47, 48). By using longitudinal automated home cage observations, we have recently focused on the genetic dissection of avoidance behavior in mice with the aim of finding more selective and effective pharmacological targets for behavioral disorders in humans.

The balance between approach and avoidance behavior is part of a behavioral strategy, which has been highly conserved across species, to obtain food or mediate social interactions while avoiding threatening situations. Such behavior is influenced by genetic variation, as shown by behavioral differences between inbred strains of mice and subsequent quantitative trait loci (QTL) analysis (49–52). Avoidance and motor activity levels are commonly studied in rodent species with traditional anxiety tests, such as the elevated plus maze and open field. However, the nature of these tests makes it difficult to differentiate between these two behavioral components, and experimenter effects can have a great impact on the behavioral outcome (53). Here, novel automated registration methods for longitudinal behavioral assessment in home cages are used to screen a panel of recently generated mouse chromosome substitution strains (CSSs) that are very powerful in QTL detection of complex traits (54). The automated home cage environment is designed to increase behavioral resolution by dissociating behavioral endophenotypes in mice (55). It assesses levels of avoidance behavior (sheltering) independent of motor activity levels (horizontal distance moved) and with minimal human interference. Longitudinal automated assessment of anxiety-related behaviors might also overcome inconsistent results, depending on subtle, short-term variations in the laboratory or test environment (56).

Furthermore, the avoidance behavior in the home cage is sensitive to benzodiazepines (55), providing predictive validity for this anxiety-related endophenotype that might relate, e.g., to mood disorders with anxious symptoms.

In contrast to the concepts of face and predictive validity, construct validity (similar to the underlying causes and mechanisms of the disease) is the most difficult to provide for animal models of psychiatric disorders, simply because of the lack of knowledge about the underlying etiological mechanisms of such complex disorders. Furthermore, for animal models of psychiatric disorders, it has been proven difficult to provide a 1:1 translation with respect to the face validity criteria. Currently, large-scale genome wide association studies (GWAS) are being performed and have in some cases revealed (unexpected) candidate genes that have low odds ratios and explain a small percentage of the variance. In light of this, genetic validity might provide a new entrance to the biology of psychiatric diseases with animal models (57). By integrating mouse and homologous human genetic mapping data, we identified a gene, adenylyl cyclase 8 (*ADCY8*), connecting mouse avoidance behavior obtained in automated home cage environments to human bipolar affective disorder (58). These findings point to novel mechanisms underlying bipolar affective disorders and open new roads for treatment and translational research of its psychiatric endophenotypes.

5. Summary

Animal models of human chronic diseases or behavioral symptoms, such as anxiety and mood disorders, should be tested in situations where long-lasting stable features ("chronic") of such behavior are manifested. Longitudinal automated home cage paradigms (or exploration from the home cage) are currently being designed as a behavioral laboratory method in order to dissociate various behavioral components. The field of behavioral neuroscience is challenged by these developments since it will allow studies on species-specific ethological relevant behavior and environmental conditions. Due to significant reduction in human interference, a major confounding factor in behavioral studies, with this method, increased sensitivity for detection of genetic or pharmacological interventions seems warranted and contributes to enhanced replication between laboratories. Integration of behavioral and physiological measures in freely behaving mice will serve as a powerful approach with diagnostic value to complement the interpretation of emotional state of rodents. Cross-species genetic studies have revealed novel translational findings that may open new roads for understanding neurobiological mechanisms underlying complex psychiatric disorders.

Novel phenotyping methods require large initial investments for development and validation time with a large contribution of descriptive analyses. Furthermore, automated acquisition of detailed behavioral observations over several days leads to large amounts of data with yet unknown depths of new information. Innovative approaches toward the analysis of these data in view of ethological species-specific relevance of behavioral components are necessary to handle, synchronize, and process these data sets efficiently. This holds especially true for scientists depending on high-throughput facilities for screening of large sets of genetically modified animals. Together, the development of these new behavioral paradigms and analysis methods as well as interpretation of the behavioral findings is expected to provide substantially improved clinical relevance for models of neurobehavioral disorders, mechanistic understanding, and novel therapeutic interventions. However, this new development will largely depend on the available critical pool of behavioral neuroscientists and their integrative view on behavior without ignoring its ecological relevance.

Acknowledgments

OS is thankful to Dr. Christian Gutzen (Biobserve) for his support in adapting the Viewer software for use with external hardware in the automated home cage system.

References

1. Hall CS, . Emotional behavior in the rat I. Defecation and urination as measures of individual differences in emotionality. J Comp Psychol 1934; 18:385–403.

2. Hall CS. Emotional behavior in the rat. III The relationship between emotionality and ambulatory activity. J Comp Psychol 1936; 22:345–352.

3. Pellow S, Chopin P, File SE, Briley M. Validation of open:closed arm entries in an elevated plus-maze as a measure of anxiety in the rat. J Neurosci Methods 1985; 14(3): 149–167.

4. Crawley JN, Goodwin FK. Preliminary report of a simple animal behavior model for the anxiolytic effects of benzodiazepines. Pharmacol Biochem Behav 1980; 13:167–170.

5. Gerlai R. Phenomics: fiction or the future? Trends Neurosci 2002; 25(10):506–509.

6. Wahlsten D, Rustay NR, Metten P, Crabbe JC. In search of a better mouse test. Trends Neurosci 2003; 26(3):132–136.

7. Koolhaas JM, De Boer SF, Buwalda B. Stress and Adaptation: Toward Ecologically Relevant Animal Models. Current directions in psychological science 2006; 15(3):109–112.

8. Wisor JP, Nishino S, Sora I, Uhl GH, Mignot E, Edgar DM. Dopaminergic role in stimulant-induced wakefulness. J Neurosci 2001; 21(5):1787–1794.

9. Gainetdinov RR, Wetsel WC, Jones SR, Levin ED, Jaber M, Caron MG. Role of serotonin in the paradoxical calming effect of psychostimulants on hyperactivity. Science 1999; 283(5400):397–401.

10. Dudley CA, Erbel-Sieler C, Estill SJ, Reick M, Franken P, Pitts S et al. Altered patterns of sleep and behavioral adaptability in NPAS2-deficient mice. Science 2003; 301(5631):379–383.

11. Drozdowicz CK, Bowman TA, Webb ML, Lang CM. Effect of in-house transport on murine plasma corticosterone concentration and blood lymphocyte populations. Am J Vet Res 1990; 51(11):1841–1846.

12. Van Ree JM, De Wied D. Behavioral approach to the study of the rat brain. Discussions in Neurosciences 1988; 5(1):11–60.

13. Chesler EJ, Wilson SG, Lariviere WR, Rodriguez-Zas SL, Mogil JS. Influences of laboratory environment on behavior. Nat Neurosci 2002; 5(11):1101–1102.

14. Crabbe JC, Wahlsten D, Dudek BC. Genetics of mouse behavior: interactions with laboratory environment. Science 1999; 284(5420): 1670–1672.

15. McIlwain KL, Merriweather MY, Yuva-Paylor LA, Paylor R. The use of behavioral test batteries: effects of training history. Physiol Behav 2001; 73(5):705–717.

16. Daan S, Pittendrigh CS. A functional analysis of circadian pacemakers in nocturnal rodents. II. The variability of the phase response curve. J Comp Physiol 1976; 106:253–266.

17. Hastings MH, Duffield GE, Ebling FJ, Kidd A, Maywood ES, Schurov I. Non-photic signalling in the suprachiasmatic nucleus. Biol Cell 1997; 89(8):495–503.

18. Kafkafi N, Benjamini Y, Sakov A, Elmer GI, Golani I. Genotype-environment interactions in mouse behavior: a way out of the problem. Proc Natl Acad Sci USA 2005; 102(12): 4619–4624.

19. Benjamini Y, Lipkind D, Horev G, Fonio E, Kafkafi N, Golani I. Ten ways to improve the quality of descriptions of whole-animal movement. Neurosci Biobehav Rev 2010; 34(8):1351–1365.

20. Benjamini Y, Fonio E, Galili T, Havkin GZ, Golani I. Quantifying the build up in extent and complexity of behavior as exemplified in the case of free exploration in mice. Proceedings of the National Academy of Science of the United States 2011. www.pnas.org/cgi/doi/10.1073/pnas.1014837108 Published online before print March 7, 2011, doi: 10.1073/pnas.1014837108 PNAS March 7, 2011.

21. Fonio E, Benjamini Y, Golani I. Freedom of movement and the stability of its unfolding in free exploration of mice. Proc Natl Acad Sci USA 2009; 106(50):21335–21340.

22. Lipkind D, Sakov A, Kafkafi N, Elmer GI, Benjamini Y, Golani I. New replicable anxiety-related measures of wall vs center behavior of mice in the open field. J Appl Physiol 2004; 97(1):347–359.

23. von Uexkull J (1957) A Stroll through the worlds of animals and men: A picture book of invisible worlds. Instinctive Behavior: The Development of a Modern Concept, ed Schiller CH (International Univ Press, New York), pp 5–80.

24. Markou A, Chiamulera C, Geyer MA, Tricklebank M, Steckler T. Removing obstacles in neuroscience drug discovery: the future path for animal models. Neuropsychopharmacology 2009; 34(1):74–89.

25. Chance MRA. An interpretation of some agonistic postures; the role of "cut-off" acts and postures. Symp Zool Soc London 1962; 8: 71–89.

26. Saper CB. The central autonomic nervous system: conscious visceral perception and autonomic pattern generation. Annu Rev Neurosci 2002; 25:433–469.

27. Rozanski A, Blumenthal JA, Davidson KW, Saab PG, Kubzansky L. The epidemiology, pathophysiology, and management of psychosocial risk factors in cardiac practice: the emerging field of behavioral cardiology. J Am Coll Cardiol 2005; 45(5):637–651.

28. Koolhaas JM, Meerlo P, De Boer SF, Strubbe JH, Bohus B. The temporal dynamics of the stress response. Neurosci Biobehav Rev 1997; 21(6):775–782.

29. Stiedl O, Misane I, Koch M, Pattij T, Meyer M, Ögren SO. Activation of the brain 5-HT2C receptors causes hypolocomotion without anxiogenic-like cardiovascular adjustments in mice. Neuropharmacology 2007; 52(3):949–957.

30. Stiedl O, Youn J, Jansen R. Cardiovascular conditioning: neural substrates. Encyclopedia of Behavioral Neurosciences, Vol. 1. Koob G, Thompson RF & Le Moal M (eds), Elsevier, Amsterdam , 226–235. 1-1-2010.

31. Stiedl O, Palve M, Radulovic J, Birkenfeld K, Spiess J. Differential impairment of auditory and contextual fear conditioning by protein synthesis inhibition in C57BL/6N mice. Behav Neurosci 1999; 113(3):496–506.

32. Stiedl O, Spiess J. Effect of tone-dependent fear conditioning on heart rate and behavior of C57BL/6N mice. Behav Neurosci 1997; 111(4):703–711.

33. Meyer M, Stiedl O. Self-affine fractal variability of human heartbeat interval dynamics in health and disease. Eur J Appl Physiol 2003; 90(3-4):305–316.

34. Van Erp AM, Kruk MR, Meelis W, Willekens-Bramer DC. Effect of environmental stressors on time course, variability and form of self-grooming in the rat: handling, social contact, defeat, novelty, restraint and fur moistening. Behav Brain Res 1994; 65(1):47–55.

35. Stiedl O, Jansen RF, Pieneman AW, Ögren SO, Meyer M. Assessing aversive emotional states through the heart in mice: implications for cardiovascular dysregulation in affective disorders. Neurosci Biobehav Rev 2009; 33(2):181–190.

36. Tovote P, Meyer M, Ronnenberg A, Ögren SO, Spiess J, Stiedl O. Heart rate dynamics and behavioral responses during acute emotional challenge in corticotropin-releasing factor receptor 1-deficient and corticotropin-releasing factor-overexpressing mice. Neuroscience 2005; 134(4):1113–1122.

37. Tovote P, Meyer M, Beck-Sickinger AG, von Hörsten S, Ögren SO, Spiess J et al. Central NPY receptor-mediated alteration of heart rate dynamics in mice during expression of fear conditioned to an auditory cue. Regul Pept 2004; 120(1-3):205–214.

38. Tovote P, Meyer M, Pilz PK, Ronnenberg A, Ögren SO, Spiess J et al. Dissociation of temporal dynamics of heart rate and blood pressure responses elicited by conditioned fear but not acoustic startle. Behav Neurosci 2005; 119(1):55–65.

39. Stiedl O, Radulovic J, Lohmann R, Birkenfeld K, Palve M, Kammermeier J et al. Strain and substrain differences in context- and tone-dependent fear conditioning of inbred mice. Behav Brain Res 1999; 104(1-2):1–12.

40. Stiedl O, Tovote P, Ögren SO, Meyer M. Behavioral and autonomic dynamics during contextual fear conditioning in mice. Auton Neurosci 2004; 115(1-2):15–27.

41. Stiedl O, Pieneman AW, Gutzen C, Schwarz S, Jansen RF. Fear conditioning in an automated home cage (DualCage) environment. Proceedings of Measuring Behavior 2010;83–84.

42. Ögren SO, Stiedl O. Passive avoidance. In: Stolerman IP, editor. Encyclopedia of Psychopharmacology, Vol. 2, Springer, Berlin. 2011: 960–966.

43. Peng CK, Havlin S, Hausdorff JM, Mietus JE, Stanley HE, Goldberger AL. Fractal mechanisms and heart rate dynamics. Long-range correlations and their breakdown with disease. J Electrocardiol 1995; 28 Suppl:59–65.

44. Meyer M, Marconi C, Ferretti G, Fiocchi R, Cerretelli P, Skinner JE. Heart rate variability in the human transplanted heart: nonlinear dynamics and QT vs RR-QT alterations during exercise suggest a return of neurocardiac regulation in long-term recovery. Integr Physiol Behav Sci 1996; 31(4):289–305.

45. Ormel J, Petukhova M, Chatterji S, Aguilar-Gaxiola S, Alonso J, Angermeyer MC et al. Disability and treatment of specific mental and physical disorders across the world. Br J Psychiatry 2008; 192(5):368–375.

46. de Mooij-van Malsen AJ, Olivier B, Kas MJ. Behavioural genetics in mood and anxiety: a next step in finding novel pharmacological targets. Eur J Pharmacol 2008; 585(2-3):436–440.

47. Gottesman II, Gould TD. The endophenotype concept in psychiatry: etymology and strategic intentions. Am J Psychiatry 2003; 160(4):636–645.

48. Kas MJ, Fernandes C, Schalkwyk LC, Collier DA. Genetics of behavioural domains across the neuropsychiatric spectrum; of mice and men. Mol Psychiatry 2007; 12(4):324–330.

49. Flint J, Valdar W, Shifman S, Mott R. Strategies for mapping and cloning quantitative trait genes in rodents. Nat Rev Genet 2005; 6(4):271–286.

50. Henderson ND, Turri MG, DeFries JC, Flint J. QTL analysis of multiple behavioral measures of anxiety in mice. Behav Genet 2004; 34(3):267–293.

51. Turri MG, Talbot CJ, Radcliffe RA, Wehner JM, Flint J. High-resolution mapping of quantitative trait loci for emotionality in selected strains of mice. Mamm Genome 1999; 10(11):1098–1101.

52. Mackay TF. Q&A: Genetic analysis of quantitative traits. J Biol 2009; 8(3):23.

53. Kas MJ, van Ree JM. Dissecting complex behaviours in the post-genomic era. Trends Neurosci 2004; 27(7):366–369.

54. Singer JB, Hill AE, Burrage LC, Olszens KR, Song J, Justice M et al. Genetic dissection of complex traits with chromosome substitution strains of mice. Science 2004; 304(5669):445–448.

55. Kas MJ, de Mooij-van Malsen AJ, Olivier B, Spruijt BM, van Ree JM. Differential genetic regulation of motor activity and anxiety-related behaviors in mice using an automated home cage task. Behav Neurosci 2008; 122(4):769–776.

56. Izidio GS, Spricigo L, Jr., Ramos A. Genetic differences in the elevated plus-maze persist after first exposure of inbred rats to the test apparatus. Behav Processes 2005; 68(2):129–134.

57. Kas MJ, Gelegen C, Schalkwyk LC, Collier DA. Interspecies comparisons of functional genetic variations and their implications in neuropsychiatry. Am J Med Genet B Neuropsychiatr Genet 2009; 150B(3):309–317.

58. de Mooij-van Malsen JG, van Lith HA, Oppelaar H, Hendriks J, de Wit M, Kostrzewa E et al. Interspecies trait genetics reveals association of adenylyl cyclase 8 with mouse avoidance behavior and a human mood disorder. Biol Psychiatry 2009; 66: 1123–1130.

Chapter 2

Using Behavioral Patterns Across Species in Mood Disorder Research

Jared W. Young and Mark A. Geyer

Abstract

Measuring motor activity has been one of the most commonly used tools to assess behavior in rodents. The macroscopic measure of behavior called motor activity is actually composed of assemblies of microscopic responses however. These microscopic responses can be measured simultaneously producing multivariate profiles that reveal differential and specific patterns of effects of pharmacological, genetic, or environmental manipulations. Thus, while stimulants increase the amount of motor activity irrespective of their varying neurotransmitter effects, they produce very different behavioral patterns of exploration. Given the utility of this approach for understanding the effects of manipulations in rodents, we have recently begun measuring behavioral patterns of exploration in humans using the same microscopic measures. Unlike rating scales, this technique has yielded information on quantitative differences of exploration in multiple psychiatric populations, including bipolar disorder, methamphetamine dependence, and schizophrenia. While the exploratory patterns of the three groups differ from controls, they also differ from each other, providing us with quantitative differences which we have used to produce more patient population-specific animal models. Thus utilizing quantitative measurement of exploration in humans may provide us with (1) more specific animal models for disease states, (2) valuable insight into neural abnormalities in patient populations, and (3) quantitative assessment on the effects of treatments.

Key words: Behavioral pattern monitor, Bipolar disorder, Mania, Motor activity, Locomotion, Mouse, Exploration patterns, Cross-species assessment

1. Background and Overview

In order to develop a greater understanding of mood disorders that result in abnormal behavior, we must first understand behavior. While there are many forms of behavior, measuring unconditioned motor activity remains the most commonly utilized tool in rodents. Motor activity does not constitute a unitary class of behavior however, but in fact is composed of multiple aspects of behavior.

Todd D. Gould (ed.), *Mood and Anxiety Related Phenotypes in Mice: Characterization Using Behavioral Tests, Volume II*, Neuromethods, vol. 63, DOI 10.1007/978-1-61779-313-4_2, © Springer Science+Business Media, LLC 2011

This macroscopic nature of motor activity means that a univariate assessment would be overly simplistic given that it consists of assemblies of microscopic responses (1). These microscopic responses could be measured separately given appropriate definitions, criteria, and measurement tools. Multivariate assessment of behavior provides measurements that may therefore be altered by orthogonal, synergistic, or competing physical systems. Differentially affecting these systems may therefore produce varying changes in these measures, resulting in different patterns of behavior. Hence, from a macroscopic level, motor activity may be viewed as chaotic, but what we call chaos can also be viewed simply as patterns we have not yet recognized.

Generating patterns of behavior can therefore provide greater information on experimental effects than examining single measurements alone. Using multivariate assessment tools has been recognized for some time by various researchers, with different approaches proposed to quantify the various components of motor activity (2–6). Some of these approaches are based on observer ratings, while others have attempted to automate the entire measurement process (7–10). The behavioral pattern monitor (BPM; Fig. 1) is a computerized activity system created to incorporate multivariate measures of exploration. Equipped with rearing sensors (both touchplates and infra-red beams) and 11 holes in the floors and walls (that serve as discrete stimuli for rodents to investigate; Fig. 2), the BPM measures activity, specific exploration, and path patterns at high levels of temporal and spatial resolution (11). This equipment produces >50 measures of exploration (Table 1) which can be used to differentiate pharmacological, genetic, and environmental influences on exploration (see below). While these measurements are disparate, many converge on hypothetical concepts. To explore the similarities between these measurements and the fundamental dimensions reflected in these multivariate

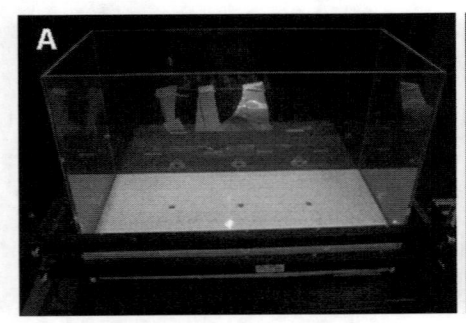 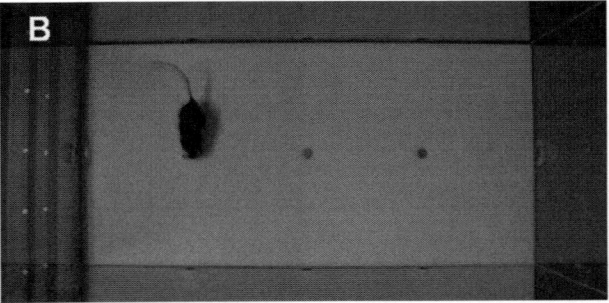

Fig. 1. The mouse behavioral pattern monitor (BPM). The mouse BPM is an activity chamber (30.5 × 61 cm). A 2.5 D view of a mouse BPM chamber in which the 11 investigatory holes (three on the near long wall, three on the far long wall, three on the floor, and one on each short wall) are visible (**a**). A top down view of mouse BPM in which the floor investigatory holes are clearly visible, as is a mouse holepoking into one of these holes (**b**).

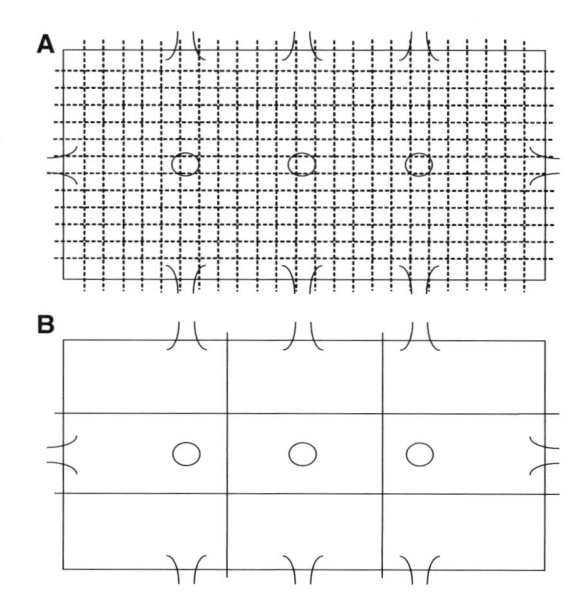

Fig. 2. Schematic of the mouse BPM. The mouse BPM records the spatial location of the mouse using a grid of 12 × 24 photobeams (*dashed lines*) located 1 cm above the floor (**a**). The chamber contains eleven 1.4 cm diameter holes (three in the floor, three on each long wall, one on each short wall), each provided with an infrared photobeam to detect investigatory holepokes (**b**). For analysis, the chamber is divided into nine unequal regions (**b**), with transitions between regions utilized as a measure of investigatory activity.

characterizations, we conducted factor analyses on the exploratory behavior of 137 rats (12). The analyses identified three primary factors (Table 1) onto which the various automated assessments of exploration loaded: the amount of activity (e.g., distance traveled, transitions, and counts); specific exploration (e.g., rearing and holepoking); and path patterns that involve novel quantitative measures based on nonlinear dynamical systems methods and fractal geometry to assess important aspects of the hierarchical and sequential organization of behavior (spatial scaling exponent d, dynamical entropy h, spatial CV, temporal CV) (2, 9, 13, 14). Together, these three factors accounted for 77% of the variance in the measures. These studies confirmed that composite variables composed of the weighted measures for each factor are highly effective in discriminating receptor subtype-specific drug effects in rats (15). While the measurements of activity levels and specific exploration are largely self-explanatory, we have provided a more in-depth explanation of path patterns below.

1.1. Linear Dynamics vs. Predefined Response Categorization

Behavior is traditionally categorized in terms of predefined response categories, even in some multivariate assessments (e.g., the OFT; (6)). Stimulant drugs disrupt normal behavioral responses however, fragmenting behavioral sequences and introducing novel elements into the normal repertoire. Thus, drug-specific categorical

Table 1
Measurements and their associated factors in the mouse behavioral pattern monitor (12)

Factor	Measure	Description
Activity measures	Counts	The cumulative number of distinct behaviors to occur during testing
	Transitions	The number of movements from one predefined area to another
	Center entries	The number of entries into the area defined as the center
	Distance traveled	Total distance traveled while exploring the chamber measured in cm
Specific exploration	Hole pokes	The total number of investigatory holepokes across all 11 holes
	Repeated poking	The number of times a hole was repeatedly poked before moving to the another hole
	Varied poking	The number of times the mouse holepoked into different holes
	Rear	The number of times the mouse reared into the air or against the side walls
	Center duration	The amount of time (s) spent in the area designated as the center of the chamber
Locomotor patterns	Spatial d	Spatial scaling exponent d measures the hierarchical and geometric organization of behavior
	Entropy h	Quantifies the sequential aspects of sequences of behavior, specifically measuring the degree to which behavior is observed along a continuum between complete order and disorder
	Spatial CV	Measures the consistency by which the animal moves from one region to another
	Temporal CV	Measures the consistency of the dwell time in each of the regions

rating scales (e.g., amphetamine stereotypy) have evolved, reflecting the fact that the arbitrary definitions of responses appropriate for normal behavior are often inadequate for the characterization of drug-manipulated behavior. Such rating scales designed for one drug (e.g., amphetamine) are often inappropriate for another drug however, even for drugs with related chemical structure such as MDMA (16). Traditional measures have fundamental problems therefore as they are (1) constrained to support inferences about the relationship of drug effects to the normal behavioral repertoire of the animal, (2) insufficient in quantifying the sequential arrangement of behavioral elements, (3) inadequate for assessing the temporal and spatial resolution of behavior, and (4) insufficient to provide comparison to other drugs. Numerous investigators acknowledge the need to quantify the sequential arrangement of behavioral elements, yet evaluation techniques have not evolved far beyond descriptively categorizing and enumerating predefined behavioral sequences. Most of the quantitative analyses are rooted in the theory of Markov processes or involve run statistics or time-series analyses. Although these approaches have confirmed that sequences of motor acts are not independent random events but show varying degrees of interdependency, they have demonstrated only limited utility, particularly in the characterization of drug effects (1). Finally, in traditional approaches, the temporal and spatial resolution used to define the measures is chosen arbitrarily. This choice is based frequently on the qualitative separation of temporal and spatial scales. As our studies indicate, however, there appears to be no distinct separation of temporal scales. Instead, "pauses" and "behavioral actions" are found on all time scales. Moreover, this separation fundamentally neglects the hierarchical nature of behavioral organization.

To quantify behavioral patterns in sequences of locomotor behavior, we began with the following premise: behavior consists of purposefully organized sequences of acts that can be observed by the motor output of an animal or human subject. Accordingly, behavioral organization can be defined as the selection, ordering, and sequencing of behavioral elements in response to external or internal stimuli to form flexible, yet stable, macroscopic patterns of behavior. The assessment of behavioral organization, as in the case with locomotor behavior, can be approached from both hierarchical and sequential points of view. Hierarchically, behavioral elements are thought to form organized behavioral components on successively larger spatial or temporal scales. Accordingly, the evaluation should quantify the scale-invariant properties of the behavioral organization. Sequentially, behavioral elements are thought to be arranged serially into organized behavioral patterns. Thus, the evaluation should quantify the sequence length-independent properties of the behavioral organization. The path pattern measures described below were originally developed for studies of rodent locomotor activity and are now being extended for studies of humans. Two of the measures, the spatial scaling exponent d, and the dynamical entropy h, quantify the hierarchical and sequential properties of behavioral organization, respectively. These measures have several advantages over traditional rating scales and other scale-dependent assessments of motor behavior. Specifically, this approach does not depend on response categories that are defined a priori, is resolution and sequence-length independent, can be extended to include time-dependent characteristics, and allows one to obtain detailed statistical evaluations.

1.1.1. Spatial d

The spatial scaling exponent, d, measures the hierarchical and geometric organization of behavior. Specifically, d is based on the principles of fractal geometry and describes the degree to which the path taken by the animal is one-dimensional or two-dimensional. Spatial d is measured by the distance traveled plotted against the number of micro-events (smallest change that can be observed) using a double-logarithmic coordinate system, generating a line of fit between these two variables (17). Spatial d typically varies between one (a straight line) and two (a filled plane), with values closer to one reflecting straight movements and values closer to two reflecting highly circumscribed, local movements. At both ends of this spectrum, the geometric pattern of movement around the BPM is highly predictable and can exhibit the same level of activity but describe almost straight line or highly circumscribed geometrical patterns.

1.1.2. Entropy h

Entropy h quantifies the sequential aspects of sequences of behavior, specifically measuring the degree to which behavior is observed along a continuum between complete order and disorder (18). Briefly, a given sequence of activity is compared to similar preceding sequences, and this comparison is conducted for varying sequence lengths. Lower values of h (low entropy) suggest highly predictable or ordered sequences of motor activity. Thus, there are commonalities in the sequences of activities being compared and future sequences could be predicted based on earlier sequences. Higher values (high entropy) suggest a greater variety, or disorder, in the level of motor activity, where the sequences being compared are not similar and are thus unpredictable from one sequence to another (see (19)).

1.1.3. Spatial CV

The spatial CV statistic describes another aspect of the spatial pattern of the animal's movement. Spatial CV measures the degree to which an animal makes the same repetitive transitions from one area to another. A transition matrix can be made of the transitions between different regions of the chamber, e.g., the corners, walls, and center. A preferential exhibition of a subset of transitions results in a high coefficient of variation (CV), reflecting the repetitive preference of certain transitions, i.e., repeatedly following the same path. A highly varied pattern transition lowers CV, i.e., different paths chosen (17, 20).

1.1.4. Temporal CV

Analogously, the length of time spent in each of the regions can also be quantified and used to calculate a CV statistic in the temporal domain. Thus, temporal CV defines the degree to which the animal remains in one area or distributes its time across multiple areas (high vs. low temporal CV respectively; (20)).

Collectively, such descriptors as spatial scaling exponent d, dynamical entropy h, spatial CV, and temporal CV assess various aspects of the hierarchical and sequential organization of behavior at a temporal macroscopic level (see e.g., Fig. 3 of paths that generate variations in these measures). This approach using nonlinear measures has provided a useful complement to the more traditional

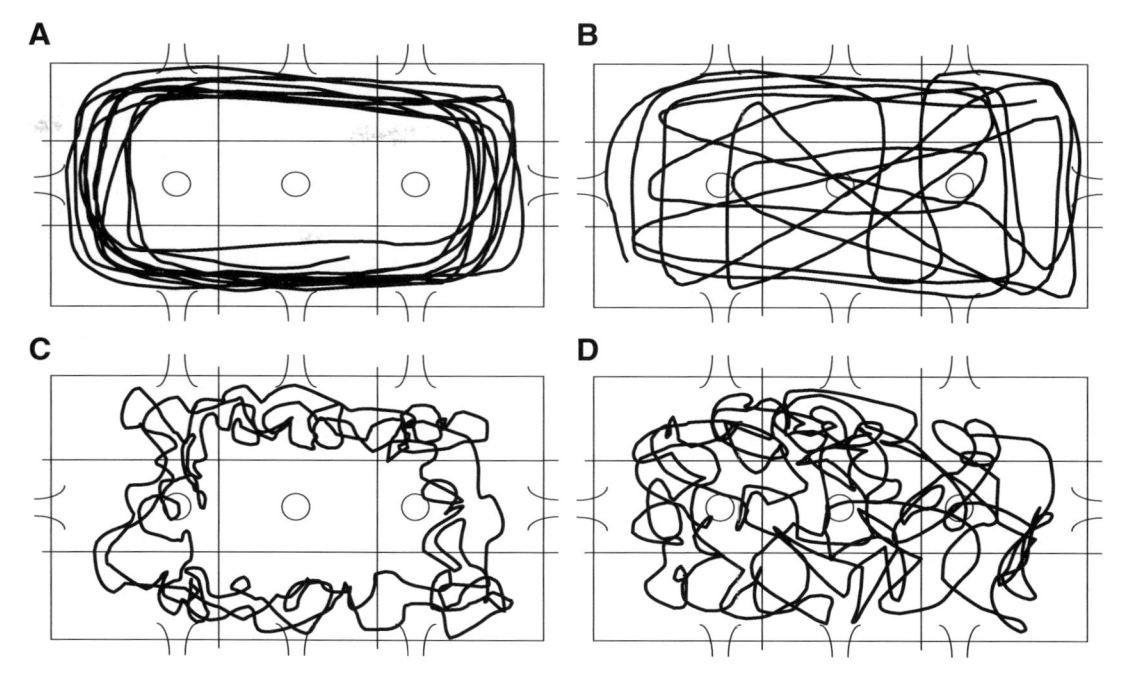

Fig. 3. Hypothetical examples of locomotor path patterns that can be measured in the mouse BPM. The spatial location of the mouse during exploration of the BPM is recorded using banks of photobeams. The path of the mouse as it moves through the BPM is continuously recorded and can be recreated for path analysis. The path of a mouse that was recorded to exhibit a low spatial d (long linear movements through space), high spatial CV (transitions made from one region to another consistently), and a low entropy h (predictable pattern of behavior) would resemble (a). The path of a mouse that was recorded to exhibit low spatial d, low spatial CV (transitions from one region to another that are random), and low entropy h would resemble (b). The path of a mouse that was recorded to exhibit a high spatial d (localized meandering movements in the chamber), high spatial CV, and low entropy h would resemble (c). The path of a mouse that was recorded to exhibit a low spatial d, spatial CV, and entropy h would resemble (d).

behavioral profiles based on a priori definitions of categorical events (10, 21).

1.2. Differentiating Stimulant Effects Using a Multivariate Approach

The multivariate approach of assessing activity, specific exploration, and path patterns has provided us with greater clarity on the differential effects of manipulations on exploratory behavior. We and others have used a multivariate approach to elucidate the behavioral characteristics and neuropharmacological mechanisms of psychoactive drugs (22, 23). Stimulants such as amphetamine, apomorphine, caffeine, 3-4-methylenedioxymethamphetamine (MDMA), nicotine, and scopolamine induce similar increases in the amount of activity (albeit dose-dependent), as measured in either the Open Field, photobeam activity chambers, or the BPM (11, 16, 24–32). Differential effects have been noted however, when even the simplest of multivariate assessments is used. For example, caffeine increases activity but also increases specific exploration as measured by rearing behavior in the Open Field and BPM, as well as holepoking in the holeboard and BPM (11, 25, 30, 33).

Nicotine had no effect on rearing behavior in the Open Field or BPM however, (11, 30), with higher doses of nicotine reducing activity and holepoking behavior (34, 35), likely due to hypothermic effects. The direct dopamine agonist apomorphine inhibits holepoking behavior (11, 36), consistent with MDMA, which releases presynaptic serotonin rather than dopamine (16). Stereotypy is observed in conjunction with both amphetamine- and apomorphine-induced hyperactivity (11, 37–39). Higher doses are required for amphetamine-induced stereotypy however, (36), and the nature of the stereotypy induced varies, with amphetamine at low doses simply producing an exaggerated wide range of activities, while apomorphine produces a more restricted behavioral repertoire (37, 40), and MDMA produced repetitive behaviors that could not be readily scored using the rating scales defined for amphetamine-treated animals (16). Multivariate assessments of behavioral patterns are therefore required, where fine discriminations can be made as to the distinctive characteristics of the behavioral profiles and thus potentially unique effects of compounds.

These stimulants also vary in their effects on path patterns. While apomorphine, MDMA, and scopolamine increased spatial CV, reflecting the generation of more repetitive patterns of movements, amphetamine increased the variety of movements, lowering spatial CV, while neither nicotine nor caffeine had any significant effect on spatial CV (11, 16). Differing effects are observed using the spatial d measure: apomorphine and scopolamine lowered spatial d, suggestive of path patterns with straighter lines; amphetamine increased spatial d at low doses and had variable effects at higher doses; neither nicotine nor caffeine affected spatial d (11, 41). Amphetamine and MDMA increased entropy h, suggesting increasingly disordered movement, while higher doses of MDMA lowered entropy h, suggesting more ordered and predictable paths (19). Temporal CV is the only path pattern measure that is consistently affected by all these psychostimulants, with amphetamine, apomorphine, scopolamine, nicotine, and caffeine all lowering temporal CV. In general, unlike the other measures of path patterns, the temporal CV measure is strongly and inversely related to the amount of activity.

Thus, despite the common psychostimulant label for all these compounds, they produce distinctive behavioral patterns that can be readily differentiated using multivariate assessments. Such differentiation should not be surprising given the different mechanisms of actions of each of these drugs. It is reassuring to know that behavioral analysis of spontaneous locomotion and exploration can differentiate between these drugs in a manner that is consistent with their differential pharmacological mechanisms. For example, amphetamine-induced blockade of norepinephrine, dopamine, and serotonin transporters increases extracellular levels of these neurotransmitters throughout the CNS leading to the myriad of effects observed, while scopolamine primarily acts as a muscarinic receptor antagonist. Hence, other drug-induced manipulation of these neurotransmitter systems can also cause hyperactivity and altered exploratory behavior and locomotor patterns.

Numerous neurotransmitter systems have been implicated in the pathophysiology of mood disorders. These systems can be investigated specifically using this multivariate approach in animals. For example, a reduction of norepinephrine in the locus coeruleus reduced specific exploration in rats, as measured by holepoking and rearing behavior (11, 42). Because either intra-ventricular or intra-hippocampal infusions of norepinephrine increase locomotor and specific exploration (43, 44), norepinephrine is likely to assist in maintaining normal exploratory behavior. Moreover, evidence suggests that β-adrenergic receptors may play a role in these noradrenergic effects (45). Such findings demonstrate the utility of the BPM because patients with mood disorders exhibit reduced volumes of brain regions that are major sources of norepinephrine in the CNS (46). Moreover, the turnover rate of norepinephrine in BD patients is increased compared to controls as measured by urinary and plasma levels, as well as postmortem analyses (47–50). Recently, we demonstrated that patients with BD mania exhibit increased activity, specific exploration, and more linear movement in the human BPM ((51) see below). Therefore, investigating the links between neuropathology in mood disorders and the mechanisms that subserve exploration in rodents may provide an avenue for the development of novel therapeutics to treat BD.

Other neurotransmitter system abnormalities may contribute to mood disorders but have received limited investigation. For example, the mAChR antagonist scopolamine increases locomotor and exploratory activity, while lowering spatial CV, temporal CV, and spatial d (11, 52). Depending upon dose and timecourse, the nicotinic acetylcholine receptor (nAChR) agonist nicotine can increase or decrease locomotor activity, presumably via actions at the α4β2 nAChR, but has limited effects on other domains of exploration. Noncompetitive NMDA antagonists, such as dizocilpine and phencyclidine, also act as stimulants (53), increasing and decreasing holepoking activity at low and high doses respectively, and inducing perseverative path patterns. Phencyclidine administration results in animals running in circles or figure of eight patterns while dizocilpine increases small localized movement activities in the corners (20). This list is not intended to be exhaustive since the primary goal of this chapter is to provide information on the cross-species translational utility of exploratory behavior. Thus, while the majority of these studies to date have been conducted in rats, we have since extended this exploratory paradigm into other species.

1.3. Cross Species Translation of the BPM

1.3.1. Rat to Mouse

The development of a mouse BPM was important on several levels: (1) since mice are more readily modified genetically, mutant disease models or knockout (KO) animals can be generated (51, 54) and utilized to complement pharmacological studies of receptor subtypes (55–58); (2) evidence for cross-species generalization increases the likelihood that findings will be of relevance across other species including humans (51, 59, 60).

Initial studies in the mouse BPM investigated which dopamine receptors contributed to the exploratory effects of MDMA using various dopamine receptor KO mice. MDMA increased activity, lowered spatial d, and increased spatial CV in wild-type mice (56) consistent with rats (16). Dopamine D1 KO mice exhibited an exaggerated responsiveness to the MDMA-induced increases in locomotor activity, while dopamine D2 KO mice exhibited a reduced amount of MDMA-induced activation (56). Although MDMA-induced hyperactivity was unaffected in dopamine D3 KO mice, these mice did not exhibit the same immediate MDMA induced-increase in perseverative locomotor path patterns (spatial CV). Altered exploration in these mice also suggested that D1 receptors may contribute to the locomotor pattern quality, i.e., linear vs. circumscribed movement (spatial d), while D2 receptors may contribute to perseverative or thigmotactic locomotor effects of MDMA (spatial CV). Thus, by utilizing genetic mutations, the mouse BPM provided information on receptors that (a) contribute to basic exploratory behaviors, and (b) mediate drug effects on exploration.

The mouse BPM has since been used to examine the contribution of the corticotropin releasing factor (CRF) receptor subtype 2 (CRF2) to alterations in exploration induced by isolation rearing during development (55). CRF2 KO mice exhibited a hypersensitivity to isolation rearing, with increased activity, holepoking, and temporal CV observed in these mice relative to isolation-reared wild-type mice (55). In another application of the mouse BPM, the hallucinogen 1-(2,5-dimethoxy-4-iodophenyl)-2-aminopropane (DOI) increases activity at low, while reducing activity at high doses. DOI-induced hyperactivity was absent in 5-HT2A KO mice however, while DOI-induced decreases in activity were attenuated with coadministration of a 5-HT2B antagonists, suggesting that these receptors mediate different aspects of DOI-induced alterations in activity (57). Other hallucinogens such as psilocin are affected by 5-HT2A, 5-HT2C, as well as 5-HT1A receptors also (61). Interactions between hallucinogens and other neurotransmitter systems can also be investigated, e.g., metabotropic glutamate receptor 5 KO mice exhibit an increased behavioral response to the 5-HT2A agonist hallucinogen DOM (2,5-dimethoxy-4-methylamphetamine) (58).

The exploratory profiles of genetic mouse models of mood disorders have also been assessed in the mouse BPM (51, 54). Using a traditional open field test, we found that mice with reduced (10% compared to WT littermates) dopamine transporters (DAT) exhibit increased activity and more linear movement (reduced spatial d) (62). Moreover, acute doses of the mood stabilizer valproate (200 mg/kg) reversed these effects (62). We have since confirmed in the mouse BPM that in addition to their locomotor hyperactivity, DAT KD mice also exhibit increased specific exploration (hole-poking) and reduced spatial d (51, 54). This pattern of exploratory behavior is consistent with mice administered the selective DAT inhibitor GBR 12909 (51, 54, 63). Given evidence that the DAT polymorphisms associated with BD result in reduced number and/or function of the DAT (64, 65), we recognized that gaining knowledge of the exploratory behavior of patients using a similar multivariate approach was paramount.

Importantly however, the mouse studies described here provide evidence for the utility and cross-species generalizability of the BPM.

1.3.2. Rodents to Humans

As reviewed here, multivariate assessment of exploratory behavior has borne a richness of results in terms of elucidating the underlying neural mechanisms of drugs in rodents. To date however, the extent of this work has not been paralleled by similar studies in humans. Consistent with early studies in rodents, initial activity studies in humans were limited to measuring motor activity only, using a wrist or leg actigraph (66–68). While these reports in psychiatric populations are consistent with expectations (increased activity in patients with attention deficit hyperactivity disorder and BD), these data are limited by their univariate assessment. Pierce and Courchesne (69) attempted a multivariate assessment of exploratory behavior in autistic children by rating videotapes of subjects left in a room with colorful and interactive objects for 8 min. The observers' ratings of decreased exploration were correlated with MRI-based measures of altered brain volumes in children with autism. These data support the hypothesis that a human exploratory paradigm would be useful in detecting behavioral deficits that are associated with brain dysfunction. The vast majority of studies on hyperactive behavior in these patient populations, however, have been limited to observer-rated and self-report scales. As discussed above, such scales: (a) may not be optimal in detecting potentially subtle alterations in activity levels; (b) are minimally effective in distinguishing psychiatric populations; and (c) do not provide the opportunity for cross-species comparisons.

In light of the paucity of quantitative multivariate assessment tools in psychiatry despite the number of disorders described as having altered motoric behavior, we recently developed a human version of the BPM. This human BPM was designed to be analogous to the rodent BPMs described above in which we would be able to rapidly and sensitively quantify activity, specific exploration, and path patterns in humans (59). The apparatus for the human BPM is described in detail below but is essentially a room that the human participant has not been exposed to and therefore, consistent with the rodent BPM, is a novel and unfamiliar environment. Eleven objects deemed to be safe, colorful, tactile, and manipulable and are placed dispersed throughout the room on items of furniture. These objects provide an analog of the exploratory holes in the walls and floor of the rodent BPM chambers. Participants are directed into the room and are asked to wait for the experimenter to return. The human BPM session has been 15 min long in our studies to date. Using a camera mounted in the ceiling and automated tracking software, we are able to generate X-Y coordinate locations for the subject in the room. The precise measurements and their collection are described below. In brief however, we are able to capture information on the activity, specific exploration, and path patterns of human subjects, consistent with that of the rodent studies described above.

To date, we have quantified the spontaneous exploration of acutely ill patients with BD mania and schizophrenia, as well as subjects dependent upon methamphetamine (MD) (51, 70–72). These three groups demonstrate strikingly different exploratory patterns of behavior, once again highlighting the importance of multivariate assessment of activity, where measurement of multiple parameters may yield distinct "signatures" of exploration that characterize and differentiate these disorders. From a diagnostic perspective, these signatures of exploration may provide an obvious and meaningful difference between acutely ill populations who are difficult to distinguish because they present with psychotic and mood symptoms (73).

Consistent with the differentiation of stimulants in rodents described above, patients with schizophrenia, BD, and MD exhibit unique patterns of exploratory behavior. For example, patients with BD exhibit hyperactivity in the first 5 min in the hBPM, with rapid habituation, patients with schizophrenia exhibit hyperactivity only in the last 5 min (51), while MD subjects do not exhibit hyperactivity at any timepoint (72). Interestingly, both patients with BD and schizophrenia exhibit increased entropy h and reduced spatial d, the latter being more prominent in patients with BD (51). These data suggest that both schizophrenia and BD patients exhibit more disorganized but straighter line movement through space when compared to controls (51), while MD subjects did not differ from control subjects (51). Both BD patients and MD subjects exhibited increased specific exploration (object interactions), while patients with schizophrenia did not differ from healthy comparison subjects (51, 71). These specific abnormal patterns and the availability of an animal version therefore provide the opportunity to dissect the abnormal neuroanatomy that underlies these changes.

Interestingly, the increased object interaction of MD subjects correlated with a number of perseverative errors they made on the Wisconsin Card Sorting Task, a cognitive task mediated by the frontal lobes (72). Subjects whose frontal lobes were lesioned exhibit inappropriate "grasping" behavior of objects that are within reach (74). There may therefore be a link between this simplistic measure of exploratory behavior and frontal functioning, which could be useful given that frontal dysfunction is associated with mood disorders (75–77). Recently, we also demonstrated that DAT KD mice exhibit increased risk taking behavior in a rodent model of the human Iowa Gambling Task, a frontally mediated behavior (78), that also correlated with increased specific exploration (holepoking) in these mice (79). The evidence for the utility of assessing behavioral patterns of exploratory behavior across species is therefore growing.

2. Equipment, Materials, and Setup

2.1. Rodent Behavioral Pattern Monitors

Both rat and mouse BPMs (Fig. 1) exist that can quantify exploratory behavior. Each chamber consists of a $30.5 \times 61 \times 38$-cm area with a hole board floor equipped with three floor holes and eight wall holes (three along each side of the long walls, and two holes in the front and back walls). Each hole (rat = 2.5 cm, mouse = 1 cm) is equipped with an infrared photobeam to detect nose poking behavior. Each chamber is illuminated from a single light source above the arena (producing 350 lux in the center, and 92 lux in the 4 corners). For the mouse BPM, a grid of 12×24 infrared photobeams 1 cm above the floor records the location of the mouse every 0.1 s (Fig. 2). For the rat BPM, a 4×8 grid records the rat's X-Y position. To measure rearing behavior, the mouse BPM has an array of 16 infrared photobeams 2.5 cm above the floor and aligned with the long axis of the chamber, while the rat BPM has touch-plates located 15.2 cm high on the walls. The position of the mouse and rat is defined across nine unequal regions (four corners: 9.375×16.875 cm, four wall regions: long, 9.375×26.25 cm, short, 11.25×16.875 cm, and a center: $11.25, 26.25$ cm $(11, 56)$). At the start of each test session, the rodent is placed in the bottom left hand corner of the chamber, facing the corner and the test session started immediately.

2.2. Human Behavioral Pattern Monitor

The human BPM takes place in a room that, at least for the initial test of a subject, is novel and unfamiliar environment to the human participant. Subjects wear an ambulatory monitoring vest that measures accelerometry in digital units at 10 Hz, stored in an onboard PDA. Currently, we have used both 2.7×4.3 and 3.5×4.9 m rooms. To date, no discernible differences between these rooms have been observed. The rooms are furnished with a desk, two bookcases, a small file cabinet, and a large file cabinet. On these furnishings are placed 11 small but visually engaging objects. These objects were chosen using the criteria that they be safe, colorful, tactile, and manipulable and thus may promote human exploration. The toys are used as an analog of the exploratory holes in the walls and floor of the rodent BPM chambers. Specific exploration is also counted for exploration of drawers in a cabinet, manipulation of blinds over the window, and the accelerometer's recording device housed in the fanny-pack around the subject's waist. Participants are escorted into the room and are asked to wait for the experimenter to return after setting up testing equipment elsewhere. The subject is then left alone in the room and so far all sessions have been 15 min long.

Data in the human BPM are gathered using three sources of measurement: (1) x-y coordinates of the subject's spatial location in the BPM, extracted from digital video recording (a video camera is embedded in the ceiling (51)); (2) experimenter ratings of exploratory activity, obtained by carefully scoring the video recording of the BPM session and measuring events such as interactions with objects (51, 71); and (3) motor activity of the subject's torso, using an accelerometer embedded in an ambulatory monitoring device that the participant wears (51, 70). It is hypothesized that similarly to the rat and mouse BPMs, three main independent factors will emerge, describing activity (counts, transitions, accelerometry), specific exploration (object interactions), and sequential organization of behavior (spatial d and spatial CV).

The digital videos of subject activity in the human BPM are subjected to frame-by-frame analysis with proprietary software (Clever Systems, Inc.1999), which generates x- and y-coordinates of the subject's location at the rate of 30 Hz. Activity can be measured by total distance traveled or transitions across nine regions of the human BPM. These regions are analogous to our definition of nine areas of the rodent BPM, the four corners, four walls, and the center (11). These data therefore enable us to obtain a distribution of amount of time spent in each region as well as to measure the number of transitions, consistent with the rodent work as movement from one region to an adjacent one. In any event, transitions between regions and dwell times within specific regions can serve as additional measures to describe different aspects of locomotor activity (51).

The video footage also enables the assessment of specific exploratory behaviors of the participant. Trained observer rating analysis is conducted on (1) total object interaction, (2) total amount of time spent with all objects, (3) multiple object interactions, (4) percent of perseverative interactions, (5) wear mask/glasses, and (6) explore drawer (for measurements and definitions see Table 2) (71).

Accelerometry data are generated using the LifeShirt System (80), an ambulatory, multisensor, continuous monitoring system that also collects data using respiratory inductive plethysmography bands, which measure pulmonary function, electrical activity of the myocardium via a three-lead EKG, and activity/posture via a two-axis accelerometer (51, 70, 81). To measure activity levels, a two-axis

Table 2
Observer rating of subjects behavior in the room – focus on object interaction (71)

Measure	Description
Object interaction	When subject makes any physical contact with the object, including when the object is in the hand, on the foot, becomes a physical extension of the body, or is used to push, poke, prod, or otherwise make physical contact with any other object or item of furniture
Mean time spent with objects	Totaling the number of seconds a subject spent interacting with objects during the 15-min period and dividing by the total object interactions
Multiple object interaction	Any instance a subject interacts with more than one object at the same time. Should a subject pick up a new object, it is scored as another multiple object interaction; however, if that same object (or one of the other objects the subject is interacting with) is put down, a new MOI is not scored
% perseverative interaction	The total number of objects a subject touched more than once divided by the total number of objects in the room
Wear mask/glasses	Subject places the mask or glasses on head (or arm, waist, leg, etc.) and releases the object completely – but the object is still on subject
Explore drawer	The instant a subject first opens a drawer to the instant they close it and release the drawer
Time spent walking	Time the subject takes one or more steps in any direction, (forward, backward, sideways, or diagonal) or when the subject shifts weight from one foot to another
Time spent sitting	Time the subject sits at the desk, cabinets, or floor

accelerometer is placed onto the shirt over the sternum, and the rectified and integrated accelerometer signal detects periods of physical activity and rest. An on-board PDA continuously encrypts and stores the patient's activity and postural physiologic data on a compact flash memory card. Accelerometry data are sampled at 10 Hz and stored numerically in digital units.

2.3. Data Acquisition The primary dependent variables of interest were locomotor activity as measured by transitions (calculated as a movement across a defined region) and center entries (cumulative entries into the center region); specific exploration as measured by holepoking, rearing, and center duration (cumulative time spent in the center); and locomotor exploratory pattern as measured by spatial d. Spatial d uses analyses based on fractal geometry to quantify the geometrical structure of the locomotor path, where a value of 2 represents highly circumscribed localized movement, 1 represents straight line distance covering movements (17).

2.4. Troubleshooting

With sufficient resolution, multivariate assessment of exploratory behavior is possible. The patterns of behavior of animals with experimental manipulation do not apparently change with chamber size (82), or whether the chamber is square, rectangular, or circular (83). Consistent with other tests of exploratory activity in rodents, rats and mice habituate with repeated exposure to the same environment, displaying less exploration when compared with previous exposures (54). Altering the environment at the time of repeated testing can reinstate exploration patterns however. The environment can be altered by altering the floor (e.g., changing from acrylic to sandpaper), or adding odorant objects (with or without odors) in the holes (encourages specific exploration) (43, 54). When assessing behavior across strains, the shape of the animal may contribute toward strain differences. For example, the lower levels of holepoking of 129 compared to C57BL mice may be due in part to the shorter nose of the former strain. GBR12909-induced increases in specific exploration can be observed in both groups however, suggesting that selective DAT inhibition increases specific exploration irrespective of the background strain (63). Finally, because of the use of infra-red beams to detect movement in the BPM, the size of the animal may make a difference in the level of activity recorded. For example, mice that are 20 g may appear less active in terms of distance traveled than mice of 45 g because the latter mouse would break more beams as it moves through space. If size differences are an issue, performance can be analyzed using weight as a co-varying factor.

2.5. Cleaning

The rodent BPM chambers should always be cleaned between testing cohorts using water and wipes (leaving no odor of a cleaning product). The pattern of cleaning should not be repetitive, e.g., do not clean in a circle, spiral figure eight etc., in order to avoid laying an odor trail of the previous animal. When cleaning between testing subjects, clean at least in all areas that the animal can conceivably touch. When time allows (at least 12 h before testing a new cohort), clean all chambers with a cleaning product and after cleaning air out as much as possible. When cleaning between cohorts, clean the whole chamber thoroughly. The human BPM must simply be kept clean and clear of any objects other than those strategically placed there.

3. Procedure

The rat, mouse, and human BPMs follow very similar procedures. The subject is introduced to the chamber with little to no guidance on what behavior is expected. In the rat and mouse BPMs, the subject is placed in the bottom left corner facing the wall and behavior is recorded immediately. In the human BPM, the subject is asked to wait while other equipment is set up and the experimenter will return soon. When the door is closed, the recording of their exploratory behavior begins. The video recording is monitored at all times to ensure the subject is not doing anything that may hurt themselves or damage anything in the room. At the end of the allotted time (rodent BPM has run anything from 15 to 180 min, human BPM 15 min to date only), the subject is removed from the chamber/room.

Exploration in the rodent BPM can be assessed under white, red, or no light conditions. Moreover, the lighting in the holding area of the rodents can be under white or red lighting. Although never directly assessed, the primary difference of white or red lighting in the rodent BPM is the amount of activity in the center of the chamber, decreased vs. increased activity respectively. To date, the human subjects have always been tested under white lighting.

4. Anticipated Results

Certain results can be expected in the rodent BPMs. Within a 60-min session, the subject will habituate to the testing chambers and the level of activity will reduce over time. The spatial scaling exponent d typically increases over time, suggesting that animals start to meander and perform localized investigatory movement as time goes on. Consistent with some of these behaviors, the rate of investigation of objects by humans does not alter over time, while spatial d does increase over time.

As described above, multiple assessments of the same subject in the BPM will result in lower activity and specific exploration levels with increased spatial d and a more rapid within-session habituation compared to BPM-naïve animals. Altering the environment (e.g., sandpapered floors or objects in the holepoke areas) can reinstate initial exploration levels in rats and mice however. Repeated testing analysis has yet to be fully quantified in the human BPM but data are being generated.

5. Discussion

In conclusion, using the BPM to provide a multivariate assessment of exploration is an important example of cross-fostering translational research. Exploratory behavior in rodents is a complex phenotype that is not sufficiently characterized by univariate approaches. Instead, measures that quantify its temporal, spatial, and dynamic organization have proven to be valuable tools to differentiate the contributions of different neural transmitter systems on locomotor and exploratory behavior. Similarly, multivariate approaches to human exploratory behavior have provided powerful insights into the different patterns exhibited by clinical populations that are often difficult to distinguish from one another during acute illness. The identification of these patterns in a cross-species task provides an opportunity to (a) develop better animal models of these disorders, (b) investigate the neural bases of these behaviors, (c) quantitatively assess current and developing treatment effects, and (d) may provide new biomarkers as targets for the development of novel therapeutics.

Acknowledgments

We would like to recognize the invaluable support from the TRIPEC (Translational Research In Psychophysiology, Exploration, and Cognition) group, including Drs. William Perry, Arpi Minassian, Brook Henry, and Martin Paulus. We would also like to thank Virginia Masten, Richard Sharp, and Mahalah Buell for their support. This work has been supported by NIH grants R01 MH071916, R21 MH085221, R01 DA002925, P50 DA026306, and R21 MH091571, as well as by the Veteran's Administration VISN 22 Mental Illness Research, Education, and Clinical Center.

References

1. Paulus M, Geyer MA (1996): Assessing the organization of motor behavior: New approaches based on the behavior of complex physical systems. In: Ossenkopp KP, Kavaliers M, Sanberg PR, editors. *Measuring movement and locomotion: From invertebrate animals to humans*. New York: Chapman & Hall.

2. Drai D, Golani I (2001): SEE: a tool for the visualization and analysis of rodent exploratory behavior. *Neuroscience and biobehavioral reviews*. 25:409–426.

3. Eilam D, Shefer G (1997): The developmental order of bipedal locomotion in the jerboa (Jaculus orientalis): pivoting, creeping, quadrupedalism, and bipedalism. *Developmental psychobiology*. 31:137–142.

4. Ossenkopp K, Kavaliers M, Sanberg PR (1996): Measuring Moevement and Locomotion: From Invertebrates to Humans. New York: Chapman & Hall.

5. Robbins TW (1977): A critique of methods available for the measurement of spontaneous locomotor activity. In: Iversen LL, Iversen SD, editors. *Handbook of Psychopharmacology*. New York, Plenum Press, 7:37–82.

6. Gould TD, Dao DT, Kovacsics CE (2009): The open field test. In: Gould TD, editor. *Mood and anxiety related phenotypes in mice: Characterization using behavioral tests*. New York: Humana Press, pp 1–20.

7. File SE, Wardill AG (1975): The reliability of the hole-board apparatus. *Psychopharmacologia*. 44:47–51.

8. Geyer MA (1990): Approaches to the characterization of drug effects on locomotor activity in rodents. In: Adler MW, Cowan A, editors. *Testing and Evaluation of Drugs of Abuse*. New York: Wiley-Liss, Inc.

9. Geyer MA, Paulus M (1996): Multivariate Analyses of Locomotor and Investigatory Behavior in Rodents. In: Ossenkopp K, Kavaliers M, Sanberg PR, editors. *Mearusing Movement and Locomotion: From Invertebrates to Humans*. New York: Chapman & Hall.

10. Paulus M, Geyer MA (1996): Assessing the Organization of Motor behavior: New Approaches Based on the Behavior of Complex Physical Systems. In: Ossenkopp K, Kavaliers M, Sanberg PR, editors. *Measuring Movement and Locomotion: From Invertebrates to Humans.* New York: Chapman & Hall.

11. Geyer MA, Russo PV, Masten VL (1986): Multivariate assessment of locomotor behavior: pharmacological and behavioral analyses. *Pharmacology, biochemistry, and behavior.* 25: 277–288.

12. Paulus MP, Geyer MA (1993): Three independent factors characterize spontaneous rat motor activity. *Behav Brain Res.* 53:11–20.

13. Eilam D, Golani I (1988): The ontogeny of exploratory behavior in the house rat (Rattus rattus): the mobility gradient. *Developmental psychobiology.* 21:679–710.

14. Kafkafi N, Lipkind D, Benjamini Y, Mayo CL, Elmer GI, Golani I (2003): SEE locomotor behavior test discriminates C57BL/6 J and DBA/2 J mouse inbred strains across laboratories and protocol conditions. *Behavioral neuroscience.* 117:464–477.

15. Krebs-Thomson K, Geyer MA (1998): Evidence for a functional interaction between 5-HT1A and 5-HT2 receptors in rats. *Psychopharmacology.* 140:69–74.

16. Gold LH, Koob GF, Geyer MA (1988): Stimulant and hallucinogenic behavioral profiles of 3,4-methylenedioxymethamphetamine and N-ethyl-3,4-methylenedioxyamphetamine in rats. *The Journal of pharmacology and experimental therapeutics.* 247:547–555.

17. Paulus MP, Geyer MA (1991): A temporal and spatial scaling hypothesis for the behavioral effects of psychostimulants. *Psychopharmacology.* 104:6–16.

18. Paulus M, Geyer MA, Mandell AJ (1991): Statistical mechanics of a neurobiological dynamical system: The spectrum of local entropies (S(a)) applied to cocaine-perturbed behavior *Physica A.* 174:567–577.

19. Paulus MP, Geyer MA, Gold LH, Mandell AJ (1990): Application of entropy measures derived from the ergodic theory of dynamical systems to rat locomotor behavior. *Proceedings of the National Academy of Sciences of the United States of America.* 87:723–727.

20. Lehmann-Masten VD, Geyer MA (1991): Spatial and temporal patterning distinguishes the locomotor activating effects of dizocilpine and phencyclidine in rats. *Neuropharmacology.* 30:629–636.

21. Paulus MP, Geyer MA (1991): A scaling approach to find order parameters quantifying the effects of dopaminergic agents on unconditioned motor activity in rats. *Progress in neuropsychopharmacology & biological psychiatry.* 15:903–919.

22. Adams LM, Geyer MA (1982): LSD-induced alterations of locomotor patterns and exploration in rats. *Psychopharmacology.* 77:179–185.

23. Flicker C, Geyer MA (1982): Behavior during hippocampal microinfusions. IV. Transmitter interactions. *Brain research.* 257:137–147.

24. Bushnell PJ (1987): Effects of scopolamine on locomotor activity and metabolic rate in mice. *Pharmacology, biochemistry, and behavior.* 26: 195–198.

25. Collins C, Laird RI, Richards PT, Starmer GA, Weyrauch S (1979): Aspirin-caffeine interaction in the rat. *The Journal of pharmacy and pharmacology.* 31:611–614.

26. Fink JS, Smith GP (1979): Abnormal pattern of amphetamine locomotion after 6-OHDA lesion of anteromedial caudate. *Pharmacology, biochemistry, and behavior.* 11:23–30.

27. Fitzgerald RE, Berres M, Schaeppi U (1988): Validation of a photobeam system for assessment of motor activity in rats. *Toxicology.* 49:433–439.

28. Gould TJ, Keith RA, Bhat RV (2001): Differential sensitivity to lithium's reversal of amphetamine-induced open-field activity in two inbred strains of mice. *Behavioural brain research.* 118:95–105.

29. Kulkarni SK, Dandiya PC (1975): Influence of chemical stimulation of central dopaminergic system on the open field behaviour of rats. *Pharmakopsychiatrie, Neuro-Psychopharmakologie.* 8:45–50.

30. Meliska CJ, Loke WH (1984): Caffeine and nicotine: differential effects on ambulation, rearing and wheelrunning. *Pharmacology, biochemistry, and behavior.* 21:871–875.

31. Paulus MP, Geyer MA (1992): The effects of MDMA and other methylenedioxy-substituted phenylalkylamines on the structure of rat locomotor activity. *Neuropsychopharmacology.* 7:15–31.

32. Sessions GR, Meyerhoff JL, Kant GJ, Koob GF (1980): Effects of lesions of the ventral medial tegmentum on locomotor activity, biogenic amines and responses to amphetamine in rats. *Pharmacology, biochemistry, and behavior.* 12:603–608.

33. Rao VS, Santos FA, Paula WG, Silva RM, Campos AR (1999): Effects of acute and repeated dose administration of caffeine and pentoxifylline on diazepam-induced mouse behavior in the holeboard test. *Psychopharmacology.* 144:61–66.

34. Marco EM, Llorente R, Perez-Alvarez L, Moreno E, Guaza C, Viveros MP (2005): The kappa-opioid receptor is involved in the

stimulating effect of nicotine on adrenocortical activity but not in nicotine induced anxiety. *Behavioural brain research*. 163:212–218.

35. Marks MJ, Stitzel JA, Romm E, Wehner JM, Collins AC (1986): Nicotinic binding sites in rat and mouse brain: comparison of acetylcholine, nicotine, and alpha-bungarotoxin. *Molecular pharmacology*. 30:427–436.

36. Makanjuola RO, Hill G, Dow RC, Campbell G, Ashcroft GW (1977): The effects of psychotropic drugs on exploratory and stereotyped behaviour of rats studied on a hole-board. *Psychopharmacology*. 55:67–74.

37. Antoniou K, Kafetzopoulos E (1991): A comparative study of the behavioral effects of d-amphetamine and apomorphine in the rat. *Pharmacology, biochemistry, and behavior*. 39:61–70.

38. Gordon D, Beck CH (1984): Subacute apomorphine injections in rats: effects on components of behavioral stereotypy. *Behavioral and neural biology*. 41:200–208.

39. Rebec GV, Bashore TR (1984): Critical issues in assessing the behavioral effects of amphetamine. *Neuroscience and biobehavioral reviews*. 8:153–159.

40. Geyer MA, Russo PV, Segal DS, Kuczenski R (1987): Effects of apomorphine and amphetamine on patterns of locomotor and investigatory behavior in rats. *Pharmacology, biochemistry, and behavior*. 28:393–399.

41. Paulus MP, Geyer MA (1993): Quantitative assessment of the microstructure of rat behavior: I, f(d), the extension of the scaling hypothesis. *Psychopharmacology*. 113:177–186.

42. Geyer MA, Gordon J, Adams LM (1985): Behavioral effects of xylamine-induced depletions of brain norepinephrine: interaction with LSD. *Pharmacology, biochemistry, and behavior*. 23:619–625.

43. Flicker C, Geyer MA (1982): Behavior during hippocampal microinfusions. I. Norepinephrine and diversive exploration. *Brain research*. 257:79–103.

44. Segal DS, Geyer MA (1976): Pre- and post-junctional supersensitivity: differentiation by intraventricular infusions of norepinephrine and methoxamine. *Psychopharmacology*. 50:145–148.

45. Geyer MA, Masten VL (1989): Increases in diversive exploration in rats during hippocampal microinfusions of isoproterenol but not methoxamine. *Physiology & behavior*. 45:213–217.

46. Baumann B, Danos P, Krell D, Diekmann S, Leschinger A, Stauch R, et al. (1999): Reduced volume of limbic system-affiliated basal ganglia in mood disorders: preliminary data from a postmortem study. *The Journal of neuropsychiatry and clinical neurosciences*. 11:71–78.

47. Manji HK, Potter W (1997): Monoaminergic Systems in Bipolar Disorder. In: Young LT, Joffe RT, editors. *Bipolar Disorder: Biological Models and their Clinical Applications*. New York: Dekker.

48. Manji HK, Quiroz JA, Payne JL, Singh J, Lopes BP, Viegas JS, et al. (2003): The underlying neurobiology of bipolar disorder. *World Psychiatry*. 2:136–146.

49. Vawter MP, Freed WJ, Kleinman JE (2000): Neuropathology of bipolar disorder. *Biological psychiatry*. 48:486–504.

50. Young LT, Warsh JJ, Kish SJ, Shannak K, Hornykeiwicz O (1994): Reduced brain 5-HT and elevated NE turnover and metabolites in bipolar affective disorder. *Biological psychiatry*. 35:121–127.

51. Perry W, Minassian A, Paulus MP, Young JW, Kincaid MJ, Ferguson EJ, et al. (2009): A reverse-translational study of dysfunctional exploration in psychiatric disorders: from mice to men. *Archives of general psychiatry*. 66:1072–1080.

52. Carey RJ (1982): Unilateral 6-hydroxydopamine lesions of dopamine neurons produce bilateral self-stimulation deficits. *Behavioural brain research*. 6:101–114.

53. Morita T, Sonoda R, Nakato K, Koshiya K, Wanibuchi F, Yamaguchi T (2000): Phencyclidine-induced abnormal behaviors in rats as measured by the hole board apparatus. *Psychopharmacology*. 148:281–288.

54. Young JW, Goey AK, Minassian A, Perry W, Paulus MP, Geyer MA (2010): The mania-like exploratory profile in genetic dopamine transporter mouse models is diminished in a familiar environment and reinstated by subthreshold psychostimulant administration. *Pharmacology Biochemistry and Behavior*. 96:7–15.

55. Gresack J, Powell S, Geyer M, Poore MS, Coste S, Risbrough V (2010): CRF2 null mutation increases sensitivity to isolation rearing effects on locomotor activity in mice. *Neuropeptides*. 44:349–353.

56. Risbrough VB, Masten VL, Caldwell S, Paulus MP, Low MJ, Geyer MA (2006): Differential contributions of dopamine D(1), D(2), and D(3) receptors to MDMA-induced effects on locomotor behavior patterns in mice. *Neuropsychopharmacology*. 31:2349–2358.

57. Halberstadt AL, van der Heijden I, Ruderman MA, Risbrough VB, Gingrich JA, Geyer MA, et al. (2009): 5-HT(2A) and 5-HT(2 C) receptors exert opposing effects on locomotor activity in mice. *Neuropsychopharmacology*. 34:1958–1967.

58. Halberstadt AL, Lehmann-Masten VD, Geyer MA, Powell SB (2011): Interactive effects of

mGlu5 and 5-HT(2A) receptors on locomotor activity in mice. *Psychopharmacology.* 215(1):81–92.

59. Young JW, Minassian A, Paulus MP, Geyer MA, Perry W (2007): A reverse-translational approach to bipolar disorder: rodent and human studies in the Behavioral Pattern Monitor. *Neuroscience and biobehavioral reviews.* 31:882–896.

60. Henry BL, Minassian A, Young JW, Paulus MP, Geyer MA, Perry W (2010): Cross-species assessments of motor and exploratory behavior related to bipolar disorder. *Neuroscience and biobehavioral reviews.* 34:1296–1306.

61. Halberstadt AL, Koedood L, Powell SB, Geyer MA (2010): Differential contributions of serotonin receptors to the behavioral effects of indoleamine hallucinogens in mice. *Journal of psychopharmacology (Oxford, England).*

62. Ralph-Williams RJ, Paulus MP, Zhuang X, Hen R, Geyer MA (2003): Valproate attenuates hyperactive and perseverative behaviors in mutant mice with a dysregulated dopamine system. *Biological psychiatry.* 53:352–359.

63. Young JW, Goey AK, Minassian A, Perry W, Paulus MP, Geyer MA (2010): GBR 12909 administration as a mouse model of bipolar disorder mania: mimicking quantitative assessment of manic behavior. *Psychopharmacology.* 208:443–454.

64. Amsterdam JD, Newberg AB (2007): A preliminary study of dopamine transporter binding in bipolar and unipolar depressed patients and healthy controls. *Neuropsychobiology.* 55:167–170.

65. Horschitz S, Hummerich R, Lau T, Rietschel M, Schloss P (2005): A dopamine transporter mutation associated with bipolar affective disorder causes inhibition of transporter cell surface expression. *Molecular psychiatry.* 10: 1104–1109.

66. Teicher MH (1995): Actigraphy and motion analysis: new tools for psychiatry. *Harvard Review of Psychiatry.* 3:18–35.

67. Teicher MH, Lawrence JM, Barber NI, Finklestein SP, Lieberman H, Baldessarini RJ (1986): Altered locomotor activity in neuropsychiatric patients. *Progress in Neuropsychopharmacology and Biological Psychiatry.* 10: 755–761.

68. Wolff EA, Putnam FW, Post RM (1985): Motor activity and affective illness. *Arch Gen Psychiatry.* 42:288–294.

69. Pierce K, Courchesne E (2001): Evidence for a cerebellar role in reduced exploration and stereotyped behavior in autism. *Biological psychiatry.* 49:655–664.

70. Minassian A, Henry BL, Geyer MA, Paulus MP, Young JW, Perry W (2009): The quantitative assessment of motor activity in mania and schizophrenia. *Journal of affective disorders.* 120:200–206.

71. Perry W, Minassian A, Henry B, Kincaid M, Young JW, Geyer MA (2010): Quantifying over-activity in bipolar and schizophrenia patients in a human open field paradigm. *Psychiatry research.* 178:84–91.

72. Henry BL, Minassian A, van Rhenen M, Young JW, Geyer MA, Perry W (2011): Effect of methamphetamine dependence on inhibitory deficits in a novel human open field paradigm. *Psychopharmacology.* 215(4):697–707.

73. Pini S, de Queiroz V, Dell'Osso L, Abelli M, Mastrocinque C, Saettoni M, et al. (2004): Cross-sectional similarities and differences between schizophrenia, schizoaffective disorder and mania or mixed mania with mood-incongruent psychotic features. *Eur Psychiatry.* 19:8–14.

74. Lhermitte F (1983): 'Utilization behaviour' and its relation to lesions of the frontal lobes. *Brain.* 106 (Pt 2):237–255.

75. Goto Y, Yang CR, Otani S (2010): Functional and dysfunctional synaptic plasticity in prefrontal cortex: roles in psychiatric disorders. *Biological psychiatry.* 67:199–207.

76. Arnone D, Cavanagh J, Gerber D, Lawrie SM, Ebmeier KP, McIntosh AM (2009): Magnetic resonance imaging studies in bipolar disorder and schizophrenia: meta-analysis. *Br J Psychiatry.* 195:194–201.

77. Koenigs M, Grafman J (2009): The functional neuroanatomy of depression: distinct roles for ventromedial and dorsolateral prefrontal cortex. *Behav Brain Res.* 201:239–243.

78. Bechara A, Damasio AR, Damasio H, Anderson SW (1994): Insensitivity to future consequences following damage to human prefrontal cortex. *Cognition.* 50:7–15.

79. Young JW, van Enkhuizen JE, Winstanley CA, Geyer MA (in press): Increased gambling behaviour in a mouse model of mania. *Journal of Psychopharmacology.*

80. Vivometrics (2002): The LifeShirt System™. Ventura, CA.

81. Henry BL, Minassian A, Paulus MP, Geyer MA, Perry W (2010): Heart rate variability in bipolar mania and schizophrenia. *Journal of psychiatric research.* 44(3):168.

82. Eilam D (2003): Open-field behavior withstands drastic changes in arena size. *Behav Brain Res.* 142:53–62.

83. Paulus MP, Geyer MA (1997): Environment and uncoditioned motor behavior influences of drugs and environmental geometry on behavioral organization in rats. *Psychobiology.* 25:324–337.

Chapter 3

Telemetry in Mice: Applications in Studies of Stress and Anxiety Disorders

Larry D. Sanford, Linghui Yang, and Laurie L. Wellman

Abstract

Telemetry in mice can be a powerful tool for assessing physiological responses to emotional stimuli. It can provide continuously recorded data over the course of the stress response and can be incorporated into established behavioral paradigms and procedures in a relatively seamless fashion. Our lab has been using telemetry in conjunction with behavioral paradigms to study the relationship between anxiety and sleep in mouse models for several years. This chapter discusses some of the current, commercially available telemetry systems, how telemetry compares to other recording methods and a number of its advantages and disadvantages. A crucial requirement for implementing telemetry in mice is successful surgery leading to fully recovered, healthy animals and ultimately the recording of the desired signals. We provide detailed surgical procedures for implanting mice for telemetry. We also provide an example experimental protocol integrating telemetry with behavioral training and sleep recording and discuss solutions for problems that may be encountered in implementing telemetry in mice.

Key words: Anxiety, Cable recording, Fear, Mouse strain differences, Sleep, Stress, Surgery, Telemetry

1. Background

Various training paradigms and behavioral assays have been developed to induce and measure emotional responses in rodents. These paradigms and assays have typically been used for examining overt behaviors during the experimental session. There can also be significant perturbations in stress and emotion-related parameters other than behavior. Common examples include heart rate and blood pressure (1–4) and body temperature (3–5). Arousal and sleep are also strongly impacted by stress and emotion, and there is a strong link between disturbed sleep and mood and anxiety disorders (6, 7).

Todd D. Gould (ed.), *Mood and Anxiety Related Phenotypes in Mice: Characterization Using Behavioral Tests, Volume II*,
Neuromethods, vol. 63, DOI 10.1007/978-1-61779-313-4_3, © Springer Science+Business Media, LLC 2011

Our lab utilizes a variety of stress paradigms as well as both behavioral methods and physiological recording to examine emotional responses in mice. Concurrent, continuous measurement of neurobiological and physiological variables typically requires animals to be surgically implanted for recording via tethers or cables or via telemetry. Thus, incorporating recordings of physiological variables into studies of emotional behavior and stress may be seen as a daunting task because of the technical issues. Compared to nonimplanted animals, there also may be additional environmental and experimental controls that have to be considered in order to obtain viable data and the required data handling and analysis procedures may be very labor intensive. The following is a discussion of some of the factors that need to be considered in using telemetry in mice and procedures we have found valuable in using telemetry in conjunction with behavioral paradigms.

2. Telemetry Basics

Telemetry has been used to record a number of variables in mice including heart rate (8–11), blood pressure (8, 10, 12), electroencephalogram (EEG) (13–17), electromyogram (EMG) (16, 17), movement (10, 11, 13), and body (9, 11) and brain (18) temperature. Based on acquired signals (e.g., EEG, EMG, and/or activity), telemetry has also been used to determine sleep-wake states in mice (13–17), rats (19–25), and other rodents (26).

One of the main arguments in favor of the use of telemetry is the reduction of stress compared to conventional measurement techniques (12, 27). This rationale for using telemetry is particularly relevant for studies which require the animal to be subjected to a stressful situation as part of an experimental protocol. For instance, in our work, we focus on the impact of aversive conditioning and stress on sleep. Telemetry allows the animals to be transferred from home cage to conditioning apparatus and back to the home cage very easily without being subjected to the potential extra stress of disconnecting and reconnecting recording cables for each manipulation. The mice also have relatively more freedom of movement within their home cages compared to animals recorded via cables. In addition, data generally can be collected continuously without the need of special animal care (27, 28).

Telemetry eliminates the need for restraint which can be a component of certain noninvasive recording procedures (e.g., tail-cuff manometry for monitoring blood pressure) and which can be a source of experimental artifact and inter-animal variability (28). The impact of restraint has been demonstrated for various measures including increases in body temperature, heart rate and blood pressure, plasma levels of epinephrine, norepinephrine, and corticosterone (for reviews see (12, 27)). Even brief restraint can produce significant alterations in subsequent behavior and physiological parameters. For instance, 5 min of manual restraint such as that required for administering microinjections can produce significant alterations in subsequent sleep that may not fully habituate across repeated sessions (29). Changes in core temperature can begin within 10 s of the onset of restraint and temperature can increase as much as 2°C during the restraint period (30, 31).

The advent of commercially available systems has allowed the use of telemetry to become relatively widespread in mice. A full discussion of the various types of telemetry systems is beyond the scope of this chapter. However, transmitters suitable for use in mice are supplied by companies such as Data Sciences International, (DSI, St. Paul, Minnesota) and MiniMitter (Bend, Oregon). DSI offers a number of implantable telemetry transmitters that can be used in mice, and collecting data with these transmitters will be the focus of the procedures discussed in this chapter. These include the PhysioTel® ETA-F10, ETA-F20, and F20-EET models that vary in physical size, the number and types of signals that can be obtained and the duration of battery life. For example, the ETA-F10 has a volume of 1.1 cc and weighs 1.6 g whereas the ETA-F20 has a volume of 1.9 cc and weighs 3.9 g. Both can record a biopotential (e.g., EEG or EMG), temperature, and provide a measure of relative activity. Battery life in the ETA-F10 is 2.0 months and in the ETA-F20 is 4.0 months. The F20-EET can record two biopotentials, temperature, and activity and has a battery life of 1.5 months. It is of approximately the same size and weight as the ETA-F20. Other types of transmitters, not listed here, are those that collect pressure information (e.g., arterial, ocular, uterine, penile, bladder, pleural, and left ventricular) or some other combination of temperature, activity, and biopotential data.

MiniMitter offers transponders to record temperature, gross motor activity, and heart rate data via telemetry. An advantage of this system is that the transponders are externally powered and thus do not require batteries. Without the requirement for battery power, the transponders can be smaller and lighter and do not require refurbishing to change the battery. For example, their smallest transponder capable of recording heart rate, temperature, and activity is the model G2 HR E-Mitter® which is 19.5×3.5 mm and weighs only 1.5 g. These transponders are not capable of recording biopotentials. However, both battery-free and battery powered models capable of recording core body temperature (32) or temperature and activity (33, 34) are sometimes implanted in animals that are also recorded via cables for signals such as the EEG and EMG.

3. Telemetry vs. Cable Recording

The major competing methodology for recording physiological signals in mice is the use of an attached cable. This approach has long been used for obtaining a variety of physiological and neural information through the surgical implantation of sensors or electrodes to record the parameter of interest. In many cases, a recording cable is attached via a plug permanently affixed to the research animal's skull, though the cable itself also may be permanently attached. The cable may be attached to the animal via a harness in certain applications. This is a valuable technique for collecting physiological information, though cables can restrict movement and potentially influence behavior in mice (35). In addition to the simple effects that cable weight and stiffness could have on behavior are the more complex effects they may have on the amount of energy and force required to overcome inertia and to provide the twisting force necessary for the mouse to initiate and continue rotational movement. These may vary with the position of the animal in its cage and with the angle of the cable (35). Relative freedom of movement can be increased by the use of a commutator that allows the cable to turn without twisting; however, proper implementation in mice can be challenging.

In a previous study (35), we found that cables of increasing weight and decreasing flexibility produced relatively linear decreases in activity. Heavier, less flexible cables also produced increases in sleep, but alterations in sleep showed more variability across cables. Perhaps most importantly, variations in cables produced a differential impact on sleep across mouse strains. The relatively small physical size of mice suggests that they may be particularly sensitive to variations in cable weight and flexibility. When unrestrained, mice can be relatively faster moving, with vertical movement such as climbing on the watering tube and cage bars, suggesting that that cables could limit their movement to a relatively greater degree than in larger animals. Cabling also could be a factor in studies of gene-altered animals, which may have motor or other deficits that affect their activity (36).

To determine how activity and sleep (nonrapid eye movement sleep (NREM) and rapid eye movement sleep (REM)) recorded via telemetry would compare to that recorded via cable, we compared results for BALB/cJ (C) and DBA/2J (D2) mice obtained using telemetry (13) with results obtained with a commercially (Plastics One Inc.) available commutator (SL6C) and lightweight cable (363-SL/6 cable, without vinyl and protective spring covering (gross weight: 4.5 g; resting weight: 1.5 g)). Approximate cable resting weights were determined by placing the free end of the cable on a scale placed at head stage height. Based on total

Table 1
Average values of representative parameters of activity and sleep recorded by a lightweight cable and telemetry in BALB/cJ and DBA/2J mice

	BALB/cJ		DBA/2J	
	Cable	Telemetry	Cable	Telemetry
Travel distance (PCRB, %)				
T	−32	−43	−19	−34
L	−27	−64	−3	−38
D	−32	−32	−23	−34
Total NREM (min)				
T	560	570	595	577
L	390	315	409	392
D	170	255	186	184
NREM duration (min)				
T	5.8	4.8	4.8	6.5
L	6.2	5.3	5.3	7.5
D	5.4	4.4	3.9	5.1
Total REM (min)				
T	58	61	70	66
L	40	26	50	43
D	18	35	19	23
REM duration (min)				
T	1.2	1.2	1.4	1.5
L	1.2	1.1	1.5	1.5
D	1.3	1.2	1.3	1.4

PCRB, percentage changes relative to presurgery baseline obtained by (baseline − treatment)/baseline*100. T, L, and D indicate total 24 h, and 12 h light and dark periods, respectively. Reprinted with permission from (35)

distance traveled (TD PCRB; considered as a percentage change relative to baseline travel distance exhibited prior to surgery), mice recorded via telemetry typically had greater reductions in activity than those recorded with a lightweight cable (Table 1). However, mice in both strains recorded with telemetry showed less reduction in activity than the 60–80% reductions in TD PCRB in recordings obtained with medium (resting weight: 2.2 g) and heavy (resting weight: 3.0 g) cables. The head pedestal, required for attaching the cable, alone (total weight about 1.6 g, including electrodes, plug, and dental cement) significantly reduced activity in the more active C mice compared to their presurgery activity levels. By comparison, pre- and postsurgery activity did not significantly differ in less active D2 mice suggesting that activity in the C strain was more sensitive to the weight of the implant alone.

Total NREM and REM were quite similar over the total 24-h recording period in C and D2 mice with both recording techniques.

Differences in total NREM across methods were 10 min for C mice and 18 min for D2 mice, and differences in total REM across methods were only 3–4 min in both strains. D2 mice also showed similar light and dark period values with both methods. Differences in REM duration between methods were no larger than 0.1 min in either strain. Compared to telemetry, NREM durations with cable recording were 0.9–1 min longer in C mice and 1.2–2.2 min shorter in D2 mice.

The lesson from this comparison is that any invasive method of recording in mice has the potential to impact behavior and neurobiological function. All things considered, lighter and less obtrusive methods are better. However, the characteristics that make inbred strains and gene-altered lines of mice good models may also interact differently with the properties of the recording methods. These factors have to be considered when determining which recording method to use.

4. Potential Disadvantages of Telemetry

The advantages of telemetry need to be carefully weighed with its disadvantages compared to other recording techniques. Compared to noninvasive recording, the use of telemetry typically requires surgical implantation of transmitters or transponders. Thus, personnel must be trained to perform the implant surgery as well as to monitor and take care of the animals during recovery. As indicated below, the recovery period can be quite long, and thus more expensive, as the animal must be maintained and cared for before useful data can be obtained.

Compared to simple cable recordings, surgical trauma for telemetry implants can be greater. In addition, most currently available transmitters are large relative to the body size of mice and can have an impact on behavior, just as cables can. The equipment required for telemetry can be more costly, both in the initial startup and over time, as battery powered transmitters must be refurbished periodically.

Lastly, when considering implementing telemetry in a lab that does not have previous experience with other types of recording (e.g., via cables or noninvasive methods), it should be recognized that telemetry (especially multichannel transmitters) can generate a wealth of data which can require time consuming and complex procedures for data handling and analysis. They may also require specialized training (such as scoring sleep, conducting a Fourier analysis of the EEG, or analyzing heart rate variability) before useful data can be obtained and/or interpreted.

5. Application of Telemetry to Behavioral Paradigms

The integration of telemetry recordings with behavioral paradigms can be relatively seamless with some limitations. The primary limitation is the restricted operating range of most current commercially available transmitters that are suitable for mice. For example, transmitters (e.g., PhysioTel® ETA-F10, ETA-F20, and F20-EET) available from Data Sciences have a maximum cage size of $33 \times 33 \times 14$ cm that is more appropriate for home cage recordings. This limitation can be circumvented by linking multiple receiver platforms. For example, multiple receivers could be combined into a single receiving system to extend the range of reception to cover a large area, such as an open field. Thus, it would be possible to integrate telemetric recordings into a variety of behavioral paradigms.

Telemetry can also easily be used to record in the period immediately after training or testing sessions when an animal has been returned to its home cage. We have used this approach to examine the poststress period in a variety of paradigms including cued (37, 38) and contextual (39, 40) fear, controllable and uncontrollable footshock (41), and the open field (40). It is also useful for paradigms in which stimuli can be presented in the home cage, e.g., fear-conditioned auditory cues (37, 38) and novel objects (42).

6. Implantation Surgery and Recovery

The primary requirement for implementing telemetry in mice is successful surgery leading to fully recovered, healthy animals and ultimately the recording of the desired signals. A major concern for surgery in mice is the size of currently available telemetry transmitters. Surgery trauma is greater for transmitters placed intraperitoneally and longer recovery periods may be needed after surgery for certain types of implants. Thus, other than the type of data one desires to collect and the appropriate transmitter to use, one of the most significant considerations is the size of the subject. DSI suggests minimal animal weights of 20 g for implantation of the ETA-F20 and F20-EET and a minimal animal weight of 17 g for the ETA-F10. These weights are good starting points; however, the success rates of implants also can vary with mouse strain and with surgical skill. Some strains (e.g., C57BL/6) are more robust, and an excellent surgeon and excellent postsurgery care can provide good results with smaller animals whereas less robust strains may require larger sizes for success.

Proper surgical technique, postsurgery care, and adequate recovery time are crucial for obtaining the best results in mice implanted for recording via telemetry. Following is a detailed protocol for the intraperitoneal implantation of a telemetry transmitter (DSI ETA-F20 or similar) using procedures we have utilized and refined for over 10 years. Implanting the transmitter subcutaneously on the back is also possible (43, 44), though in our experience this type of implant is more prone to problems in long-term studies that are carried out over the course of several weeks or months. Over time, the skin, which typically is thin in mice, above the transmitter can degrade though this may not be a significant concern for short-term experiments.

We use isoflurane as the surgical anesthetic and we also cover the cranial implant with dental acrylic which results in an exposed mound. Others may prefer to use injectable anesthetics and to close the skin over the cranial mound for either aesthetic reasons or to lessen the potential for infection at the incision site. There may also be variations in acceptable anesthesia, analgesia, antibiotic regimens, or other procedures across institutions. However, we have had minimal problems with infection at the incision sites. We also often pair telemetry recordings with intracranial injections (either intracerebroventricular (ICV) or locally into specific brain regions) which necessitates a cranial mound to support the implanted injection cannula(e). A description of the integration of this surgical procedure with that required for the telemetry implant is included below though others have performed the implantation of an ICV cannula as a separate surgical procedure (15).

The procedures described in this section should be easily used or adapted by researchers with experience in animal surgery. They also should be a good starting point for researchers with minimal experience who are encouraged to obtain training in basic surgical skills. Note that the surgery requires two people: the surgeon who conducts the surgery and aseptic procedures and an assistant who monitors the animal and performs tasks such as mixing dental cement. Table 2 provides a list of instruments and supplies required to complete the procedures described below.

6.1. Preoperative Procedures

Preoperatively, the mouse should be prepared for surgery by providing it with analgesic. For each mouse, we make available ad lib access to ibuprofen (30 mg/kg; 4.75 ml children's Motrin® (or similar) in 500 mL water) in its water bottle for 24–48 h prior to surgery and continuing for at least 72 h postsurgery. Antibiotics (gentamicin 5–8 mg/kg IM and potassium penicillin 100,000 IU/kg) are given subcutaneously preoperatively and may be used additionally if infection occurs. If intracranial cannulae are to be implanted, dexamethasone (0.4 mg (0.2 mL total dosage)) may be administered preoperatively to reduce brain swelling.

Table 2
Instruments and supplies used for implanting a telemetry transmitter (ETA-F20 or similar)

General instruments and supplies	Scale, cautery, UV light, water blanket, surgical drill, #9 drill bit, #14 drill bit, 10 mL syringe, 1 mL syringes, 3 M ESPE Scotchbond dual cure dental adhesive liquid system, dental acrylic, tape, sutures, scalpel blade (#10 or similar), sterile saline
For surgeon	Lab coat, surgical gloves, mask, scrub, sterile towels
Surgical area	Electric clippers, betadine soaked gauze, 70% alcohol soaked gauze
Drugs	Antibiotic ointment, Paralube® ocular ointment, gentamycin (0.005 mg/gWT), potassium penicillin (100 IU/gWT)
Anesthesia system	Vaporizer, induction chamber, isoflurane, oxygen
Autoclave pack	Drapes, 2×2 gauze, cotton swabs, stainless steel tube (trocar), needle point forceps, curved point forceps, micro-dissecting forceps, micro-dissection scissors, dull scissors, scalpel handle, needle driver (holder)
Cold sterilized items	Screw driver, scissors, anchor screws, 14 gauge needle
Surgical board	Covered with 4×4 gauze
Forms	Anesthesia monitor log, animal health action and immediate recovery/pain forms
Extras for cannula implants	18 gauge needles, gel foam, dexamethasone (0.4 mg), dental pick (in the autoclave pack), stereotaxic apparatus, cannulae and dummy cannulae (in cold sterile)

For the surgical implantation of electrodes and transmitter, the mouse is initially anesthetized with isoflurane (5% induction, 2% maintenance (note: the requirement for maintaining anesthesia may vary depending on how the animal responds)) in an induction chamber. In our case, this is a plastic cylindrical chamber (diameter 12 cm and height 11 cm). Following attainment of surgical anesthesia, the head and abdomen sites are clipped. Eye lubricant (Paralube® sterile ocular ointment) is then applied to the eyes.

6.2. Operative Procedures

After it reaches a surgical plane of anesthesia, the animal is placed on a surgical board in the prone position and the head is cleaned using gauze soaked in Betadine® (10% povidone-iodine) followed by cleaning using gauze soaked in 70% alcohol. Afterward, an incision approximately 1 cm in length is made using microdissection scissors. A scalpel blade is used to scrape away the membrane of connective tissue covering the skull. A cautery may be used to stop excessive bleeding. A drop of phosphoric acid is placed on the skull and a cotton swab is used to "scrub" the skull (avoiding the skin) to ensure all connective tissue has been removed. The skull is then rinsed twice with saline. The head incision is covered with sterile gauze, and the animal is turned to the supine position.

During surgery, operative tissues are kept moist with applications of saline warmed to approximately normal body temperature 37°C (98°F).

The abdominal skin is cleaned using gauze soaked in Betadine® and in 70% alcohol as described above. An incision is made in the skin (approximately 2.0 cm in length) using scissors, and blunt dissection is used to separate the skin from the underlying muscle. An incision is then made in the muscle along the linea alba (white center line). The transmitter is placed in the abdominal cavity, and the wires are routed through a separate hole in the muscle with the aid of a 14-ga. needle. The transmitter is then sutured to the abdominal wall as the muscle is closed. The abdominal incision is covered with sterile gauze and the animal is turned on its side.

A small stainless steel tube (this must be large enough such that the wires from the transmitter can easily slide through it) is used as a trocar to create a subcutaneous path from the head incision to the abdominal incision. Marcaine® (0.5%) is administered along the trocar tract. The wires are routed through the tube and it is removed. The animal is returned to the supine position for closure of the abdominal skin incision. This is done using a simple interrupted suture pattern. Once closed, antibiotic ointment is placed on the incision along with sterile gauze.

The animal is returned to the prone position and three holes are drilled in the skull. One hole (posterior to lambda) is made with a #14 drill bit for placing an anchor screw (stainless steel; 0–80×3–32). Two additional holes (one in the left anterior quadrant and the other in the right posterior quadrant around bregma) (A:1.0, L:1.0; P:3.0, L:3.0) are made with a #9 drill bit for the lead wires from the transmitter. The anchor screw is inserted and left raised off the skull. This screw is placed to aid in permanently affixing the implant to the skull. The wires are cut to length (use dull scissors as the wire ruins surgical scissors) and the insulation is removed approximately 1.0 mm from the cut end. The wire tips are bent into a "Z" shape and then placed into the hole such that the "Z" is completely under the skull. The wires are secured to the skull using 3 M dual cure dental glue (activated by UV light (Patterson Brand, Model TCL490 Plus, visible light curing unit)). Dental acrylic is placed over the wires and exposed skull to finish the head cap.

Cannula(e) for the local or ICV administration of drugs or other compounds can be placed during the same surgical procedure with modifications to account for stereotaxic placement of the cannulae. For ICV microinjections, a 1-mm hole is drilled 1.0 mm lateral and 0.5 mm posterior to the Bregma. The dura is then pierced with a dental pick. Afterward, the tip of a 26-gauge stainless steel infusion cannula (e.g., PlasticsOne, part number 315 G with custom length for use in mice) is stereotaxically placed 2.0 mm below the skull surface in the lateral ventricle. Gelfoam is carefully inserted around the cannula to prevent dental cement from touching the brain. The cannula is then secured to the skull with dental cement and a stylus (dummy cannula) is inserted to maintain patency. Simple variations of this procedure can be used to place cannulae for local delivery of drugs in specific brain regions. Note also, that the position of the EEG and anchor screws may need to be moved to accommodate certain cannula placements.

6.3. Postoperative Period

Once an animal is taken off anesthesia, it is returned to its home cage and kept warm by an under cage (water circulating) heating pad. We place the animal on a clean paper towel to prevent the abdominal incision from coming in contact with the bedding material while also ensuring that the animal does not aspirate the bedding while coming out of anesthesia. We also administer warmed sterile saline subcutaneously (0.5–1.0 mL depending on body weight) to compensate for intraoperative fluid loss. Initially and every 15 min for at least 2 h postoperatively, a record is made of the animal's coloring, returning reflexes, and pain level (posture, vocalization, etc.). The time it took the animal to wake up and ease of recovery is also recorded.

In addition, we have found that the following procedures are helpful for ensuring successful implantations and recovery. We feed the animal highly palatable food (e.g., Grapenut® cereal) for 3 or more days both pre- and postsurgery. Presurgery feeding allows the mice to become accustomed to the novel food. We also place the transmitter carefully so that it does not impinge on the bladder or interfere with the diaphragm and then densely sutured the incision in the abdomen. A topical antibiotic is applied to the incisions.

6.4. Postsurgery Recovery Period

An adequate postsurgery recovery period is essential for obtaining data that are not potentially affected by the recovery process. At present, we allow a minimum of 21 days for the mice to recover from an intraperitoneal implant prior to beginning baseline sleep studies and typically 28 days prior to beginning any experimental manipulation. Others report periods ranging from 28 days (14) up to 8 weeks (45) prior to beginning experiments. These may seem like long recovery periods, but we have found that the longer recovery periods provide more consistent data across animals. Fully recovered and stabilized animals are important to insure that stressful manipulations or behavioral tests used to induce or assess emotional state are not interacting with the recovery process.

Studies using telemetry in mice report a large range of recovery times allowed before experiments are started, and the time required for postsurgery stabilization may vary with the parameter of interest and surgical site. The best course of action is to empirically verify stabilization prior to beginning studies. For instance, telemetrically recorded heart rate in mice with subcutaneously implanted transmitters reportedly stabilized 4 days after implantation (compared to 24 days postsurgery), but heart rate variability required at least 10 days to stabilize (44). Recovery periods may also vary with species. Rats with subcutaneous transmitter showed stabilized NREM and total sleep amounts within 2–3 days (compared to sleep at 15 days postsurgery) whereas REM amounts may require 7 or more days (20).

7. Sample Protocol for Examining the Effects of Controllable and Uncontrollable Stress

7.1. Subjects

We primarily use inbred strains in our studies and have conducted most of our work in two commonly used mouse strains (C57BL/6J (B6); BALB/cJ (C)). These strains differ significantly in behavioral measures of anxiety, and in the effect of shock training and fearful cues and contexts on sleep. Behavioral data suggest that B6 mice are a less "anxious" or reactive strain, and that C are a more "anxious" or reactive strain (46). Indeed, C mice have been suggested to exhibit pathological anxiety (47) whereas B6 mice show an intermediate phenotype in most behaviors (36) and are often used in strain comparisons.

7.2. Experimental Procedure

The paradigm we use for studies of the effects of controllable and uncontrollable stress on sleep is a variant of the yoked escapable and inescapable footshock training similar to that used in studies of learned helplessness in mice (48).

1. The mice are implanted for recording via telemetry and allowed to completely recover as described above. When the animals are not on study, the transmitter is inactivated to preserve battery life. The recording room is kept on a 12:12 light:dark cycle and ambient temperature is maintained at $24.50 \pm 0.5°C$. Cages are changed 2 days prior to recording onset and then timed such that they do not impact subsequent phases of the study. Two days are allowed to provide sufficient time for the mice to construct new nests (the mice are provided with a new Nestlet® (AnCare, Bellmore, NY) at each cage change).

2. A minimum of two days of uninterrupted, baseline sleep is recorded for each mouse. For recording, the mice are housed in individual cages with food and water available ad libitum. Individual home cages are placed on a DSI telemetry receiver (RPC-1), and the transmitter is activated with a magnetic switch. Data acquisition is conducted using a configuration file containing preentered information regarding the transmitter, its calibration, and type of signals to be collected. Setting up

the configuration is carried out using instructions in the DSI system manual.

3. After baseline recording is completed, the mice are trained with escapable and inescapable footshock using a yoked control design. Training is conducted in a shuttlebox with a grid floor for presenting shock (Coulbourn Instruments, Model E10-15SC). The mice are randomly assigned to one of three conditions (escapable shock; inescapable shock; no shock). Mice receiving escapable shock are presented with 20 shocks. To escape shock, the mice must learn to enter the nonoccupied chamber, which terminates the shock. Mice in the inescapable shock condition receive the same amount of shock. Mice in the no shock condition are placed in the shuttlebox and allowed to freely move from one chamber to the other with no shock presentations.

4. Training takes place in a room separate from the colony/ recording room. The mice are gently transported in their home cages to the training room and placed in one side of the shuttlebox. The mice are allowed to freely explore the shuttlebox for 5 min after which they are presented with 20 footshocks (0.5 mA, 5.0 s maximum duration) at 1.0 min intervals. Five minutes after the last shock, the mice are returned to their home cages. Training is conducted during the same hour if multiple sessions are conducted across days. The chamber is thoroughly cleaned with diluted alcohol prior to each conditioning session.

5. Training parameters are controlled by computer. We use Coulbourn Graphic State software running on a personal computer to control the administration and timing of footshock to ensure that mice in the escapable and inescapable footshock conditions receive the same duration of footshock. Footshock is produced via Coulbourn Precision Regulated Animal Shockers (Model E13-14) and administered via grid floors in the shuttleboxes.

6. Following training, the mice are returned to the colony/ recording room and are monitored for the subsequent 20 h. For experimental paradigm changes or studies in new strains, recording is continued for an additional 24 h to follow the time course of the impact of the stress and to determine whether baseline sleep is reestablished on the posttraining day.

7. Six to seven days following training, the mice are returned to the shuttlebox and allowed to freely explore for the same amount of time as required for the training sessions. The animals are videotaped during this time for subsequent scoring of behavioral freezing. These tests take place at the same circadian time as the shock training.

8. Following exposure to context, the mice are recorded for 20 h on the day of exposure and for an additional 24 h, if needed.

7.3. Data Collected

1. *Escape Data*: We record the number of successful escapes (the number of times an animal enters the safe chamber prior to the full duration of shock being given) and the duration of shock received. A consistent improvement in escape latencies (a reduction in the duration of received shock) across training demonstrates that the animal has learned the escape response.

2. *Freezing*: Freezing is scored in 5-s intervals over the course of the initial 5-min period prior to shock training and the context reexposure session. From these data, the percentage time spent in freezing is calculated (FT%: freezing time/observed time $\times 100$) for each animal for each observation period.

3. *Telemetry Data*: Telemetric data include the EEG (and EMG with the appropriate transmitter model), a relative measure of activity and core body temperature. These signals can be used to provide relevant information on their own. For example, the EEG can be subjected to a Fourier analyses and the temperature measure can be used to assess stress-induced hyperthermia. With appropriate training, they can also be used to derive basic sleep parameters such as total sleep time, number of NREM and REM episodes, number of arousals, and duration of NREM and REM episodes.

8. Experimental Design and Recording Considerations

8.1. Husbandry and Housing

Simple husbandry requirements such as routine changes of an animal's home cage can be stressful. In fact, transferring an animal to a clean cage and the introduction of novel objects (even those such as used for environmental enrichment) into the cage can be used as mild stressors for experimental purposes and can produce significant alterations in activity (49, 50), heart rate and blood pressure (51), and sleep and wakefulness (49, 50, 52–55). Animals also do not appear to habituate to these procedures (51, 56) and responses can vary across inbred strains (49, 50) and gene-altered mice (53–55). Thus, it is very important that the timing of behavioral procedures and recording session take into account the amount of time since the last cage change.

8.2. Effects on Activity

As discussed above, mice intraperitoneally implanted with transmitters may show significantly decreased home-cage activity compared to presurgical levels (13). Intraperitoneally implanted transmitters have also been reported to decrease running wheel activity (9, 10). This suggests that while current telemetry recording devices may allow a greater freedom of movement for certain activities, the overall level may still be lower than prior to implantation. The reduction in activity should be carefully considered when using implanted animals in behavioral tests as many tests of emotion have a significant motor component (36, 46, 57, 58). Unfortunately, the amount of impact on activity can vary with mouse strain (13) adding a potential confound that must be considered.

8.3. Survival Rates

Adult male mice, of appropriate size, are highly tolerant of the implant surgery. In earlier work we had a mortality rate of 9.3% between 3 and 9 days after surgery (13). These rates are similar to the survival rate of over 90% reported for telemetric implants used for recording blood pressure and heart rate in B6 mice (10). With good surgical technique and postsurgical care, it is possible to increase the number of viable implants and our current mortality rate is around 7%.

For intraperitoneal implants, proper placement and securement of the transmitter is crucial to prevent ileus and potential problems with bladder and diaphragm function. Interestingly, D2 mice had higher mortality (16.7%) that could be associated with the strain's naturally lower activity, which could possibly lead to more chances of having secondary abdominal complications (13).

8.4. Signal Quality

Sources for decreased signal quality can come from a failing battery, degradation in the interface between the animal and the transmitter, signal dropout when the animal moves outside the range of the receiver, or environmental factors.

Problems with the battery and implant are best prevented rather than remedied. For the battery, turn off the transmitter when it is not in use and keep accurate notes as to how long it has been in use. By comparing these notes to the projected life of the battery, one can obtain a reasonable estimate of whether a battery will last through a study. Using these simple steps, we routinely implant transmitters in multiple animals before having them refurbished. With respect to the animal interface, repair may be possible in certain cases. However, firmly affixing the lead wires as described in the surgical section should provide adequate signal quality for the life of most studies.

If signal dropout occurs, the most likely source is that the animal cage or test area is too large and allows the animal to move outside the range of the receiver. Environmental sources of signal interference may include "cross-talk" between two or more receivers and their respective transmitters or electromagnetic interference. Detailed instructions for testing for and dealing with these problems should be included in system manuals. Solutions may be as simple as using a smaller cage for signal dropout or moving receivers farther apart for cross-talk. They may also be more complicated such as shielding the receivers and cage or making other changes to protect the system from electromagnetic interference. Fortunately, these types of problems can typically be solved permanently unless there is a subsequent change in the recording environment.

Our lab has two telemetry systems currently running (one in a fully shielded room and the other in an unshielded room) that have provided noise free recordings for several years.

8.5. Conclusion

Telemetry can provide physiological recordings to complement data obtained in behavioral tests. It can provide continuously recorded data over the course of the stress response. It also has the advantage in that telemetered animals can be tested using established behavioral paradigms and procedures in a relatively seamless fashion. Drawbacks include the potential need for lengthy surgery recovery times and the fact that the implanted transmitter may significantly alter activity levels. However, with proper consideration of these factors, telemetry can be a valuable tool in the use of mouse models in research on mood and anxiety disorders.

Acknowledgments

This work was supported by NIH research grants MH61716 and MH64827.

References

1. Hoppe CC, Moritz KM, Fitzgerald SM et al (2009) Transient hypertension and sustained tachycardia in mice housed individually in metabolism cages. Physiol Res 58:69–75

2. Lee DL, Leite R, Fleming C et al (2004) Hypertensive response to acute stress is attenuated in interleukin-6 knockout mice. Hypertension 44:259–263

3. Meijer MK, Sommer R, Spruijt BM et al (2007) Influence of environmental enrichment and handling on the acute stress response in individually housed mice. Lab Anim 41:161–173

4. Meijer MK, Spruijt BM, van Zutphen LF et al (2006) Effect of restraint and injection methods on heart rate and body temperature in mice. Lab Anim 40:382–391

5. Vinkers CH, van Oorschot R, Olivier D et al (2009) Stress-induced hypothermia in the mouse. In: Gould TD (ed) Mood and anxiety related phenotypes in mice, Humana Press, New York

6. Szelenberger W, Soldatos C (2005) Sleep disorders in psychiatric practice. World Psychiatry 4:186–190

7. Sateia MJ (2009) Update on sleep and psychiatric disorders. Chest 135:1370–1379

8. Carlson SH, Wyss JM (2000) Long-term telemetric recording of arterial pressure and heart rate in mice fed basal and high NaCl diets. Hypertension 35:E1–5

9. Johansson C, Thoren P (1997) The effects of triiodothyronine (T3) on heart rate, temperature and ECG measured with telemetry in freely moving mice. Acta Physiol Scand 160:133–138

10. Mills PA, Huetteman DA, Brockway BP et al (2000) A new method for measurement of blood pressure, heart rate, and activity in the mouse by radiotelemetry. J Appl Physiol 88:1537–1544

11. Spani D, Arras M, Konig B et al (2003) Higher heart rate of laboratory mice housed individually vs in pairs. Lab Anim 37:54–62

12. Kramer K, Voss HP, Grimbergen JA et al (2000) Telemetric monitoring of blood pressure in freely moving mice: a preliminary study. Lab Anim 34:272–280

13. Tang X, Sanford LD (2002) Telemetric recording of sleep and home cage activity in mice. Sleep 25:691–699

14. Morrow JD, Vikraman S, Imeri L et al (2008) Effects of serotonergic activation by 5-hydroxytryptophan on sleep and body temperature of C57BL/6J and interleukin-6-deficient mice are dose and time related. Sleep 31:21–33

15. Olivadoti MD, Opp MR (2008) Effects of i.c.v. administration of interleukin-1 on sleep and body temperature of interleukin-6-deficient mice. Neuroscience 153:338–348

16. Dugovic C, Shoblock JR, Shelton J et al (2010) Enhanced Rem Sleep And Resistance To Antidepressants In Mice Lacking The Relaxin-3 Peptide SLEEP (Abstracts) 33:A48

17. Wurts Black S, Morairty S, Iacopetti C et al (2010) Gammahydroxybutyrate And R-Baclofen Promote NREM Sleep In Hypocretin/Ataxin-3 And Wild-Type Mice SLEEP (Abstracts) 33:A4

18. Bejanian M, Jones BL, Syapin PJ et al (1991) Brain temperature and ethanol sensitivity in C57 mice: a radiotelemetric study. Pharmacol Biochem Behav 39:457–463

19. Sanford LD, Yang L, Liu X et al (2006) Effects of tetrodotoxin (TTX) inactivation of the central nucleus of the amygdala (CNA) on dark period sleep and activity. Brain Res 1084: 80–88

20. Tang X, Yang L, Sanford LD (2007) Sleep and EEG spectra in rats recorded via telemetry during surgical recovery. Sleep 30:1057–1061

21. Tang X, Yang L, Sanford LD (2007) Individual variation in sleep and motor activity in rats. Behav Brain Res 180:62–68

22. Hamrahi H, Stephenson R, Mahamed S et al (2001) Selected Contribution: Regulation of sleep-wake states in response to intermittent hypoxic stimuli applied only in sleep. J Appl Physiol 90:2490–2501

23. Ben V, Bruguerolle B (2000) Effects of bilateral striatal 6-OHDA lesions on circadian rhythms in the rat: a radiotelemetric study. Life Sci 67:1549–1558

24. Uchida M, Suzuki M, Shimizu K (2007) Effects of urocortin, corticotropin-releasing factor (CRF) receptor agonist, and astressin, CRF receptor antagonist, on the sleep-wake pattern: analysis by radiotelemetry in conscious rats. Biol Pharm Bull 30:1895–1897

25. Neuhaus HU, Borbely AA (1978) Sleep telemetry in the rat. II. Automatic identification and recording of vigilance states. Electroencephalogr Clin Neurophysiol 44:115–119

26. Herold N, Spray S, Horn T et al (1998) Measurements of behavior in the naked mole-rat after intraperitoneal implantation of a radiotelemetry system. J Neurosci Methods 81: 151–158

27. Kramer K, Kinter LB (2003) Evaluation and applications of radiotelemetry in small laboratory animals. Physiol Genomics 13:197–205

28. Kramer K, Kinter L, Brockway BP et al (2001) The use of radiotelemetry in small laboratory animals: recent advances. Contemp Top Lab Anim Sci 40:8–16

29. Tang X, Yang L, Sanford LD (2007) Interactions between brief restraint, novelty and footshock stress on subsequent sleep and EEG power in rats. Brain Res 1142:110–118

30. Krarup A, Chattopadhyay P, Bhattacharjee AK et al (1999) Evaluation of surrogate markers of impending death in the galactosamine-sensitized murine model of bacterial endotoxemia. Lab Anim Sci 49:545–550

31. Clement JG, Mills P, Brockway B (1989) Use of telemetry to record body temperature and activity in mice. J Pharmacol Methods 21: 129–140

32. Meerlo P, Westerveld P, Turek FW et al (2004) Effects of gamma-hydroxybutyrate (GHB) on vigilance states and EEG in mice. Sleep 27: 899–904

33. Szentirmai E, Kapas L, Sun Y et al (2010) Restricted feeding-induced sleep, activity, and body temperature changes in normal and pre-proghrelin-deficient mice. Am J Physiol Regul Integr Comp Physiol 298:R467–477

34. Huitron-Resendiz S, Sanchez-Alavez M, Wills DN et al (2004) Characterization of the sleep-wake patterns in mice lacking fatty acid amide hydrolase. Sleep 27:857–865

35. Tang X, Orchard SM, Liu X et al (2004) Effect of varying recording cable weight and flexibility on activity and sleep in mice. Sleep 27: 803–810

36. Crawley JN (1999) Behavioral phenotyping of transgenic and knockout mice: experimental design and evaluation of general health, sensory functions, motor abilities, and specific behavioral tests. Brain Res 835:18–26

37. Sanford LD, Fang J, Tang X (2003) Sleep after differing amounts of conditioned fear training in BALB/cJ mice. Behav Brain Res 147: 193–202

38. Sanford LD, Tang X, Ross RJ et al (2003) Influence of shock training and explicit fear-conditioned cues on sleep architecture in mice: strain comparison. Behav Genet 33:43–58

39. Sanford LD, Yang L, Tang X (2003) Influence of contextual fear on sleep in mice: a strain comparison. Sleep 26:527–540

40. Tang X, Xiao J, Liu X et al (2004) Strain differences in the influence of open field exposure on sleep in mice. Behav Brain Res 154:137–147

41. Sanford LD, Yang L, Wellman LL et al (2010) Differential effects of controllable and uncontrollable footshock stress on sleep in mice. Sleep 33:621–630

42. Tang X, Xiao J, Parris BS et al (2005) Differential effects of two types of environmental novelty

on activity and sleep in BALB/cJ and C57BL/J mice. Physiol Behav 85:419–429

43. Schuler B, Rettich A, Vogel J et al (2009) Optimized surgical techniques and postoperative care improve survival rates and permit accurate telemetric recording in exercising mice. BMC Vet Res 5:28

44. Thireau J, Zhang BL, Poisson D et al (2008) Heart rate variability in mice: a theoretical and practical guide. Exp Physiol 93:83–94

45. Arras M, Rettich A, Cinelli P et al (2007) Assessment of post-laparotomy pain in laboratory mice by telemetric recording of heart rate and heart rate variability. BMC Vet Res 3:16

46. Tang X, Orchard SM, Sanford LD (2002) Home cage activity and behavioral performance in inbred and hybrid mice. Behav Brain Res 136:555–569

47. Belzung C, Griebel G (2001) Measuring normal and pathological anxiety-like behaviour in mice: a review. Behav Brain Res 125:141–149

48. Anisman H, Merali Z (2009) Learned helplessness induced in mice. In: Gould TD (ed) Mood and anxiety related phenotypes in mice, Humana Press, New York

49. Tang X, Liu X, Yang L et al (2005) Rat strain differences in sleep after acute mild stressors and short-term sleep loss. Behav Brain Res 160:60–71

50. Tang X, Xiao J, Parris BS et al (2005) Differential effects of two types of environmental novelty on activity and sleep in BALB/cJ and C57BL/6J mice. Physiol Behav 85:419–429

51. Duke JL, Zammit TG, Lawson DM (2001) The effects of routine cage-changing on cardiovascular and behavioral parameters in male Sprague-Dawley rats. Contemp Top Lab Anim Sci 40:17–20

52. Schiffelholz T, Aldenhoff JB (2002) Novel object presentation affects sleep-wake behavior in rats. Neurosci Lett 328:41–44

53. Hunsley MS, Palmiter RD (2003) Norepinephrine-deficient mice exhibit normal sleep-wake states but have shorter sleep latency after mild stress and low doses of amphetamine. Sleep 26:521–526

54. Mochizuki T, Crocker A, McCormack S et al (2004) Behavioral state instability in orexin knock-out mice. J Neurosci 24:6291–6300

55. Parmentier R, Ohtsu H, Djebbara-Hannas Z et al (2002) Anatomical, physiological, and pharmacological characteristics of histidine decarboxylase knock-out mice: evidence for the role of brain histamine in behavioral and sleep-wake control. J Neurosci 22: 7695–7711

56. Balcombe JP, Barnard ND, Sandusky C (2004) Laboratory routines cause animal stress. Contemp Top Lab Anim Sci 43:42–51

57. Finger FW (1972) Measuring behavioral activity. In: Myers RD (ed) Methods in psychobiology, Academic Press, New York

58. Tang X, Sanford LD (2005) Home cage activity and activity-based measures of anxiety in 129P3/J, 129X1/SvJ and C57BL/6J mice. Physiol Behav 84:105–115

Chapter 4

Modeling Mouse Anxiety and Sensorimotor Integration: Neurobehavioral Phenotypes in the Suok Test

Elisabeth Dow*, Valerie Piet*, Adam Stewart*, Siddharth Gaikwad, Jonathan Cachat, Peter Hart, Nadine Wu, Evan Kyzar, Eli Utterback, Alan Newman, Molly Hook, Kathryn Rhymes, Dillon Carlos, and Allan V. Kalueff

Abstract

Animal behavioral tests are useful tools for modeling complex human brain disorders. The Suok test (ST) is a relatively new behavioral paradigm that simultaneously examines anxiety and neurological/vestibular phenotypes in rodents. The novelty and instability of the ST apparatus induces anxiety-related behavior in mice, whereas the elevation of the horizontal rod allows for the assessment of motor and neurological phenotypes. This chapter discusses the utility of the ST in detecting mouse anxiety, habituation, exploration, motorisensory deficits, and the interplay between these domains. With a growing number of laboratories using this model, a detailed protocol for the ST behavioral analysis (with a focus on video-tracking tools and novel applications) is also provided.

Key words: Mice, Behavioral models, Anxiety, Stress, Exploration, Ethological analysis, Vestibular phenotypes, Stress-evoked sensorimotor disintegration

1. Introduction

Experimental animal models are widely used to improve our understanding of complex psychiatric disorders, and to screen the effects of various pharmacological, genetic, and behavioral manipulations (1–8). As will be shown in several chapters in this book, mice frequently display neurobehavioral similarities with humans. This supports the utility of murine models for anxiety research (9, 10), including both the improvements in existing tests and the establishment of new paradigms (11–13).

---

*These authors contributed equally.

Todd D. Gould (ed.), *Mood and Anxiety Related Phenotypes in Mice: Characterization Using Behavioral Tests, Volume II,* Neuromethods, vol. 63, DOI 10.1007/978-1-61779-313-4_4, © Springer Science+Business Media, LLC 2011

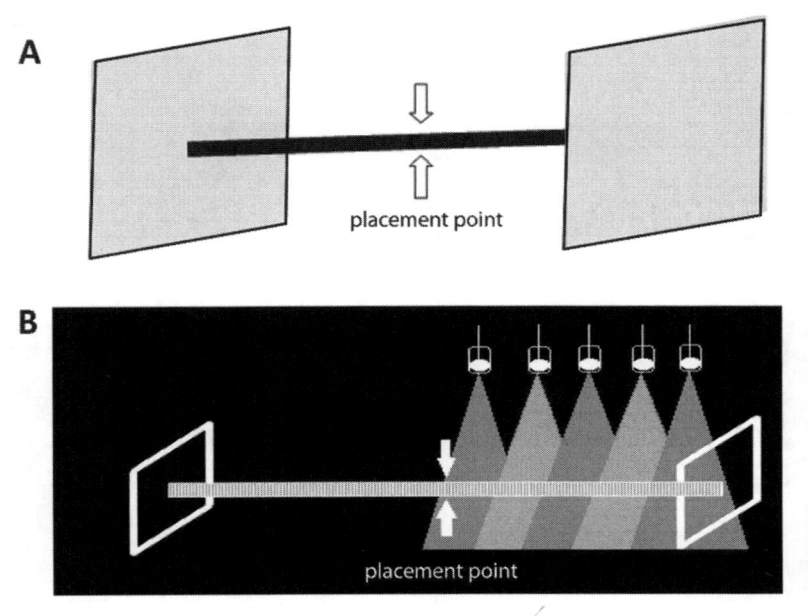

Fig. 1. Murine Suok test apparatus: the regular Suok test (**a**) and its light-dark version (**b**).

The Suok test (ST, Fig. 1) is a recently introduced behavioral model that applies ethological analysis to examine mouse and rat anxiety (5, 14, 15). The novelty and utility of this paradigm arise from its ability to simultaneously assess rodent anxiety, vestibular phenotypes, and motor performance, as well as their complex interplay, such as stress-evoked sensorimotor disintegration (SSD) (2, 16–19). Although SSD is a common clinical phenomenon, its pathogenesis remains largely unknown (17, 20). The ST's rationale and construct validity come from a well-known ability of unprotected, open, and elevated areas to evoke anxiety and panic (acrophobia) as well as vestibular symptoms (vertigo, dizziness) in both clinical patients (21–25) and in normal human subjects (26–29). The concept of SSD is further supported by anxiolytic drugs' ability to reduce vestibular deficits in humans (19, 30, 31) and by animal data on the comorbidity between vestibular and anxiety phenotypes (see (17) for a detailed review).

Compared to other anxiety tests, the ST enhances the dimensionality of mouse data, serving as a conceptual combination of the elevated plus maze, open field (OFT), and horizontal beam tests (32, 33). Representing a long, elevated horizontal rod with a Plexiglas wall on either end (Fig. 1a), the mouse ST simultaneously assesses lateral (e.g., horizontal locomotion) and vertical (e.g., head dipping, falls) behaviors (5, 15, 32–34). At the same time, the ST is a typical novelty-based paradigm, similar to the elevated plus maze and OFT, where anxiety is evoked and examined based on the classical approach-avoidance theory (35). While the ST novelty couples with the instability of the apparatus to induce animal anxiety, the elevated testing surface is used to assess rodent balance and motor performance (similar to the traditional beam test (22, 36–38)) by the number of falls and hind leg slips (32, 33). The light-dark ST version (Fig. 1b), which utilizes animals' natural aversion to a novel and brightly lit environment, further enhances the model by adding an additional stressor (5).

Basic methodology of rodent ST behavioral testing and its validity have been discussed previously in detail (5, 15, 32–34). With a growing number of laboratories using the ST for different rodent applications (e.g., (5, 14, 39, 40)), this chapter aims to provide an update on this model and its utility for mouse behavioral phenotyping. We will specifically emphasize the ST ability to target multiple behavioral domains, and how this can be enhanced by the use of modern video-tracking technology. The latter not only enables the correction of manual observations but also generates additional indices reflecting velocity, immobility, high mobility, and distance traveled. The developing utility of the ST to study basic cognitive functions (e.g., habituation) as well as other aspects of mouse novelty-evoked responses (e.g., homebase behaviors) will also be discussed.

2. Equipment, Materials, and Setup

Various inbred, outbred, selectively bred, and genetically modified (mutant or transgenic) mice may be used in the ST to observe anxiety, motor function, and neurological phenotypes. When selecting a mouse model, the strain difference in activity and emotionality are important to consider. For example, BALB/cJ mice generally exhibit high anxiety, whereas C57BL/6J and NMRI have low baseline anxiety levels. Activity levels and novelty seeking also differ markedly between strains. For example, 129 S1/SvImJ mice generally display low activity, the NMRI strain has moderate activity, while both BALB/cJ and C57BL/6J strains are usually highly active. Similarly, 129 S1/SvImJ and BALB/cJ mice are neophobic, and C57BL/6J mice show high novelty-seeking behavior (9, 41, 42). Factors such as age, weight, sex, estrous cycle stage, and husbandry should also be considered when designing ST experiments. In addition, the most updated and detailed nomenclature for mouse strains must be used (see Mouse Phenome Project for mouse strains: http://phenome.jax.org, and Mouse Genome Informatics for genetically modified mice: http://www.informatics.jax.org).

The equipment required for the regular or light-dark ST is simple, inexpensive, and sufficient to assemble the apparatus and collect data. The typical mouse ST apparatus is a 1–2-m aluminum tube ~2 cm in diameter, elevated to a height of 20–25 cm above a cushioned test surface (Fig. 1a). The rods for both ST versions can easily be purchased from home utility stores, costing approximately $10 per rod. The rod is demarcated into 10-cm sectors to allow quantification of distance moved by the mouse. Two Plexiglas walls (50×50×1 cm) are fixed on either end of the aluminum tube to prevent mice from leaving the test apparatus, and paper towels or cloths placed directly underneath the rod act as protective cushions (to prevent injuries during falls and enable efficient clean up between subjects). Use 70% percent ethanol to clean the aluminum rod between sessions. To avoid the potentially confounding effects of bright lights (42), the experimental room must not be brightly illuminated (in our studies at Tulane University, 700–900 lux appears to be appropriate for mouse ST).

The light-dark ST apparatus, identical to the regular ST test, includes 4–6 light bulbs (60 W) fixed ~40–50 cm above one-half of the rod, providing the only light source in the dark experimental room (Fig. 1b). The few additional pieces of equipment for data collection are easily attainable, and include a manual observation template, timer, light meter, and video-recorder. The template generates a per-minute distribution of behavioral endpoints (see further) for the quick detection of temporal trends, such as habituation. For video-tracking mouse ST behavior, special software packages are required. For example, our laboratory uses Noldus Ethovision XT7 (Wageningen, the Netherlands) and Clever Sys LocoScan (Reston, VA).

The light meter (e.g., Sper Scientific, Scottsdale, AZ) is a hand-held device that measures lighting of the ST apparatus. To ensure proper lighting (e.g., 700–900 lux) for the regular ST test, take 10–15 measures for three points on the ST apparatus (in the center and on either end). If necessary, adjust the light source or the ST apparatus location to ensure homogenous illumination.

3. Procedure

3.1. Acclimation

The acclimation entails transporting mice from their holding room to the experimental room 1 h prior to behavioral testing, and leaving subjects undisturbed to minimize their transfer anxiety. If the mice are obtained from a commercial vendor or another laboratory, allow at least a 2–3-week acclimation period before testing, to reduce transportation stress.

3.2. Suok Test Procedure

Mice must be tested in the ST during their normal waking cycle, to avoid interference with circadian rhythms. When performing a battery of tests, consider how the effects of these prior tests may confound the mouse ST performance and drug sensitivity. At the beginning of each trial and after each fall from the apparatus, place mice at the center of the rod (0 cm) with snout facing either end (or, in the light-dark modification, orient the animal facing the dark end). If necessary, subjects can be gently supported by hand during initial placement, to avoid falls caused by incorrect positioning. Note that if video-tracking is used, place mice back to the point where they fell off, to prevent artificial inflation of the endpoint "distance traveled" when the software analyzes the videos. To minimize detection problems, allow ~5 s to pass at the start of each recording before placing the subject into the test arena (see Troubleshooting 1).

3.3. Behavioral Testing and Analyses

While a typical ST experiment is a short 5–6-min trial, its duration can be altered at the discretion of the experimenter, depending on experimental needs (e.g., we recently applied an extended 20-min trial to examine mouse ST exploratory behavior in depth). A digital camera mounted in front (or on top) of the test apparatus, combined with video-tracking software, will enable the collection of accurate behavioral data. If video-tracking software is used, the camera should be positioned ~50 cm away from the apparatus. During the observational period, the experimenter usually sits and records mouse behavior ~2 m away from the apparatus. The observers must refrain from making noise or moving, as this may alter animal behavior. Also, intra- and inter-rater reliability should be assessed for consistency (desired level is ~0.85 or more) by Spearman rank correlation coefficient.

During each trial, the following behavioral measures are recorded manually or using video-tracking software: (a) horizontal exploration activity, which includes latency to leave central zone, number of segments visited (four paws), time spent moving, velocity, average inter-stop distance (distance traveled divided by number of stops) distance traveled, number of stops, time spent immobile; (b) vertical exploration (number of vertical rears and wall leanings); (c) directed exploration (number of head dips and side looks); (d) risk assessment behavior (stretch-attend postures); (e) vegetative responses (latency to defecate, number of fecal boli and urination spots); and (f) motor behavioral parameters (number of missteps or hind-leg slips and falls) (see Fig. 2 for details). Note that tail position may also be a useful index (usually elevated and erect if anxiety is high). The value of each "latency" endpoint will equate to total observation time if the animal does not show the respective behavior. At the end of each testing session, mice are returned to a holding room, and the ST apparatus should be wiped with 70% ethanol, to remove olfactory cues that may affect the behavior of sequential subjects.

3.4. Data Analysis   *Statistics.* The ST behavioral data can be analyzed with the Wilcoxon–Mann–Whitney U-test for comparing two groups (parametric Student's t-test may be used if data is normally distributed), or analysis of variance (ANOVA) for >2 groups, including one-way ANOVA with repeated measures (time), and n-way ANOVA for

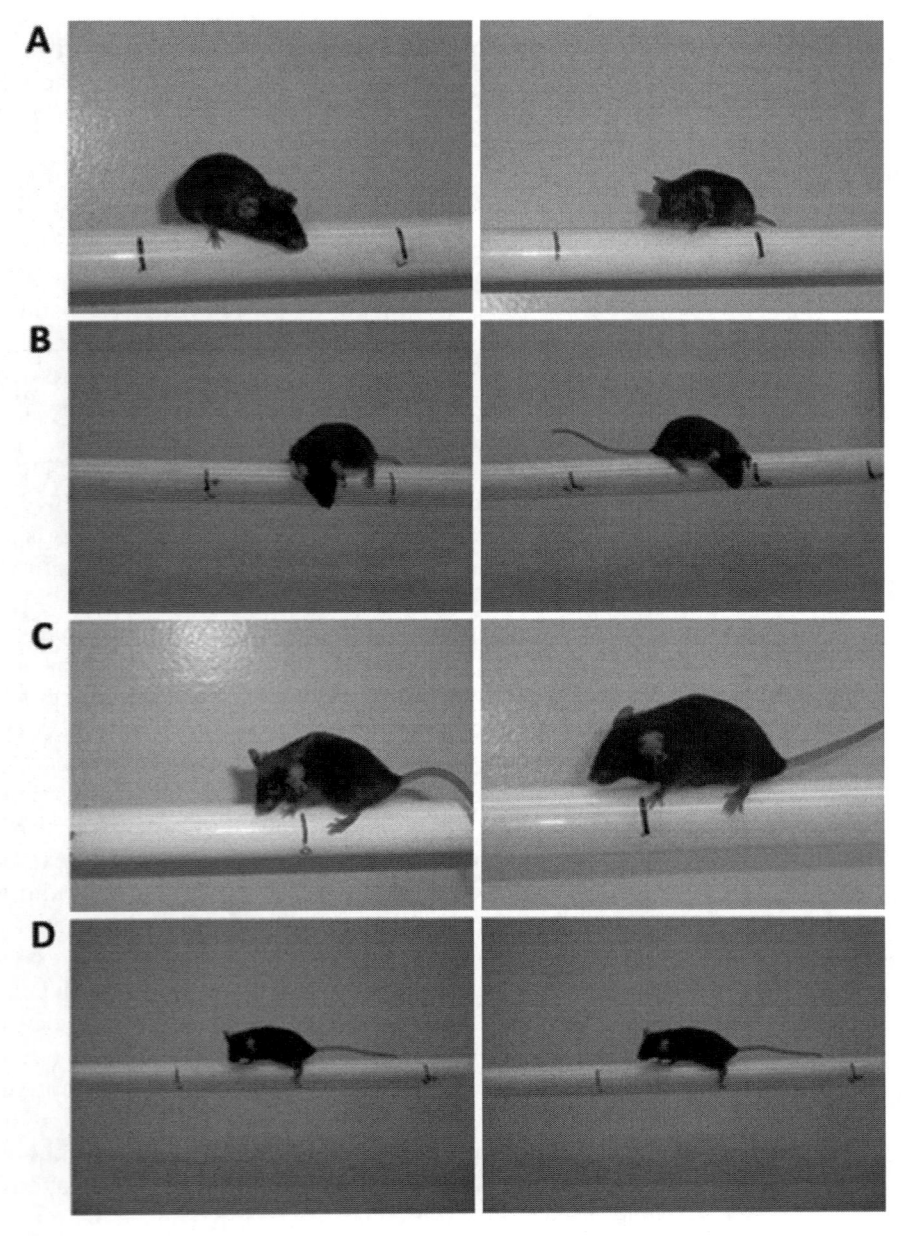

Fig. 2. Typical mouse behaviors observed in the Suok test: (a) side looks, (b) head dips, (c) freezing, (d) hind leg slips, (e) "anxious tail" position, (f) stretch-attend posture, (g) grooming behavior.

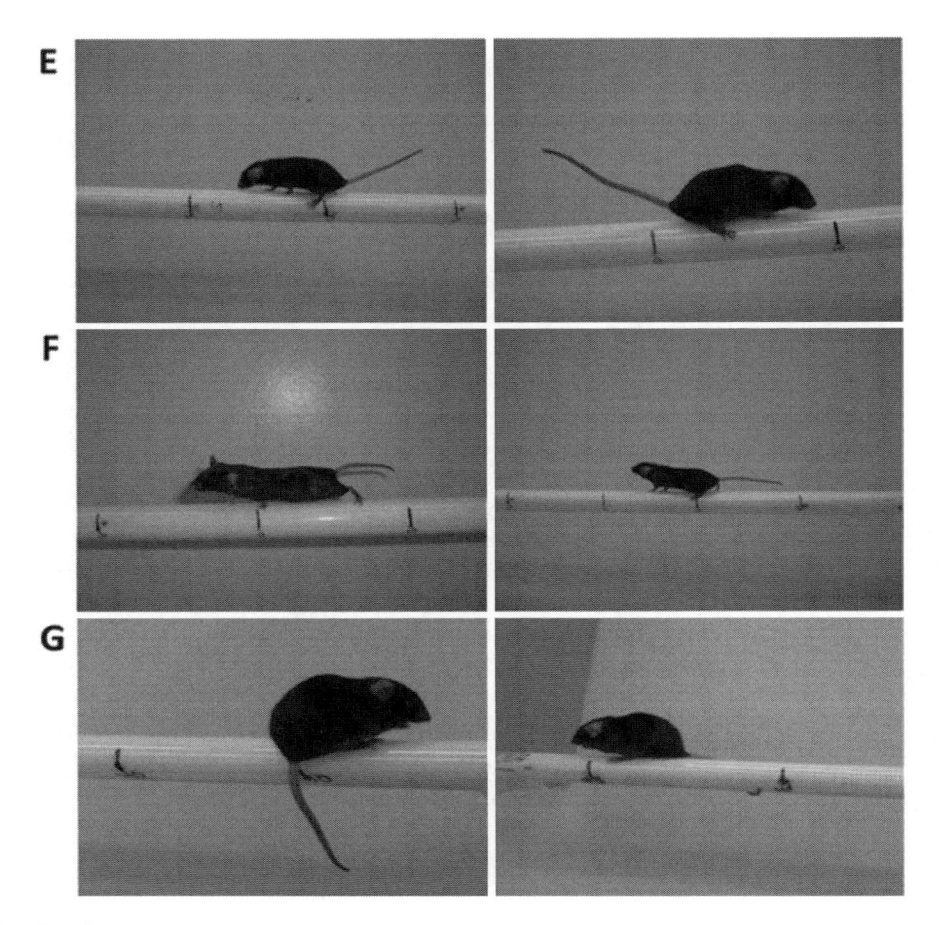

Fig. 2. (continued)

more complex studies (e.g., including treatment, genotype, sex, and/or stress), followed by an appropriate post-hoc test, such as Bonferroni adjustment, Dunn, Dunnett, or Tukey tests.

Video analysis: The ST videos can be analyzed and its endpoints (e.g., distance traveled, velocity, and time spent moving) calculated using an automated video-tracking system. Before analyzing videos, frames including the researcher must be removed to avoid skewing data. Generally, researchers stay out of camera sight, away from the ST apparatus during testing. However, at the beginning of each session or if the animal falls, they must be close to the apparatus and may briefly appear in the videos. If the frames are not removed from the video recording, researcher's body parts could be "detected" as mice (see Troubleshooting 2). A video-editing program, such as Windows Movie Maker, may be used to remove such frames.

After videos have been edited appropriately, they may be analyzed using video-tracking programs, such as Noldus Ethovision XT7. To properly acquire videos, first establish a rectangular arena for the experiment, with the boundaries of the arena formed by the bottom of the rod, including ~5 cm past each end (to include Plexiglas end walls), and a line ~10 cm above the rod. Limiting the size of the arena (by excluding the area between the test surface and the underside of the rod) ameliorates detection setting problems and reduces rogue endpoints. To determine which detection settings work best, evaluate the three detection settings, "Static Subtraction," "Differencing," and "Dynamic Subtraction" in concurrence with playing a video. When tracking using Noldus Ethovision XT7, yellow shading will cover the subject as it moves around the arena. On the Experiment Settings screen, set the program to track all morphological endpoints, including tail, center, and nose. After acquisition, remove any rogue detection points and interpolate missing data. If there are apparent errors, readjust detection settings and reacquire videos before exporting data for behavioral analyses.

The behavioral data generated by video-tracking complements the manual observation endpoints. Recommended indices to calculate include total distance moved, mean velocity, absolute and mean turn angle, turning rate (absolute and mean angular velocity), turning bias (relative and mean angular velocity), absolute and mean meandering, duration and frequency of movement, and duration and frequency of elongation. All of these behavioral endpoints reflect different aspects of the mouse ST performance and are common for many other behavioral paradigms and tests. Endpoints only attainable through video-tracking (e.g., velocity and movement) can quantify whether the subject moves in short, quick bouts or longer, more cautious movements. Calculations of turning rate and bias describe the nature of circular exploratory movement (turning movements with a higher velocity may represent potentially interesting phenotypes; see further).

3.5. Time Required

The acclimation period typically requires 1 h prior to the ST procedure. However, if the initial level of mouse anxiety is very high, using a longer acclimation time and/or handling each animal (e.g., for 5 min per day for 3–4 days prior to ST) may reduce potential anxiety related to experimental procedures. Animal testing in the ST requires approximately 9 min per animal (6 min of testing and 2–3 min of clean-up of apparatus). Depending on the amount of data collected, analysis for manual observations may take approximately 1 day, and an additional 2–4 days may be needed to analyze video-generated data.

4. Anticipated Results

In general, the ST is highly sensitive to behavioral differences in mouse anxiety. For example, the model correctly detects major differences between strains' behavioral phenotypes (e.g., anxiety and motor functioning) and state or trait behaviors (3, 5). A typical experiment examining baseline anxiety in BALB/cJ, NMRI, and C57BL/6J strains is shown in Fig. 3. Note that BALB/cJ mice, an

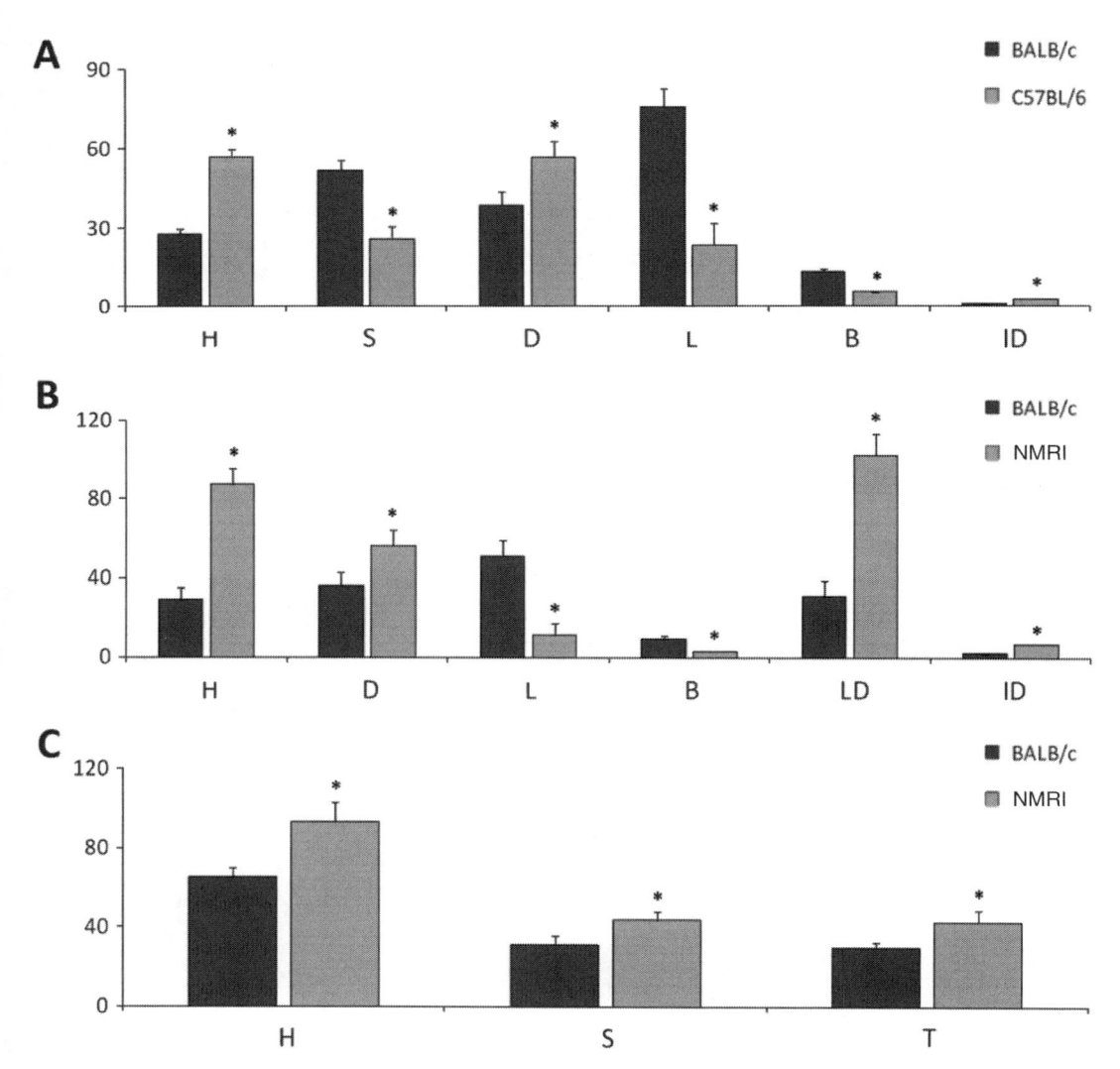

Fig. 3. Representative behavioral responses of male NMRI, BALB/cJ, and C57BL/6J mice in regular (**a, b**) or light-dark (**c**) Suok test for 5 min (graphs are based on data published previously by our group (5)). (**a, b**) *H* horizontal activity (segments); *S* stops; *D* head dips; *O* orientation (side-directed exploration); *L* latency to leave center; *B* defecation boli; *LD* latency to defecate; *ID* average inter-stop distance. (**c**) *H* horizontal activity in the light; *S* sectors visited in light; *T* time in light; values expressed as percentages. *$P < 0.05$ (U-test) between strains.

innately anxious strain, exhibit predictably more anxiety and less exploratory behavior than both NMRI and C57BL/6J strains. Increased anxiety was demonstrated by shorter inter-stop distance, increased stops and fecal boli, whereas exploratory behavior was signified by higher latencies to leave the center, less horizontal activity, and fewer head dips (Fig. 3). BALB/cJ mice show preference for the dark area of the light-dark ST, assessed by significantly fewer stops and less time spent in light, consistent with their higher trait anxiety (Fig. 3).

The ST sensitivity to evoked anxiety has been demonstrated in a recent experiment where C57BL/6J mice were roughly handled (ten strokes of backward petting) for 1 min (Fig. 4). The stressed mice displayed predictably higher anxiety, as indicated by more falls and decreased exploratory behavior (increased duration of stops and a lower total distance moved). Similar results were obtained using other psychological stressors in mice, such as pretest exposure to a rat, which is a strong stressor as rats are natural

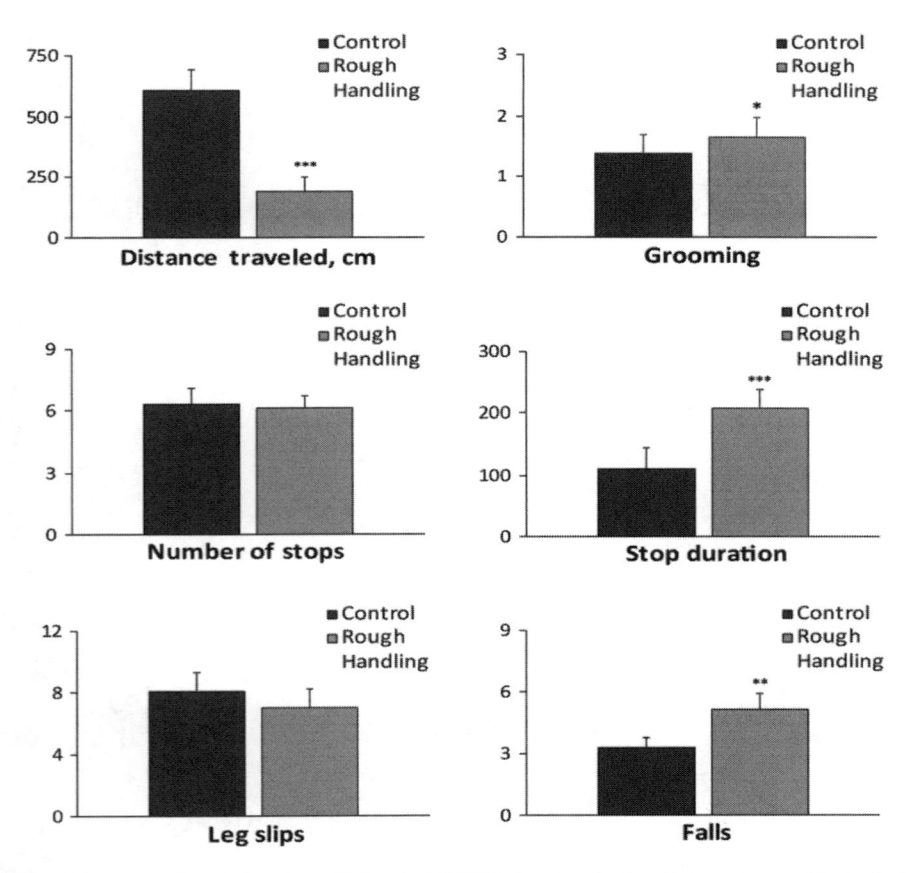

Fig. 4. Behavioral responses of control and roughly handled C57BL/6J male mice ($n=20$ in each group) tested in the regular Suok test. Handled mice exhibited a significantly higher number of falls, a longer stopping duration and a shorter distance traveled, suggesting their increased anxiety. $*P<0.05$, $**P<0.01$, $***P<0.005$ (U-test).

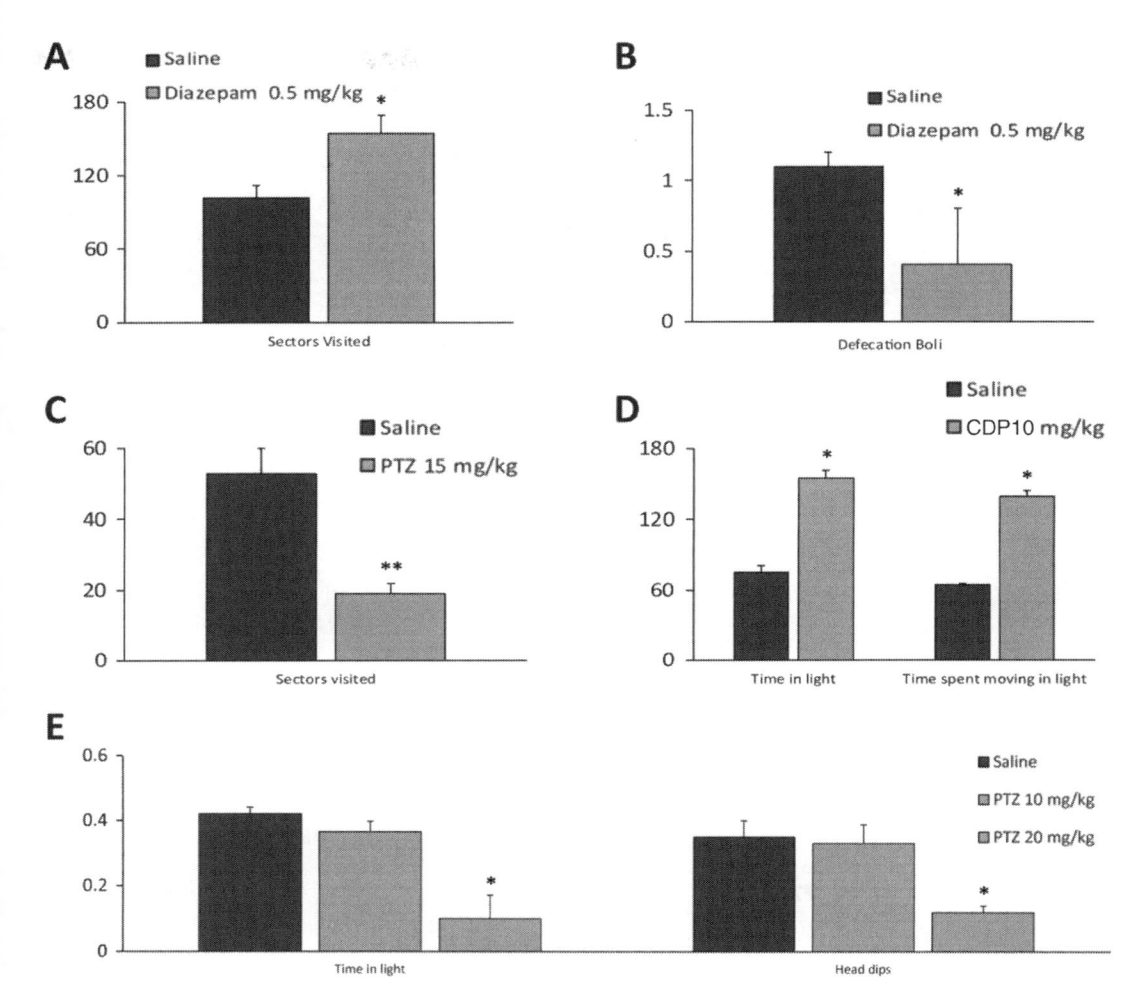

Fig. 5. Behavioral responses of male BALB/cJ mice to diazepam, chlordiazepoxide (CDP) and pentylenetetrazole (PTZ) in the regular (**a–c**) and light-dark (**d–e**) Suok tests. Diazepam increased exploration and lowered the number of defecation boli. PTZ increased anxiety in both tests by decreasing sectors visited, head dips and time spent in light, and showing decreased motor functioning by increasing the falls and misstep. CDP decreased anxiety by increasing time spent and movement in light. Graphs are based on data previously published by our group (5). *$P < 0.05$, **$P < 0.01$ (U-test).

predators of mice. Rat-exposed mice exhibited increased anxiety and impaired balance compared to a nonexposed control group (33).

In addition to genetic strain differences and experimental stressors, the ST is also sensitive to pharmacogenic anxiety (32). A typical experiment assessing the ST responses to various pharmacological agents is shown in Fig. 5. In this study, the anxiolytic drug diazepam increased exploration and lowered the number of fecal boli. In the light-dark ST version, the anxiolytic drug chlordiazepoxide (CDP) decreased anxiety by increasing time spent and movement in light. By contrast, the anxiogenic drug pentylenetetrazole (PTZ) increased anxiety in both the regular and light-dark ST (Fig. 5) and

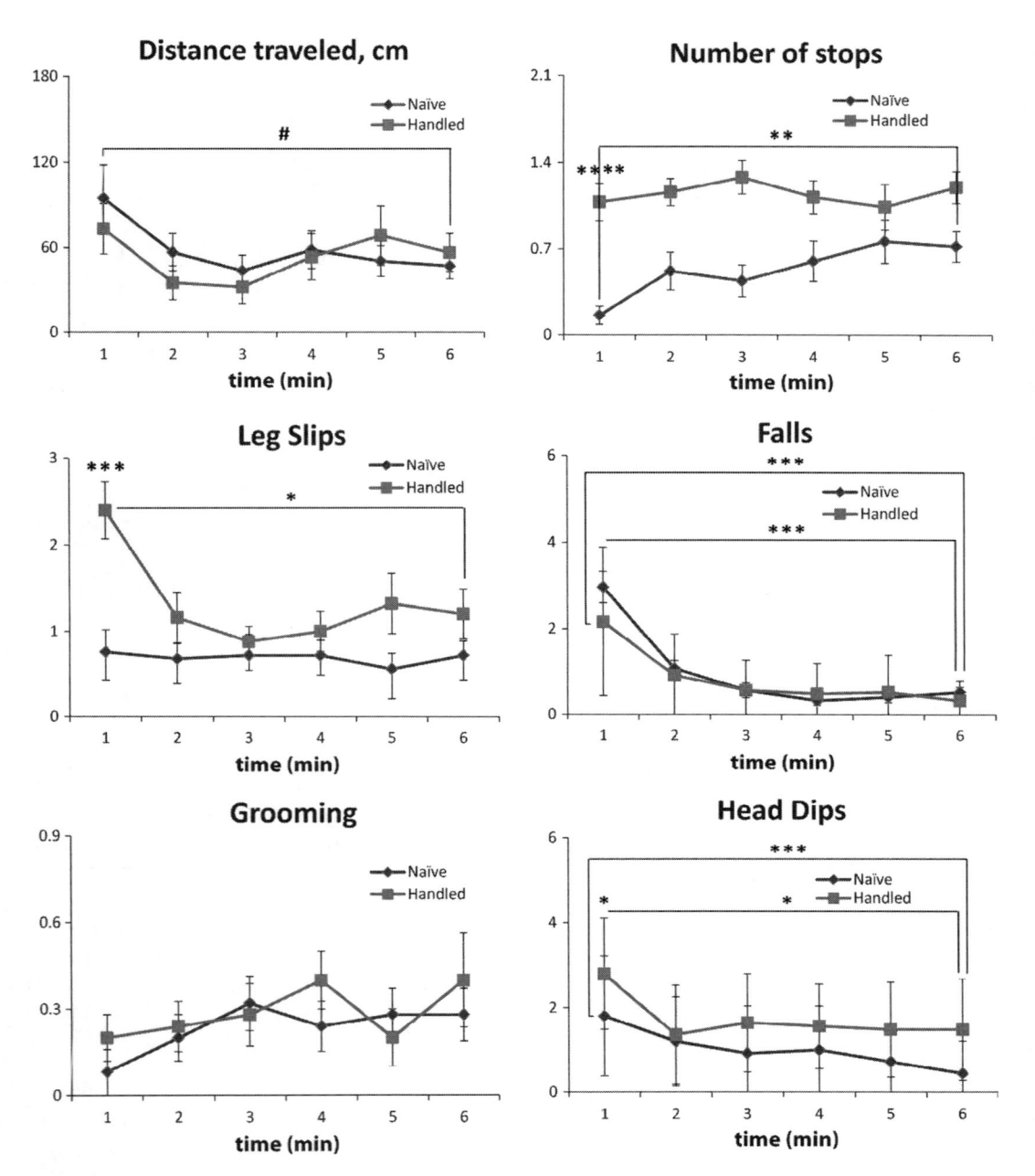

Fig. 6. Habituation of Suok test behaviors in male C57BL/6J mice. Control (naïve) mice traveled less distance over the course of the 6-min trial. Note that acutely stressed mice show slightly impaired habituation as compared to control mice, consistent with the known negative effect of acute stressors on rodent spatial working memory (57–59). Min 1 data between groups was compared using paired U-test. Min 1 vs. min 6 within each group was compared using unpaired U-test. *Asterisks* on *top* of *horizontal line* denote difference between respective min 1 and min 6. *Asterisks* on *top* of min 1 data denote difference between initial (min 1) anxiety in stressed (handled) vs. naïve control mice. *$P < 0.05$, **$P < 0.005$, ***$P < 0.0005$, #$P = 0.05$–0.1, trend (U-test).

also impaired mouse motor function by increasing the falls and missteps (43). Taken together, these findings support the utility of the ST for screening a wide spectrum of pharmacological agents in rodents.

In addition to producing quantifiable data, video-tracking software can provide an accurate visual summary of murine ST traces (Fig. 6–9). Center-point tracking shows overall (global) distance moved, as some subjects may never leave the center, show preference for certain areas of the rod, or utilize the entire apparatus. However, the tail and nose-point tracking, in our opinion, better detect exploratory behavior. For example, a head dip is represented in a side view trace by a nose-point line below the center-point trace. As shown in these traces, the nose and tail-traces often form circular patterns, indicating head dips and vertical explorations that occur in more of a sweeping manner. Top view traces can also be generated by positioning the video recorder above the test rod. Unlike side view traces, top view traces can visually represent and detect exploration on either side of the ST apparatus (Fig. 8), which appear as rotating or swiveling maneuvers.

Finally, video-tracking software can produce "density maps," which show the overall frequency of time spent over the length of the ST apparatus. As shown in Fig. 9, the density of behavior is not homogenous over the ST rod's length, as the mouse clearly prefers locations in the center (initial placement point) or close to the walls of the apparatus (thigmotaxis; see further).

5. Additional Potential Applications

Within-trial habituation is an important phenotype (observed in mouse behavioral tests), reflecting rodent spatial working memory (44–46). Our recent experiments reveal the ST's utility for examining mouse habituation. As shown in Fig. 6, roughly handled (stressed) mice demonstrate poorer habituation for distance traveled, head dips, and number of stops (vs. robust habituation curves in their controls). While control mice traveled less distance over the course of the trial, stressed mice traveled approximately the same distance each minute. Similarly, control mice performed less head dips per minute, while the stressed group had a more gradual decline (Fig. 6).

Although leg slips and falls are nonexploratory behaviors (and, therefore, do not reflect habituation), the negative slope of their graphs suggests the occurrence of some kind of aversive learning. An alternative explanation of these temporal phenotypes may also be due to reduced activity (e.g., an increased number of stops and decreased overall distance traveled, see Fig. 6) since if subjects

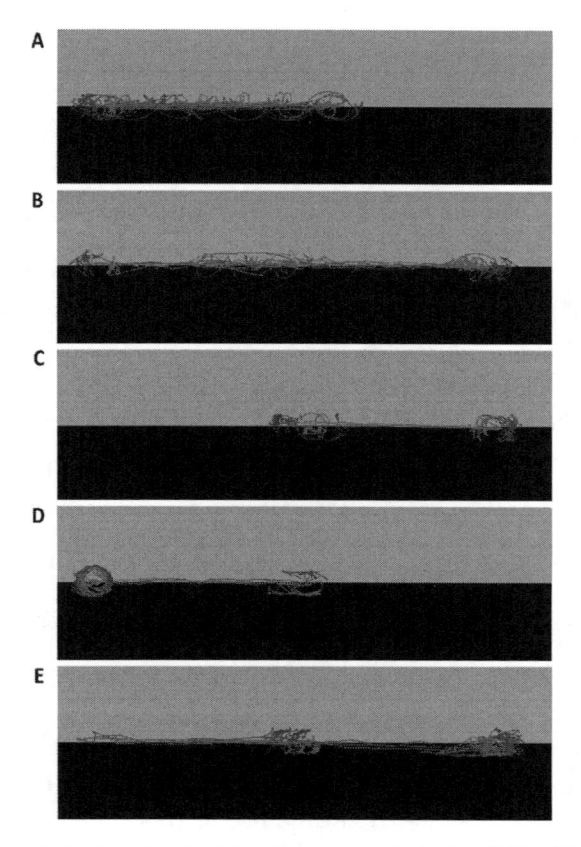

Fig. 7. Representative top-view Suok test traces generated using Noldus Ethovision XT7 video-tracking software. As explained in the text, Ethovision XT7 can track the nose, center, and tail points of subjects, to produce traces. The traces presented here were saved from the software and superimposed onto a *gray* and *black* background, to indicate the location of the test apparatus. (**a**) Trace in which the subject failed to leave the center, circular rings around the center point by the nose and tail points indicate that the mouse spun around to explore the novel environment; (**b**) traces in which the subject performed moderate exploratory behavior on one side only. This trace shows the mouse swiveled at regular intervals across the left side of the rod. (**c, d**) This mouse performed exploratory behavior on one side only, but most of the behavior was localized to the center and left endpoints. (**e, f**) These animals performed exploratory behavior over the entire rod. The lack of full circles in these traces shows that these mice did not perform as much swiveling behavior as in previous (**a, c**).

move less distance and stop more frequently, they are less likely to fall or slip. Whether this signifies altered habituation, different processing of sensory information, or both, it is an interesting direction for further studies (47, 48), also suggesting that the ST has the potential for screening various mnemotropic drugs.

While the behavioral effects of antidepressants have not been examined in the ST, the well-known ability of selective serotonin reuptake inhibitors' (SSRI) to improve balance and reduce anxiety

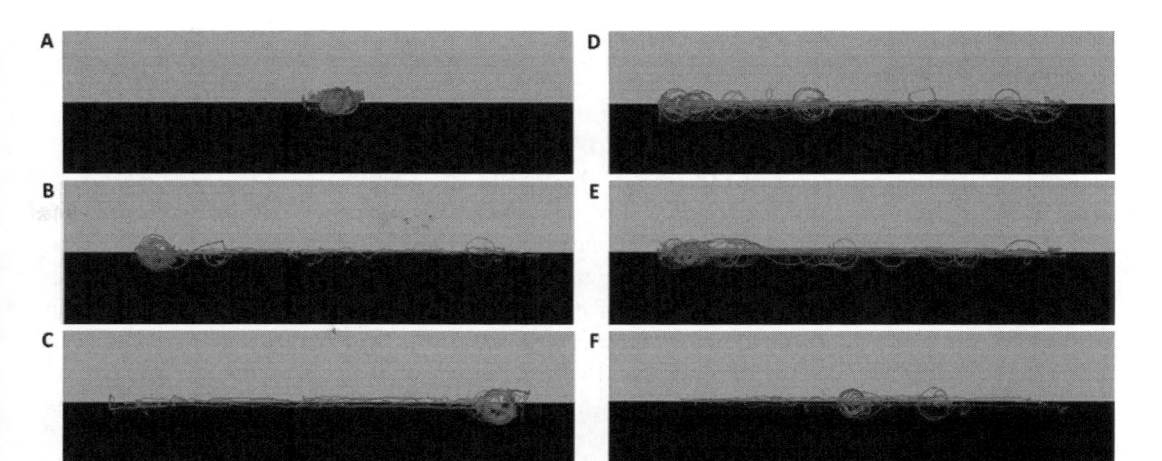

Fig. 8. Representative side-view Suok test traces generated using Noldus Ethovision XT7 video-tracking software. (a) subject failed to leave the center, showing extensive rotational exploratory behavior at the center point; (b) subject utilized the entirety of the test rod, spending more time on the left side of the test; (c) subject utilized the entirety of the apparatus, performing more consistent exploratory behaviors; (d, e) these mice utilized the entire of the apparatus, exhibiting vertical exploratory behaviors in certain nonregular intervals; (f) subject showed more horizontal exploratory behavior than vertical.

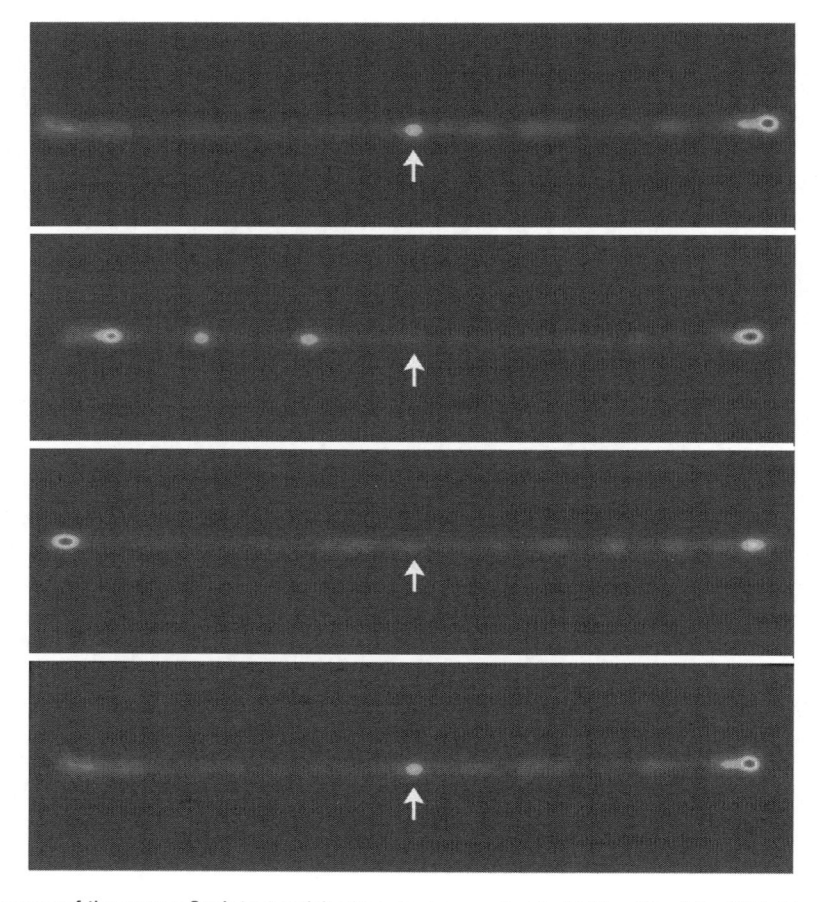

Fig. 9. Density maps of the mouse Souk test activity (*top* view) generated by Noldus Ethovision XT7 video-tracking software. Concentrated *red/yellow* color would indicate a large percentage of time spent in a particular zone on the apparatus (*white arrow* indicates the placement point).

in both humans and animals (48–50) implies the ST's potential sensitivity to these drugs. Furthermore, the ST is likely to be sensitive for novel drugs targeting the vestibular system, agents affecting SSD and anxiety, as well as some other drug classes, such as hallucinogens. For example, the sensitivity to a hallucinogenic lysergic acid diethylamide (LSD) has already been demonstrated in a mouse ST (4). Recent rodent studies from other laboratories have identified additional potential applications of the ST. For example, the test showed superior (vs. OFT) sensitivity to behavioral effects of long-term alcoholization (14), and high sensitivity to behavioral effects of bioflavonoids' on stress-related behavioral activity (51) in rats, collectively suggesting that the rodent ST may also be applied to study a wide spectrum of drug abuse-related phenomena, such as long-term behavioral alteration, withdrawal-evoked anxiety and SSD.

Another potential novel application of the ST is the analysis of homebase formation. Homebase formation is an adaptive behavioral strategy used by rodents to facilitate spatial orientation and exploration (52–55). In a novel environment, animals establish one or two "safe" zones where they spend most of their time and frequently visit, while exploring their environment. Rodent homebases tend to be established near vertical surfaces and show higher grooming and rearing activity (56). Our observation of ST-induced behaviors presents an innovative opportunity for studying rodent homebase formation. For example, we observed the mouse ability to form preferred loci in the ST apparatus, (Fig. 9), demonstrating that mice spent considerably more time at 2–3 nonrandom locations, usually near the side walls or at the center drop point (Fig. 9).

6. Troubleshooting

Several practical recommendations, briefly summarized here, may enable more reliable and reproducible behavioral data in the mouse ST experiments.

1. When initially placing the mouse on the bar (or after a fall), orient the mouse with the snout facing either end. Support the animal during initial placement to avoid a fall due to poor positioning. If a mouse falls off the testing rod, place the animal back on the rod with minimal disturbance, to the same spot from where it fell (if the mouse is returned to a different location, a video-tracking program will artificially inflate total distance traveled by the mouse).

2. When using video-tracking software, minimize the amount of time researchers spend within camera range. For example, reduce the time spent in frames by having one individual stationed

near the ST apparatus to quickly return mice to the rod, and the other ready to pause the experiment timer. Alternatively, a careful editing of video files will help solve the problem. To edit videos using this program, open a new project file and import one video at a time. Remove all the video segments in which the mouse has fallen off the test apparatus or a researcher is in frame; alternating between various zoom settings may increase accuracy. Save the video in DV-AVI format (the Windows Movie Maker version of AVI files supported by video-tracking software).

3. Setting the detection arena tightly around the testing rod can minimize confounds in the video tracking process. If raw points are still being detected, attempt to reduce the complexity of the entire screen shot. Try to buffer bright lighting with white paper and create a surface of white paper behind the testing platform, to increase contrast and object detection.

4. Testing sessions around 5–6 min are usually sufficient for the ST. This testing time is desirable as it is sensitive to anxiety, yet long enough to produce significant habituation responses (Fig. 6). However, this amount of time may not be sufficient if mice with impaired motor or vestibular function are used. For example, several initial minutes may be lost from repeatedly returning the falling mouse to the rod. To retain experimental time, pause the experimental timer during each fall or run the experiment for a longer duration (e.g., 10–20 min). Pausing the experimental timer can also help synchronize manual observation data with edited tracking videos. Analysis of home-base-like behavior may require an even longer observation time, as suggested by early OFT studies investigating rodent homebase formation (56).

5. High levels of transfer anxiety may lead to poor initial retention on the testing apparatus. To prevent this problem, gently support the animals by hand for ~5 s to facilitate a better grip. If the animal continues to display high transfer anxiety, exclude it from the experiment (record, however, the % of such animals in each group). In addition, improved animal husbandry in the holding areas and the use of a dimly lit experimental room can reduce initial anxiety levels.

6. Depending on the overall motor ability of the experimental mice, the type of experimental rod can be altered. For mice with severely impaired vestibular function, masking tape along the surface of the rod, wider or wooden rods for a better grip, and (in extreme cases) a flattened surface similar to a narrow meter stick, can be used. In this case, the control mice would also fall and slip less, producing a habituation curve with less amplitude. If mice continue to struggle with balance or motor

abilities, assess motor and vestibular functions separately, as these behaviors may be due to a neuromuscular or motor coordination problem unrelated to vestibular deficits or anxiety.

7. Low motor or vertical activity may be a strain-specific phenotype. Less active mouse strains will produce lower activity overall, and may not be suitable for this model. Likewise, hyperactive strains generally display less nonhorizontal exploration and may have difficulties with balance. A narrower apparatus will encourage the animal to show its horizontal activity, enabling other behavioral responses.

8. Performance on the ST is strongly determined by physical factors, such as body size and weight (larger animals have predictably more difficulty). Only use animals of similar age, size, and weight to reduce possible confounds and accurately compare between groups.

9. If the study involves a battery of behavioral tests, consider the potential effects of test batteries on ST performance. For example, because the ST utilizes rather strong anxiety evoked by height and novelty, administer less stressful tests before subjecting animals to the ST. Acclimate animals for at least 7 days before or between STs to reduce habituation confounds. Likewise, this model may not be suitable for long-term follow-up studies, since mice quickly habituate to the apparatus (Fig. 6). However, the ST habituation itself may provide a readily testable mouse model with an additional (cognitive) dimension.

7. Conclusion

Overall, the ST simultaneously examines anxiety, vestibular, and neuromuscular deficits by combining an unstable, elevated rod with novelty. Anxiolytic or anxiogenic drugs predictably modulate mouse ST exploration, risk assessment, and vegetative behaviors. The model is also sensitive to anxiety-evoked vestibular/balancing deficits (such as SSD), as anxiogenic drugs increase the number of falls and missteps, while anxiolytic agents generally improve balance (4, 6). Some basic cognitive (e.g., habituation) phenotypes may easily be assessed in this model. A light-dark ST modification may also be employed to further examine these domains. The test combines an economical experimental apparatus (Fig. 1) with well-defined behavioral endpoints (Fig. 2). Representing a useful behavioral paradigm for mouse neurophenotyping, it can be strengthened by applying video-tracking and data-mining software.

Acknowledgment

The study was supported by Tulane University Intramural funds, Provost's Scholarly Enrichment Fund, Newcomb Fellows Grant, and NARSAD YI Award. This chapter is dedicated to the memory of Eli Utterback.

References

1. Gold, R., C. Linington, and H. Lassmann, *Understanding pathogenesis and therapy of multiple sclerosis via animal models: 70 years of merits and culprits in experimental autoimmune encephalomyelitis research*. Brain, 2006. 129(Pt 8): p. 1953–71.

2. Avni, R., et al., *Mice with vestibular deficiency display hyperactivity, disorientation, and signs of anxiety*. Behav Brain Res, 2009. 202(2): p. 210–7.

3. Kalueff, A.V., A. Minasyan, and P. Tuohimaa, *Behavioural characterization in rats using the elevated alley Suok test*. Behav Brain Res, 2005. 165(1): p. 52–7.

4. McKinney, W.T., *Overview of the past contributions of animal models and their changing place in psychiatry*. Semin Clin Neuropsychiatry, 2001. 6(1): p. 68–78.

5. Kalueff, A.V., et al., *The regular and light-dark Suok tests of anxiety and sensorimotor integration: utility for behavioral characterization in laboratory rodents*. Nat Protoc, 2008. 3(1): p. 129–36.

6. Kalueff, A.V., M. Wheaton, and D.L. Murphy, *What's wrong with my mouse model? Advances and strategies in animal modeling of anxiety and depression*. Behav Brain Res, 2007. 179(1): p. 1–18.

7. Lang, P.J., M. Davis, and A. Ohman, *Fear and anxiety: animal models and human cognitive psychophysiology*. J Affect Disord, 2000. 61(3): p. 137–59.

8. van den Buuse, M., et al., *Importance of animal models in schizophrenia research*. Aust N Z J Psychiatry, 2005. 39(7): p. 550–7.

9. Belzung, C. and G. Griebel, *Measuring normal and pathological anxiety-like behaviour in mice: a review*. Behav Brain Res, 2001. 125(1-2): p. 141–9.

10. Lang, P.J., M.M. Bradley, and B.N. Cuthbert, *Emotion, motivation, and anxiety: brain mechanisms and psychophysiology*. Biol Psychiatry, 1998. 44(12): p. 1248–63.

11. Rodgers, R.J., et al., *Animal models of anxiety: an ethological perspective*. Braz J Med Biol Res, 1997. 30(3): p. 289–304.

12. Blanchard, R.J., et al., *The characterization and modelling of antipredator defensive behavior*. Neurosci Biobehav Rev, 1990. 14(4): p. 463–72.

13. Blanchard, D.C., et al., *Human defensive behaviors to threat scenarios show parallels to fear- and anxiety-related defense patterns of non-human mammals*. Neurosci Biobehav Rev, 2001. 25(7–8): p. 761–770.

14. Filatova, E.V., et al., *Influence of individual features on the formation of ethanol preference in Wistar male rats*. Academic Reports of Russian Academy of Sciences, 2010. 430: p. 562–564.

15. Kalueff, A.V., A. Minasyan, and P. Tuohimaa, *Behavioural characterization in rats using the elevated alley Suok test*. Behav Brain Res, 2005. 165(1): p. 52–57.

16. Venault, P., et al., *Balance control and posture in anxious mice improved by SSRI treatment*. Neuroreport, 2001. 12(14): p. 3091–4.

17. Kalueff, A.V., K. Ishikawa, and A.J. Griffith, *Anxiety and otovestibular disorders: linking behavioral phenotypes in men and mice*. Behav Brain Res, 2008. 186(1): p. 1–11.

18. Zheng, Y., et al., *Effects of bilateral vestibular deafferentation on anxiety-related behaviours in Wistar rats*. Behav Brain Res, 2008. 193(1): p. 55–62.

19. Asmundson, G.J., D.K. Larsen, and M.B. Stein, *Panic disorder and vestibular disturbance: an overview of empirical findings and clinical implications*. J Psychosom Res, 1998. 44(1): p. 107–20.

20. Balaban, C.D. and J.F. Thayer, *Neurological bases for balance-anxiety links*. J Anxiety Disord, 2001. 15(1-2): p. 53–79.

21. Emmelkamp, P.M. and M. Felten, *The process of exposure in vivo: cognitive and physiological changes during treatment of acrophobia*. Behav Res Ther, 1985. 23(2): p. 219–23.

22. Lorivel, T. and P. Hilber, *Motor effects of delta 9 THC in cerebellar Lurcher mutant mice*. Behav Brain Res, 2007. 181(2): p. 248–53.

23. Rothbaum, B.O., et al., *Virtual reality graded exposure in the treatment of acrophobia: A case*

80   E. Dow et al.

report. Behavior Therapy, 1995. 26(3): p. 547–554.

24. Menzies, R.G. and J.C. Clarke, *The etiology of acrophobia and its relationship to severity and individual response patterns.* Behav Res Ther, 1995. 33(7): p. 795–803.

25. Davey, G.C., R. Menzies, and B. Gallardo, *Height phobia and biases in the interpretation of bodily sensations: some links between acrophobia and agoraphobia.* Behav Res Ther, 1997. 35(11): p. 997–1001.

26. Yardley, L., et al., *Effects of anxiety arousal and mental stress on the vestibulo-ocular reflex.* Acta Otolaryngol, 1995. 115(5): p. 597–602.

27. Wada, M., N. Sunaga, and M. Nagai, *Anxiety affects the postural sway of the antero-posterior axis in college students.* Neurosci Lett, 2001. 302(2-3): p. 157–9.

28. Bolmont, B., et al., *Mood states and anxiety influence abilities to maintain balance control in healthy human subjects.* Neurosci Lett, 2002. 329(1): p. 96–100.

29. Viaud-Delmon, I., A. Berthoz, and R. Jouvent, *Multisensory integration for spatial orientation in trait anxiety subjects: absence of visual dependence.* Eur Psychiatry, 2002. 17(4): p. 194–9.

30. Erez, O., et al., *Balance dysfunction in childhood anxiety: findings and theoretical approach.* J Anxiety Disord, 2004. 18(3): p. 341–56.

31. Nagaratnam, N., J. Ip, and P. Bou-Haidar, *The vestibular dysfunction and anxiety disorder interface: a descriptive study with special reference to the elderly.* Arch Gerontol Geriatr, 2005. 40(3): p. 253–64.

32. Kalueff, A.V., et al., *Pharmacological modulation of anxiety-related behaviors in the murine Suok test.* Brain Res Bull, 2007. 74(1–3): p. 45–50.

33. Kalueff, A.V. and P. Tuohimaa, *The Suok ("ropewalking") murine test of anxiety.* Brain Res Brain Res Protoc, 2005. 14(2): p. 87–99.

34. Kalueff, A.V., et al. *The developing utility of the suok test in anxiety pharmacology and behavioral research.* in *10-th Jubilee Multidisciplinary International Conference of Biological Psychiatry "Stress and Behavior".* 2007. St-Petersburg, Russia.

35. Rodgers, R.J. and N.T. Johnson, *Factor analysis of spatiotemporal and ethological measures in the murine elevated plus-maze test of anxiety.* Pharmacol Biochem Behav, 1995. 52(2): p. 297–303.

36. Cummings, B.J., et al., *Adaptation of a ladder beam walking task to assess locomotor recovery in mice following spinal cord injury.* Behav Brain Res, 2007. 177(2): p. 232–41.

37. Dluzen, D.E., et al., *Evaluation of nigrostriatal dopaminergic function in adult +/+ and +/–* *BDNF mutant mice.* Exp Neurol, 2001. 170(1): p. 121–8.

38. Furman, J.M. and R.G. Jacob, *A clinical taxonomy of dizziness and anxiety in the otoneurological setting.* J Anxiety Disord, 2001. 15(1-2): p. 9–26.

39. Tubaltseva, I., et al. *The effects of quercetin on behavioral parameters of stressed rats in the suok-test.* in *10-th Jubilee Multidisciplinary International Conference of Biological Psychiatry.* 2007. St-Petersburg, Russia.

40. Filatovaa, E.V., et al., *The influence of social conditions on the development of ethanol preference in rats.* Dokl Biol Sci, 2010. 430: p. 23–5.

41. Roman, E. and L. Arborelius, *Male but not female Wistar rats show increased anxiety-like behaviour in response to bright light in the defensive withdrawal test.* Behav Brain Res, 2009. 202(2): p. 303–7.

42. Hogg, S., *A review of the validity and variability of the elevated plus-maze as an animal model of anxiety.* Pharmacol Biochem Behav, 1996. 54(1): p. 21–30.

43. Quinn, L.P., et al., *A beam-walking apparatus to assess behavioural impairments in MPTP-treated mice: pharmacological validation with R-(–)-deprenyl.* J Neurosci Methods, 2007. 164(1): p. 43–9.

44. Chapillon, P. and P. Roullet, *Habituation and memorization of spatial objects' configurations in mice from weaning to adulthood.* Behavioural Processes, 1997. 39(3): p. 249–256.

45. Salomons, A.R., et al., *Behavioural habituation to novelty and brain area specific immediate early gene expression in female mice of two inbred strains.* Behav Brain Res, 2010. 215(1): p. 95–101.

46. Leussis, M.P. and V.J. Bolivar, *Habituation in rodents: a review of behavior, neurobiology, and genetics.* Neurosci Biobehav Rev, 2006. 30(7): p. 1045–64.

47. Mogg, K., et al., *Effect of short-term SSRI treatment on cognitive bias in generalised anxiety disorder.* Psychopharmacology (Berl), 2004. 176(3-4): p. 466–70.

48. Bolivar, V. and L. Flaherty, *A region on chromosome 15 controls intersession habituation in mice.* J Neurosci, 2003. 23(28): p. 9435–8.

49. Boulenger, J.P., et al., *Baseline anxiety effect on outcome of SSRI treatment in patients with severe depression: escitalopram vs paroxetine.* Curr Med Res Opin. 2010. 26(3): p. 605–14.

50. Szabo, S.T., C. de Montigny, and P. Blier, *Progressive attenuation of the firing activity of locus coeruleus noradrenergic neurons by sustained administration of selective serotonin reuptake inhibitors.* Int J Neuropsychopharmacol, 2000. 3(1): p. 1–11.

51. Griebel, G., et al., *The free-exploratory paradigm: an effective method for measuring neophobic behaviour in mice and testing potential neophobia-reducing drugs.* Behav Pharmacol, 1993. 4(6): p. 637–644.

52. Eilam, D. and I. Golani, *Home base behavior in amphetamine-treated tame wild rats (Rattus norvegicus).* Behav Brain Res, 1990. 36(1-2): p. 161–70.

53. Eilam, D. and I. Golani, *Home base behavior of rats (Rattus norvegicus) exploring a novel environment.* Behav Brain Res, 1989. 34(3): p. 199–211.

54. Mintz, M., et al., *Sharing of the home base: a social test in rats.* Behav Pharmacol, 2005. 16(4): p. 227–36.

55. Wallace, D.G., M.M. Martin, and S.S. Winter, *Fractionating dead reckoning: role of the compass, odometer, logbook, and home base establishment in spatial orientation.* Naturwissenschaften, 2008. 95(11): p. 1011–26.

56. Stewart, A., et al., *Phenotyping of Zebrafish Homebase Behaviors in Novelty-Based Tests,* in *Zebrafish Neurobehavioral Protocols,* A. Kalueff and J. Cachar, Editors. 2010: Humana Press, New York.

57. Voigt, J.P. and E. Morgenstern, *Pentylenetetrazole kindling impairs learning in mice.* Biomed Biochim Acta, 1990. 49(1): p. 143–5.

58. Angelucci, M.E., et al., *The effect of caffeine in animal models of learning and memory.* Eur J Pharmacol, 1999. 373(2-3): p. 135–40.

59. Grecksch, G., A. Becker, and C. Rauca, *Effect of age on pentylenetetrazol-kindling and kindling-induced impairments of learning performance.* Pharmacol Biochem Behav, 1997. 56(4): p. 595–601.

Chapter 5

Assessment of Social Approach Behavior in Mice

Orsolya J. Kuti and Damon T. Page

Abstract

Assays of social approach behavior involve measuring the social investigatory behavior of a subject mouse toward a stimulus mouse. Such assays represent a phenotyping tool that may be applied to mouse models of mood and anxiety disorders, as well as other neuropsychiatric disease, to understand how genetic and environmental risk factors impact brain development and behavior. Here we discuss a measure of social approach behavior in mice using a three-chamber apparatus: we start with an overview of the approach, we then present a protocol that utilizes computer-assisted analysis of social approach data, and finally, we provide information on troubleshooting this protocol. The aim of this chapter is to provide investigators with the information necessary to start testing mice for social approach behavior and to adapt the protocol to their particular needs.

Key words: Social interaction, Social behavior, Social approach, Anxiety, Autism, Schizophrenia, Mouse, Three-chamber apparatus

1. Background and Historical Overview

Social behaviors are increasingly studied in mice as a controlled means to understand the etiology of human mood and anxiety disorders, as well as disorders such as autism spectrum disorder, schizophrenia, and Williams syndrome (1). Mice are social animals that live in colonies and exhibit complex relationships such as dominance hierarchies, gendered interactions, and aggression (2) and are able to recognize conspecifics and strangers through a variety of sensory modalities such as olfaction, vision, and touch. Social behaviors may be studied across the lifespan of the mouse, including: ultrasonic vocalizations in neonates, play in juveniles, adult behaviors such as social approach, social recognition, preference for social novelty, aggression, dominance, mating behaviors and vocalizations, and maternal behaviors such as pup retrieval and nursing (3–13).

Todd D. Gould (ed.), *Mood and Anxiety Related Phenotypes in Mice: Characterization Using Behavioral Tests, Volume II*, Neuromethods, vol. 63, DOI 10.1007/978-1-61779-313-4_5, © Springer Science+Business Media, LLC 2011

Social approach can be studied in both male and female mice and utilizes the natural tendency of a mouse to approach and investigate an unfamiliar conspecific through following, sniffing, and grooming (2, 14). Mice with altered social approach behavior may spend less time engaged in investigatory behavior and this can be interpreted as a deficit in sociability. As a measure of innate behavior, social approach is highly reproducible within mouse strains (15), while strain-specific changes in social behavior can be detected (16, 17). Genetic mutations that alter neural development have been shown to influence social approach (7, 9, 18–25), as have pharmacological agents (5, 26–29) and environmental modifiers (30).

Tests of social interaction were originally developed as an ethologically relevant method of studying anxiety in rodents without the use of food/water deprivation or punishment (31). At the most basic level, measurement of social interaction involves the observation of a subject mouse and a stimulus mouse interacting, noting the investigatory behaviors that result. To dissociate social interaction initiated by the subject mouse as opposed to investigation initiated by the stimulus mouse, current protocols for studying social approach behavior generally use a three-chamber apparatus where a subject mouse is given a choice between interacting with a tethered or caged stimulus mouse and an inanimate object (11, 29, 32). By using gender-matched mice and a neutral location (the testing apparatus), mating motivated behavior and territorial aggression are minimized.

Several methods for collecting and analyzing data from the social approach assay may be employed. An experimenter may observe the trial, recording time spent in each chamber of the apparatus during the trial (26, 33). The testing apparatus may be fitted with photobeam break detectors that measure chamber crossings and the amount of time spent in each chamber (34). Video recordings may be taken of the trial and subsequently analyzed to determine time spent in each chamber by an experimenter or a computer program (18, 20, 35). Software packages such as Ethovision (Noldus), SMART Triwise Video Tracking (Harvard Apparatus), or VideoMot2 (TSE) allow the user to track locomotion within chambers and identify specific aspects of mouse behavior such as rearing and head orientation and can measure interactions between multiple mice; some (Ethovision) are able to detect behaviors such as grooming, sniffing, fighting, or tail rattling. These packages can be of great utility; however, they may be beyond the budgetary reach of some researchers. Open source software, such as Matlab, ImageJ, and WinTrack, can be customized by users proficient in computer programming and used to analyze social approach video files (22–25).

Three-chamber social approach assays may be set up and carried out in most laboratories by one or two experimenters. Presented here is a method for analyzing social approach data recorded in a three-chamber apparatus, using computer scripts written in ImageJ, an image analysis program freely available from the National Institutes of Health (NIH).

2. Equipment, Materials, and Setup

2.1. General Note on Methodology

The approach detailed here describes a protocol for testing social approach behavior that has worked well in our hands (24, 36); however, numerous protocols have successfully used the three-chamber apparatus for testing social approach behavior, as well as social recognition and novelty preference, and the reader is directed to these to learn of the range of methods and testing parameters employed by various laboratories (18, 20, 22–26, 33–35).

2.2. Behavior Room and Personnel

A room used for testing social approach behavior should be in a quiet location with minimal traffic and should not be used for testing any other species of animal, as mice will react fearfully to the odors of rats, ferrets, and many other experimental animals, and this may influence social approach behavior (37, 38). The social approach box and recording device should be placed in a position in the room far from drafts or air flow, where mice have limited perception of personnel moving around the room. The apparatus should be evenly lit across all chambers; this may be tested using a light meter and light sources may be adjusted as necessary. It is desirable to have separation between the testing area where the apparatus is located, the location where the subject mice are held before and after testing, and the location where the stimulus mice are held before and after testing, to limit distracting odors. It is a necessary control to observe the behavior of a cohort of control mice in an empty (no stimulus mouse) behavior apparatus before beginning any testing, to ensure that no side preference within the three-chamber apparatus is observed. It is additionally important to alter the orientation of the apparatus relative to the testing room and record movement of the mice to verify that subject mice do not display positional preference. If subject mice display differential exploration in chambers 1 and 3 during acclimation, or if, in the presence of a stimulus mouse, the time spent in chamber 1 vs. chamber 3 varies significantly for the control subject mouse population based on apparatus orientation, then a new testing room should be found. Personnel conducting experiments should ideally be consistent for the length of a study and should be the same personnel as those who habituated the mice to handling.

2.3. Apparatus

A social approach apparatus can be purchased from a number of retailers or can be built by hand. As an example, one may use opaque acrylic sheets (US Plastics, Lima, OH, USA) bonded using acrylic solvent (GE Polymershapes, Huntsville, NC, USA) to form an open top box with three chambers, as shown in Fig. 1. A commonly used

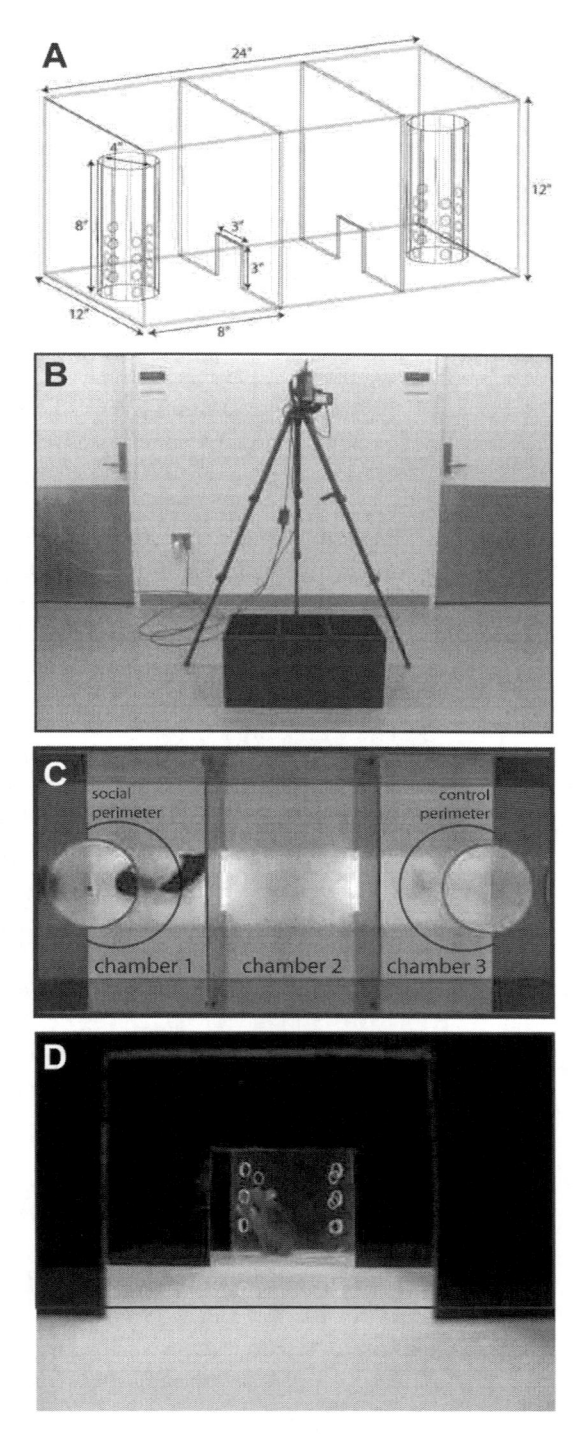

Fig. 1. Social approach apparatus setup. (a) Schema of social approach apparatus, showing dimensions. (b) Placement of video camera for recording social approach behavior. (c) *Top down* view of social approach box, with stimulus mouse in chamber 1. Subject mouse shown is interacting within the social perimeter. (d) View of stimulus mouse in chamber 1. Figure (a–c) and legend adapted from (36).

box size is $24''L \times 12''W \times 12''H$, divided into three chambers ($8''L \times 12''W$) by acrylic sheets with $3'' \times 3''$ square openings cut into them to allow test mice to move between chambers. The floor of the apparatus should contrast with the coat color of the mouse; in experiments using strains with dark or black coats, a white floor is recommended. If lines with a range of coat colors will be tested, a red floor may provide the best overall contrast. In describing the chambers of the behavior apparatus, we number chambers 1–3 from left to right for reference. Chambers 1 and 3 hold cages and chamber 2 holds no cage and serves as a passage between chambers 1 and 3. The cage in chamber 1 holds the social stimulus mouse; the cage in chamber 3 is left empty or may contain a nonsocial object as a control. The cages may be wire cups purchased commercially, or cages built from clear acrylic cylinders measuring $8''H \times 4''D$ (US Plastics, Lima, OH, USA) drilled with holes to allow for olfactory, visual, and tactile investigation of the stimulus mouse by the subject mouse. The cages should be large enough for an adult mouse to move about comfortably and heavy enough to prevent a stimulus mouse from moving them during normal exploratory behavior.

2.4. Animals

For assaying social approach behavior in adult mice, test subjects should be at least 6–8 weeks old. A large number of inbred and outbred strains have been tested in social approach behavior, and strain-specific behavior differences are common (8, 17, 33); therefore, care should be taken when using transgenic animals to maintain consistency in genetic background. Unfamiliar strain-, gender-, and age-matched stimulus mice are typically used in standard social approach assays. All mice should be group-housed throughout testing, on a 12-h light/dark cycle, with *ad libitum* access to food and water. Though there has been no observed effect of testing social approach in the light vs. dark cycle (39), it is advisable that behavioral testing and handling take place during the same time of the day for a given experiment.

Before testing social approach behavior in a new mouse line, these animals should preferably first be screened for deficits in locomotor and sensory function, especially in the domain of olfaction. Any deficits in these areas represent a potential confound that needs to be taken into account in the design of experiments for the mouse line under consideration.

3. Procedure

3.1. Acclimation

As previously described (36), mice being tested should be acclimated to the behavior room, apparatus, and handling by personnel several days before experiments start. Mice may be moved to the behavior room in their home cages around the time of day that the experiment will be taking place. Mice should be acclimated to handing to a point where they can be removed from the home cage with minimal stress. It is generally sufficient for mice to be handled by personnel for approximately 5–10 min a day for several days prior to testing. If mice will be identified by viewing ear tags or tattoos, mice should also be acclimated to this aspect of handling. If injection of drug or vehicle will be part of the experiment, mice should be acclimated to the restraint necessary to humanely administer an injection.

Mice must also be acclimated to the behavior apparatus before testing. This may be accomplished at the same time that acclimation to handling is taking place. Test mice should be placed into the clean three-chamber social approach box (without cages) and allowed to explore the chambers for 5–10 min. Mice to be used as a stimulus should also be handled and acclimated to the holding cage for a similar amount of time.

On the day of an experiment, mice should be moved to the behavior testing area in their home cages and allowed to acclimate for one to several hours.

3.2. Social Approach

Immediately prior to testing, each subject mouse should receive a final acclimation to the social approach apparatus. An experimenter should place the subject mouse into chamber 2 of the clean empty social approach box (no cages), allowing it to explore all three chambers for 5 min. Video recordings taken during this time can be analyzed to test whether mice showed chamber preference. At the end of 5 min, the subject mouse will be removed, and a previously acclimated stimulus mouse placed into a cage in chamber 1 while an empty control cage is placed in chamber 3 of the social approach box. The subject mouse is replaced in chamber 2 and video recording is taken as the subject mouse explores the apparatus for 10 min. At the end of the 10-min testing period, the subject and stimulus mice are removed and the apparatus and cages cleaned thoroughly with a cleaner and disinfectant such as Quatricide PV (Pharmacal, Naugatuck, CT, USA) to remove odors and allowed to dry before being used to test another mouse.

3.3. Video Acquisition

Video recordings of social approach trials can be taken with a digital camcorder (e.g., Sony Handycam) held directly above the apparatus at a height of 3″ using a tripod (e.g., Manfrotto). Correct calibration of camera zoom is essential for downstream analysis of videos. To do this, the camera zoom is set so that the floor of the social approach apparatus fits precisely into a 120×60-pixel box in the exported video (see below). Videos are recorded at 320×240 pixels, 30 frames per second, using Quicktime Pro software (Apple) run on a Apple laptop computer. To reduce file size to approximately 325 KB, images are exported at 160×120 pixels, 3 frames per second for subsequent analysis, using options in Quicktime's export settings.

4. Data Analysis and Anticipated Results

4.1. Processing of Data

1. Install ImageJ (from the NIH, installation instructions are included) to data processing computer. ImageJ is available for Microsoft Windows and Linux operating systems; however, the scripts have only been verified using Mac (Fig. 2).

2. Install custom scripts downloaded from http://www.frontiersin.org/behavioralneuroscience/paper/10.3389/neuro.08/048.2009/by moving these seven scripts into the folder ImageJ\plugins\Macros.

3. Load video files using the File > Open function. At this stage, users are given the option of converting video files to grayscale, this is advised.

4. Run scripts in order to analyze video. Scripts can be accessed from the Plugins > Macros dropdown menu.

   - The "Step 1 field selection" script file defines a 120×60-pixel box. Place box to precisely fit the floor of the social approach apparatus. If the video camera or box position were to accidentally shift during the course of an experiment such that floor no longer precisely fits the 120×60 pixel box, the movie file can be adjusted postrecording using Image > Adjust > Size (to correct a problem with zoom) or Image > Rotate > Arbitrarily (to correct a problem with rotation).

   - The "Step 2 blocking" script blocks out the area of the cages and partitions of the apparatus to remove signal from these areas in downstream analysis.

   - The "Step 3 chamber 1 ROI" script uses ImageJ's thresholding, noise reduction, and variance filters (more information about these can be found at http://rsb.info.nih.gov/ij/docs/menus/image.html) to generate a signal from the mouse that can be analyzed against a background with minimal noise. The output from the "chamber 1 ROI" script is a Z-axis plot of mean gray value signal within the region of interest corresponding to chamber 1 for the recording time. Copy and paste Z-axis plot to data sheet using the "copy" function.

   - The "Step 4 chamber 2 ROI" obtains the same information from chamber 2, Z-axis plot should be transferred to data sheet.

   - The "Step 5 chamber 3 ROI" obtains the same information from chamber 3, Z-axis plot should be transferred to data sheet.

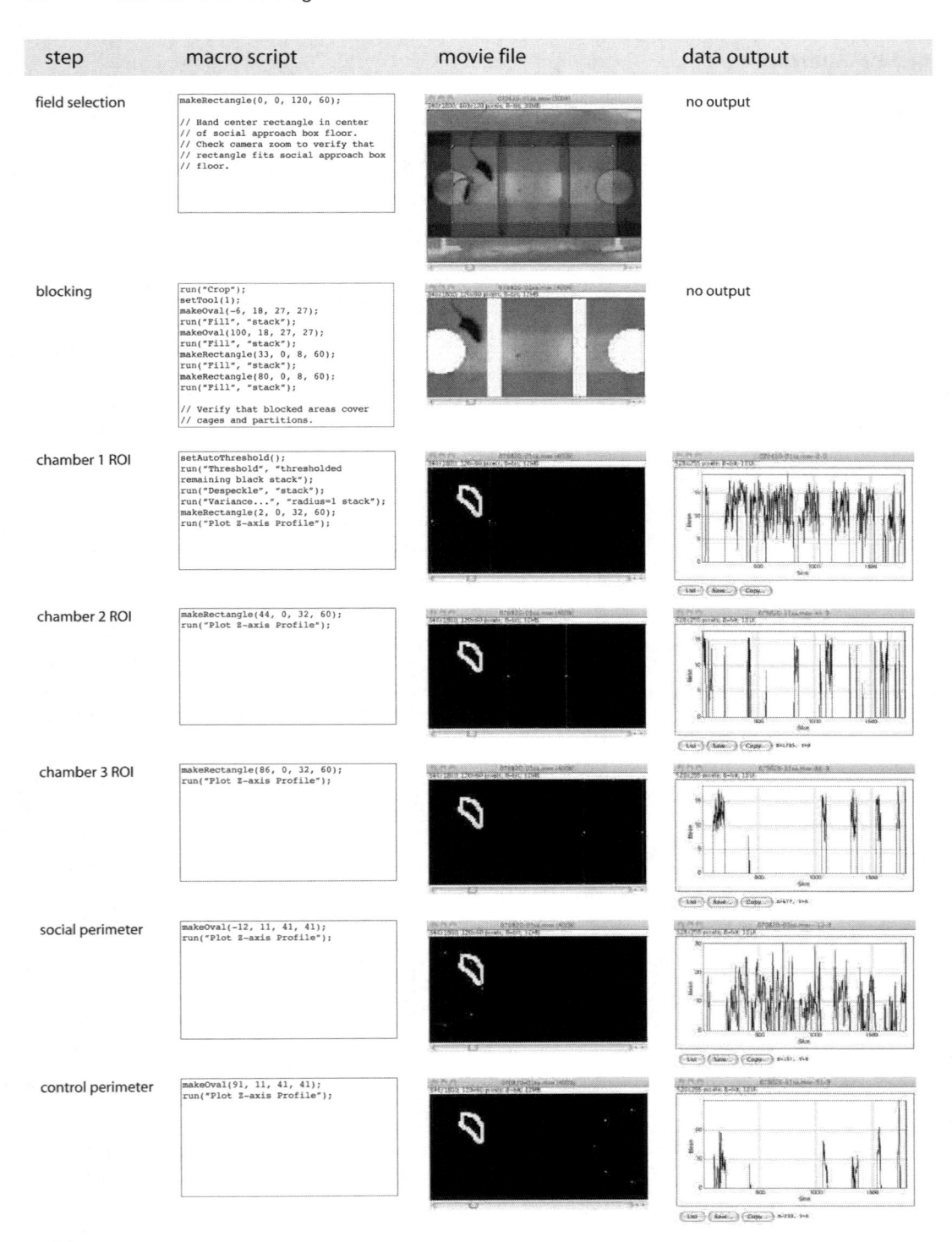

Fig. 2. Procedure for analyzing social approach movie files using ImageJ computer scripts. Scripts are run in order shown, data output shows pixel change in ROI over each frame of video. Figure and legend adapted from (36).

- The "Step 6 social perimeter" and "Step 7 control perimeter" scripts generate Z-axis grey value plots for regions of interest corresponding to a perimeter centered on the cage in chamber 1 (social perimeter) or chamber 3 (control perimeter), which is 1.5 times larger than the circumference of the cage. This allows for quantifying the time spent in close proximity to the cage, where social interaction is likely to take place, as opposed to time spent in the chamber but not in close proximity to the cage. Transfer data for each to data sheet.

5. Review data in data sheet to check for gaps and conflicts as described below. After hand-checking for errors, calculate percent time spent in each chamber in Excel. We do this by dividing the number of frames containing signal for a given ROI by the total number of frames for the assay (this should be 1,800 frames for a 10-min movie exported at 3 frames per second) and multiplying by 100.

4.2. Statistics

Analysis of social approach behavior may be carried out using one way ANOVA, comparing the number of frames containing mouse signal between ROIs. Where a significant effect is seen, *post hoc* comparison is performed using Tukey HSD test. The significance level is set at $P < 0.05$.

5. Experimental Variables and Troubleshooting

5.1. Variability in Social Approach Behavior

Social approach behavior is sensitive to environmental stressors, such as aversive sounds, odors, or handling, which can occur in a laboratory testing environment. Mice that are stressed or fearful may be observed freezing, hiding, or darting during a social approach trial, creating unusually long chamber stays or a large number of chamber crossings. If a cohort of control mice exhibits abnormal social approach behavior, then it is necessary to examine whether the mice were exposed to an uncontrolled environmental stressor and, if so, use data collected from another cohort of mice once that variable is removed.

Variability may be reduced through controlling the environment of the mouse being tested. Mice should be acclimated to the behavior apparatus as described above, noting that time spent acclimating may change between mouse strains. Subject mice should not be housed individually, as isolated housing leads to various behavioral effects including aggression and anxiety that may alter baseline social behavior. Levels of home cage enrichment should be kept consistent among cohorts of mice for a given experiment. Inconsistent handling between personnel can be a source of uncontrolled variability; it is advisable to have one person carry out all mouse handling for a given experiment to remove this variable. While it is possible to successfully carry out trials using intraperitoneal or subcutaneous injected drugs, we have noted that if the latency between injection and social approach testing is less than 15 min, wild-type C57BL/6 animals injected with control vehicle (saline) will sometimes show altered patterns of behavior in the apparatus, with increased chamber crossings or freezing being the most notable effect.

5.2. Verification of Data and Analysis

Scripts for "chamber 1 ROI," "chamber 2 ROI," and "chamber 3 ROI" are designed to minimize the occurrence of gaps (resulting from no mouse signal in any chamber region of interest in a frame, a "disappearing mouse") and conflicts (resulting from signal from the mouse occurring in two-chamber regions of interest in the same frame, "two mice"). It is important to keep in mind, however, that both gaps and conflicts occur at a low rate. Therefore, the data for all regions of interest should be hand-checked for accuracy by reviewing raw data for errors and verified by comparing to the original video. In cases where a gap is present, caused by the signal for the mouse briefly dropping below threshold, a notation is made to count signal in the gap areas. If a conflict is present, caused by signal for the mouse appearing in ROIs for two chambers at the same time (e.g., as the mouse moved between chambers), a chamber entry is recorded, and signal from the exiting chamber is not counted if at least the head and shoulders of the subject mouse are in the newly entered chamber.

6. Discussion

Social approach behavioral assays offer a controlled, reproducible method to investigate the tendency of a mouse to engage in social investigation (Fig. 3). Using open source software to assist in the analysis of social approach data increases the ease with which data can be analyzed, and the modular format of the scripts allows users to customize social approach by writing additional programs to measure behaviors. We expect that users of this protocol will modify the scripts and experimental parameters presented here to optimize for their particular needs. By writing custom scripts for different applications, users can study a range of social behaviors using one apparatus.

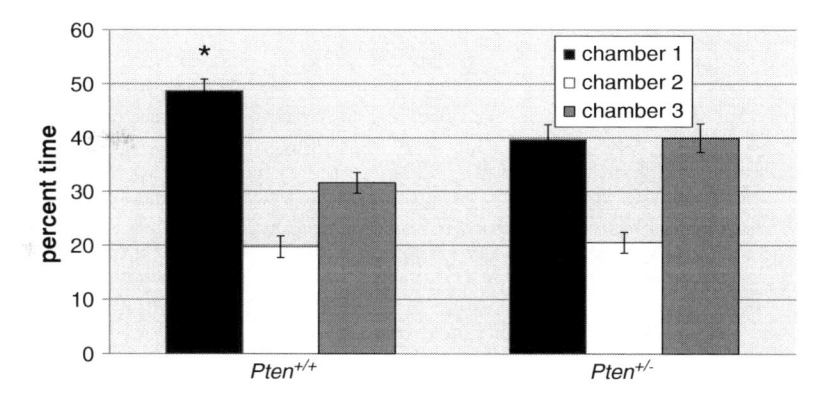

Fig. 3. Characterization of social approach behavior in *Pten*^{+/−} mice. Data showing percent time spent in chamber containing a cage that holds a stimulus mouse (chamber 1), empty chamber (chamber 2), or chamber containing a cage with no stimulus mouse (chamber 3). All mice tested were 12 weeks of age. In females, *Pten*^{+/+} mice show a significant preference for chamber 1 over chamber 3, whereas this preference is not seen in *Pten*^{+/−} mice, indicating a deficit in sociability. *$P < 0.05$, ANOVA within group comparison between chamber 1 and chamber 3. $n = 12$ females for each genotype. *Error bars* indicate SEM. Figure and legend adapted from (24).

Acknowledgments

Research described here was supported by grants from the Nancy Lurie Marks Family Foundation, the Simons Foundation, and the Autism Consortium. Many thanks to Mriganka Sur for space, support and useful discussions regarding this work.

References

1. Crawley JN (2004) Designing mouse behavioral tasks relevant to autistic-like behaviors. Ment Retard Dev Disabil Res Rev 10:248–258.

2. Blanchard RJ, Blanchard DC (2003) Bringing natural behaviors into the laboratory: A tribute to paul maclean. Physiol Behav 79:515–524.

3. Blanchard DC, Griebel G, Blanchard RJ (2001) Mouse defensive behaviors: Pharmacological and behavioral assays for anxiety and panic. Neurosci Biobehav Rev 25:205–218.

4. Ferguson JN, Aldag JM, Insel TR, Young LJ (2001) Oxytocin in the medial amygdala is essential for social recognition in the mouse. J Neurosci 21:8278–8285.

5. Griebel G, Belzung C, Perrault G, Sanger DJ (2000) Differences in anxiety-related behaviours and in sensitivity to diazepam in inbred and outbred strains of mice. Psychopharmacology (Berl) 148:164–170.

6. Laviola G, Terranova ML, Sedowofia K, Clayton R, Manning A (1994) A mouse model of early social interactions after prenatal drug exposure: A genetic investigation. Psychopharmacology (Berl) 113:388–394.

7. Long JM, LaPorte P, Paylor R, Wynshaw-Boris A (2004) Expanded characterization of the social interaction abnormalities in mice lacking dvl1. Genes Brain Behav 3:51–62.

8. Mondragon R, Mayagoitia L, Lopez-Lujan A, Diaz JL (1987) Social structure features in three inbred strains of mice, c57bl/6j, balb/cj, and nih: A comparative study. Behav Neural Biol 47:384–391.

9. Moretti P, Bouwknecht JA, Teague R, Paylor R, Zoghbi HY (2005) Abnormalities of social interactions and home-cage behavior in a mouse model of rett syndrome. Hum Mol Genet 14:205–220.

10. O'Tuathaigh CM, Babovic D, O'Sullivan GJ, Clifford JJ, Tighe O, Croke DT, Harvey R, Waddington JL (2007) Phenotypic characterization of spatial cognition and social behavior

in mice with 'knockout' of the schizophrenia risk gene neuregulin 1. Neuroscience 147: 18–27.

11. Pomerantz SM, Nunez AA, Bean NJ (1983) Female behavior is affected by male ultrasonic vocalizations in house mice. Physiol Behav 31:91–96.

12. Rodriguiz RM, Chu R, Caron MG, Wetsel WC (2004) Aberrant responses in social interaction of dopamine transporter knockout mice. Behav Brain Res 148:185–198.

13. Wersinger SR, Caldwell HK, Christiansen M, Young WS, 3 rd (2007) Disruption of the vasopressin 1b receptor gene impairs the attack component of aggressive behavior in mice. Genes Brain Behav 6:653–660.

14. Wills GD, Wesley AL, Moore FR, Sisemore DA (1983) Social interactions among rodent conspecifics: A review of experimental paradigms. Neurosci Biobehav Rev 7:315–323.

15. Pearson BL, Defensor EB, Blanchard DC, Blanchard RJ (2010) C57bl/6j mice fail to exhibit preference for social novelty in the three-chamber apparatus. Behav Brain Res 213:189–194.

16. Moy SS, Nadler JJ, Perez A, Barbaro RP, Johns JM, Magnuson TR, Piven J, Crawley JN (2004) Sociability and preference for social novelty in five inbred strains: An approach to assess autistic-like behavior in mice. Genes Brain Behav 3:287–302.

17. Moy SS, Nadler JJ, Young NB, Perez A, Holloway LP, Barbaro RP, Barbaro JR, Wilson LM, Threadgill DW, Lauder JM, Magnuson TR, Crawley JN (2007) Mouse behavioral tasks relevant to autism: Phenotypes of 10 inbred strains. Behav Brain Res 176:4–20.

18. Blundell J, Tabuchi K, Bolliger MF, Blaiss CA, Brose N, Liu X, Sudhof TC, Powell CM (2009) Increased anxiety-like behavior in mice lacking the inhibitory synapse cell adhesion molecule neuroligin 2. Genes Brain Behav 8:114–126.

19. Crawley JN, Chen T, Puri A, Washburn R, Sullivan TL, Hill JM, Young NB, Nadler JJ, Moy SS, Young LJ, Caldwell HK, Young WS (2007) Social approach behaviors in oxytocin knockout mice: Comparison of two independent lines tested in different laboratory environments. Neuropeptides 41:145–163.

20. Kwon CH, Luikart BW, Powell CM, Zhou J, Matheny SA, Zhang W, Li Y, Baker SJ, Parada LF (2006) Pten regulates neuronal arborization and social interaction in mice. Neuron 50:377–388.

21. Labrie V, Lipina T, Roder JC (2008) Mice with reduced nmda receptor glycine affinity model some of the negative and cognitive symptoms of schizophrenia. Psychopharmacology (Berl) 200:217–230.

22. Nakajima R, Takao K, Huang SM, Takano J, Iwata N, Miyakawa T, Saido TC (2008) Comprehensive behavioral phenotyping of calpastatin-knockout mice. Mol Brain 1:7.

23. Nakatani J, Tamada K, Hatanaka F, Ise S, Ohta H, Inoue K, Tomonaga S, Watanabe Y, Chung YJ, Banerjee R, Iwamoto K, Kato T, Okazawa M, Yamauchi K, Tanda K, Takao K, Miyakawa T, Bradley A, Takumi T (2009) Abnormal behavior in a chromosome-engineered mouse model for human 15q11-13 duplication seen in autism. Cell 137:1235–1246.

24. Page DT, Kuti OJ, Prestia C, Sur M (2009) Haploinsufficiency for pten and serotonin transporter cooperatively influences brain size and social behavior. Proc Natl Acad Sci USA 106:1989–1994.

25. Tanda K, Nishi A, Matsuo N, Nakanishi K, Yamasaki N, Sugimoto T, Toyama K, Takao K, Miyakawa T (2009) Abnormal social behavior, hyperactivity, impaired remote spatial memory, and increased d1-mediated dopaminergic signaling in neuronal nitric oxide synthase knockout mice. Mol Brain 2:19.

26. Brigman JL, Ihne J, Saksida LM, Bussey TJ, Holmes A (2009) Effects of subchronic phencyclidine (pcp) treatment on social behaviors, and operant discrimination and reversal learning in c57bl/6j mice. Front Behav Neurosci 3:2.

27. Cutler MG (1991) An ethological study of the effects of buspirone and the 5-ht3 receptor antagonist, brl 43694 (granisetron) on behaviour during social interactions in female and male mice. Neuropharmacology 30:299–306.

28. File SE, Hyde JR (1979) A test of anxiety that distinguishes between the actions of benzodiazepines and those of other minor tranquilisers and of stimulants. Pharmacol Biochem Behav 11:65–69.

29. Landauer MR, Balster RL (1982) A new test for social investigation in mice: Effects of d-amphetamine. Psychopharmacology (Berl) 78:322–325.

30. Shi L, Fatemi SH, Sidwell RW, Patterson PH (2003) Maternal influenza infection causes marked behavioral and pharmacological changes in the offspring. J Neurosci 23:297–302.

31. File SE, Seth P (2003) A review of 25 years of the social interaction test. Eur J Pharmacol 463:35–53.

32. Winslow JT (2003) Mouse social recognition and preference. Curr Protoc Neurosci Chapter 8:Unit 8 16.

33. Brodkin ES, Hagemann A, Nemetski SM, Silver LM (2004) Social approach-avoidance behavior of inbred mouse strains towards dba/2 mice. Brain Res 1002:151–157.

34. Nadler JJ, Moy SS, Dold G, Trang D, Simmons N, Perez A, Young NB, Barbaro RP, Piven J, Magnuson TR, Crawley JN (2004) Automated apparatus for quantitation of social approach behaviors in mice. Genes Brain Behav 3:303–314.

35. Tabuchi K, Blundell J, Etherton MR, Hammer RE, Liu X, Powell CM, Sudhof TC (2007) A neuroligin-3 mutation implicated in autism increases inhibitory synaptic transmission in mice. Science 318:71–76.

36. Page DT, Kuti OJ, Sur M (2009) Computerized assessment of social approach behavior in mouse. Front Behav Neurosci 3:48.

37. Hebb AL, Zacharko RM, Dominguez H, Laforest S, Gauthier M, Levac C, Drolet G (2003) Changes in brain cholecystokinin and anxiety-like behavior following exposure of mice to predator odor. Neuroscience 116:539–551.

38. Takahashi LK, Nakashima BR, Hong H, Watanabe K (2005) The smell of danger: A behavioral and neural analysis of predator odor-induced fear. Neurosci Biobehav Rev 29:1157–1167.

39. Yang M, Scattoni ML, Zhodzishsky V, Chen T, Caldwell H, Young WS, McFarlane HG, Crawley JN (2007) Social approach behaviors are similar on conventional versus reverse lighting cycles, and in replications across cohorts, in btbr t+ tf/j, c57bl/6j, and vasopressin receptor 1b mutant mice. Front Behav Neurosci 1:1.

Chapter 6

The Mouse Defense Test Battery: A Model Measuring Different Facets of Anxiety-Related Behaviors

Guy Griebel and Sandra Beeské

Abstract

Defensive behaviors of lower mammals constitute a significant model for understanding human emotional disorders. They generally occur in response to a number of threatening stimuli, including predators, attacking conspecifics, and dangerous objects or situations. Such behaviors can readily be studied in wild rats, wild mice, or in several laboratory mice, which show a complete defensive repertoire in response to danger. Here we describe the mouse defense test battery (MDTB), which measures flight, freezing, defensive threat and attack, and risk assessment in response to an unconditioned predator stimulus, and postthreat (conditioned) defensiveness to the test context. The MDTB represents a significant improvement over other animal models for evaluating drugs active against emotional disorders since it is capable of responding to and differentiating anxiolytic drugs of different classes through specific profiles of effect on different measures.

Key words: Mouse defense test battery, Anxiety, Panic, Defense repertoire, Risk assessment, Flight, Defensive aggression

1. Background and Historical Overview

Defensive behaviors occur in response to a number of threatening stimuli, including predators, attacking conspecifics, and dangerous objects or situations. Such behaviors can readily be studied in wild rats which show a complete defensive repertoire in response to danger. In contrast, in laboratory rats, defensive threat and attack behaviors in response to predators have been much reduced through systematic selection for docility by breeders (1). However, the disadvantages of using wild rats as subjects in laboratory research are obvious. For example, it is clear that the difficulty and cost in obtaining and maintaining these animals are greater than for laboratory rats.

Todd D. Gould (ed.), *Mood and Anxiety Related Phenotypes in Mice: Characterization Using Behavioral Tests, Volume II,* Neuromethods, vol. 63, DOI 10.1007/978-1-61779-313-4_6, © Springer Science+Business Media, LLC 2011

There are reasons to believe that the laboratory mouse has not been so severely selected on the basis of its defensive behaviors. The smaller size of the mouse and its reduced potential to inflict serious wounds, plus the ease of handling mice with a tail pickup, have enabled greater tolerance of defensive attack behavior in this species and, indeed, domesticated mice often show biting behavior to human handling (2). Thus, it has been demonstrated that mice from four lines, three inbred (BALB/c, C57BL/6, and DBA/2) and one outbred (Swiss), show intense defense reactions when confronted with an approaching threat stimulus (laboratory rat). They display initial flight, followed by risk assessment (RA) and defensive vocalization, and biting occur when escape is blocked (3). The concept of RA has emerged from the work of Blanchard and colleagues (4). These authors defined RA in terms of orientation toward present or potential threat, often followed by specific approach responses. They demonstrated that RA is associated with gathering of information concerning threat sources. Together, these defense patterns closely resemble those of wild rats, suggesting that mice of these strains do not show the reductions in flight and defensive threat/attack that are typical of laboratory rats. Such findings clearly indicate that the laboratory mouse may be a suitable subject for studies concerned with defensive behaviors.

However, it was not clear in these initial studies whether the responses displayed by the mice were specific to the encounter with a laboratory rat. The idea that defensive reactions may be elicited by any approaching stimulus was addressed by studying the influence of various stimuli on defensive reactions of Swiss mice (5). Briefly, this study demonstrated that when compared to mice approached by a leather glove, animals confronted with an anesthetized or a conscious rat displayed potentiated flight responses and defensive threat/attack reactions, while RA behavior was generally similar in all three conditions. Furthermore, escape attempts following removal of the stimulus were higher in the rat conditions compared to the leather glove group. In this latter case, however, responses displayed by the leather glove group mice were also higher than those observed in a group which was not exposed to any stimulus, indicating that the leather glove stimulation also elicited defense reactions, albeit at a lower level. Taken together, these results demonstrated that a rat stimulus elicits higher levels of flight reactions and defensive threat/attack responses than a leather glove stimulus, thereby suggesting that this experimental situation is appropriate for investigating antipredator defense.

Factor analyses are commonly used to describe the relationship between different variables and, consequently, to identify specific indices or factors such as anxiety and locomotor activity. Thus, the question whether the different defensive responses elicited in the MDTB provide different measures of the same state or measure distinct states of defensiveness, fear, or anxiety has been approached by performing a factor analysis of the various behavioral defense reactions observed in the battery. The factor analysis identified four main independent factors (6). Factor 1 included cognitive aspects of defensive behaviors that appear to be related to the process of acquiring and analyzing information in the presence of threatening stimuli (i.e., risk assessment). Flight responses heavily loaded on Factor 2. Several defensive threat/attack reactions (i.e., upright postures and biting) highly loaded on Factor 3, indicating that this Factor reflects more affective-orientated defense reactions. Finally, Factor 4, which includes escape attempts in the absence of the rat, relates to contextual defensiveness. Together, this pattern is consistent with the idea that defense reactions of mice exposed to a threat stimulus may relate to different emotional states, and perhaps that they may model different aspects of human anxiety.

To address this hypothesis further, a variety of different clinically effective and marketed anxiolytic agents have been tested in the MDTB (see below). Results suggest that certain defensive behaviors may be considered particularly relevant in modeling specific aspects of anxiety disorders. For example, the observation that there is a rather good correspondence in terms of drug effects between the clinical outcome in panic disorder and generalized anxiety disorder and the ability to modify flight and risk assessment responses, respectively, suggests that the latter behaviors may be considered particularly relevant in modeling some aspects of panic disorder or generalized anxiety disorder. Moreover, previous reports have suggested that there may be an isomorphism between risk assessment in the MDTB and several behaviors often described in generalized anxiety disorder patients (7) such as apprehensive expectation, and vigilance and scanning, involving hyperattentiveness (8). In addition, the observation that panic disorder patients usually report an urgent desire to flee from where the attack is occurring (8) has led to suggest that panic symptoms are due to pathological, spontaneous activation of neuronal mechanisms underlying flight reactions (9, 10). As such, flight behaviors in the mouse defense test battery may model certain aspects of panic (11).

2. Equipment, Materials, and Setup

2.1. Animals

The Swiss strain of mouse is recommended for use in this protocol, because of the animals' high levels of defensiveness (see also point 4 below). Mice are 10–12-week-old at the beginning of the experiment. The stimulus subject is a Long Evans male rats (400–500 g). The use of another strain of rat as a stimulus has not been studied extensively. Some preliminary findings tend to suggest that the Long Evans strain is the most suitable as it elicits more stronger defensive behaviors in mice than Wistar or Sprague–Dawley rats.

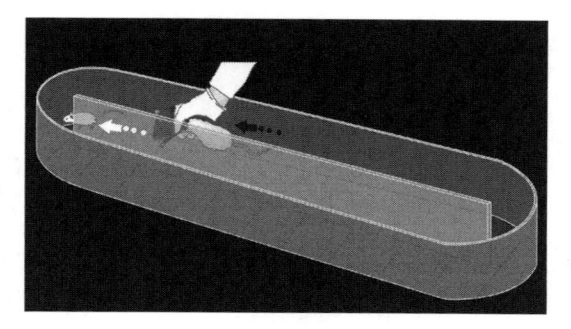

Fig. 1. Runway apparatus.

Both mice and rats are housed singly in polycarbonate cages in a room maintained under a 12-h light/dark cycle with light onset at 6 a.m. Although it is possible to use a live rat as stimulus, it is highly recommended, for obvious ethical reasons, to use an anesthetized (e.g., with 40 mL/kg of pentobarbital) or a freshly terminated (killed by CO_2 inhalation) rat.

2.2. Test Environment

Experiments are performed in a quiet, darkened test room away from disturbance and under red light to minimize visual contact with the experimenter.

2.3. Setup

Two video cameras are mounted vertically above the runway apparatus and connected to a television screen located in an adjacent room. They allow the live recording of pre- and posttest activities. Assessments of defensive responses are made from the recordings with the observer unaware of the original pretreatment.

2.4. Apparatus

The test is conducted in an oval runway, 0.40 m wide, 0.30 m high, and 6.0 m in total length, consisting of two 2 m straight segments joined by two 0.4 m curved segments and separated by a median wall ($2.0 \times 0.30 \times 0.06$) (Fig. 1). The apparatus is elevated to a height of 0.80 m from the floor to enable the experimenter to easily hold the rat, while minimizing the mouse's visual contact with him. All parts of the apparatus are made of black Plexiglas. The floor is marked every 20 cm to facilitate distance measurement. Experiments are performed under red light between 1 p.m. and 5 p.m.

3. Procedure

3.1. Subjects and Drug Administration

Mice are brought into a holding area immediately adjacent to the test room at least 1 h before testing. Animals must be randomly allocated to the various drug groups and injected at the interval appropriate to the route of injection of a particular drug.

3.2. Familiarize Test Subjects to the Test Arena: The Pretest (Minutes 1–3)

Place a mouse in the middle of the runway apparatus. Allow 3 min of free exploration and count line crossings, wall rears, wall climbs, and jump escapes (Fig. 2a).

3.3. Rat Avoidance Test (Minutes 4–6)

Immediately after the 3-min familiarization period, introduce the hand-held stimulus rat at one end of the runway, 2 m from the subject. Bring it up to the subject at a speed of approximately 0.5 m/s, initiating approach only if the subject is at a standstill with its head oriented towards the hand-held rat. Consequently, intervals between trials are variable but never exceeded 15 s. Terminate approach when contact with the subject is made or the subject runs away from the approaching rat. If the subject flees, record avoidance distance (the distance from the rat to the subject at the point of flight). Remove the rat from the apparatus between each trial so that there is no visual contact between the threat stimulus and the subject. Repeat for a total of five approaches (Fig. 2b).

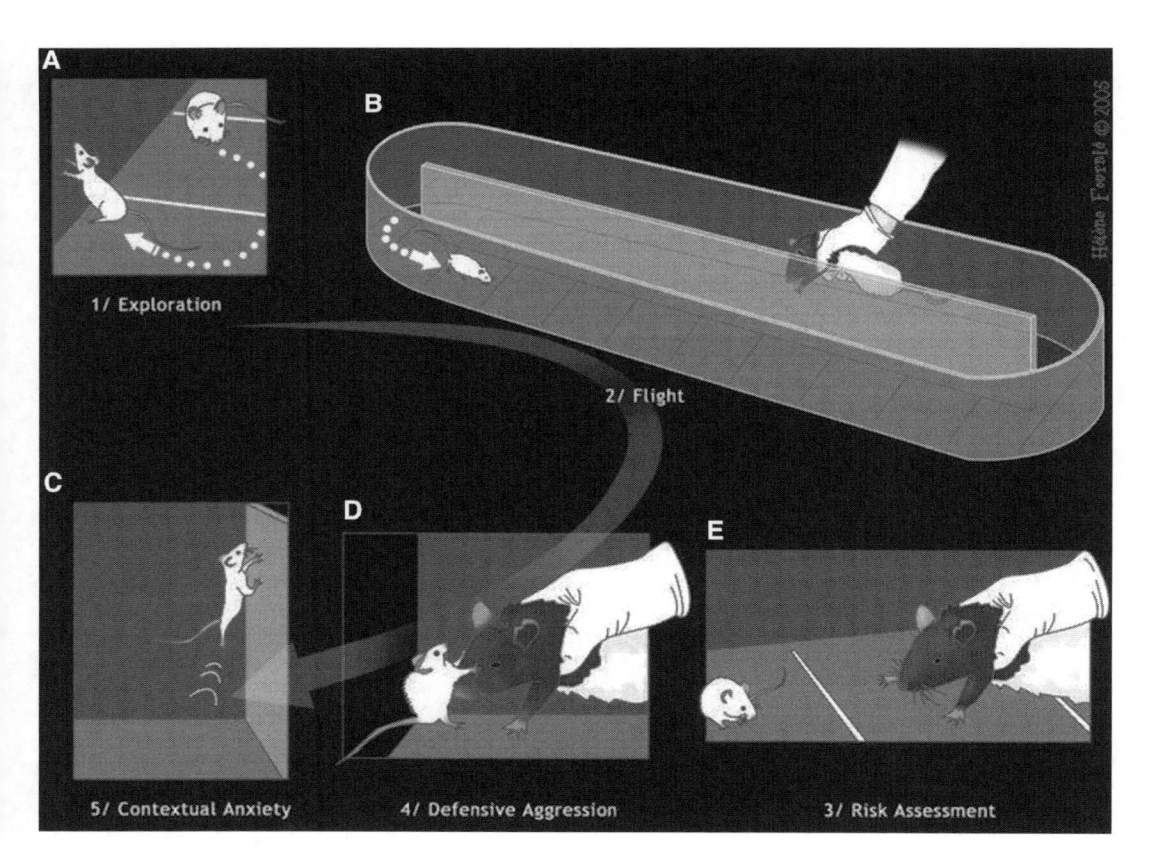

Fig. 2. The different behaviors displayed by mice in the runway apparatus before (exploration), during (flight, risk assessment, and defensive aggression), and after (escape attempts elicited by contextual anxiety) exposure to a Long Evans rat.

3.4. Chase/Flight Test (Minutes 7–8)

Introduce hand-held rat at a distance of 2 m from the subject and initiate chase only when subject is at a standstill with its head oriented toward the rat. Bring rat up to the subject at a speed of approximately 2 m/s. Chase is terminated when the subject has traveled a distance of 15 m. During chase, maintain a constant distance of 20 cm between the two animals. Consequently, if subject stops fleeing before traveling the full 15 m, stop the chase too in order to avoid contact between the two animals; resume by moving the hand-held rat quickly from left to right in front of the subject to elicit flight. Record the following parameters: flight speed (measured when the subject is running straight), number of stops (pause in movement), orientations (subject stops, then orients the head toward the rat), and reversals (subject stops, then runs in the opposite direction). Remove the rat after the chase is completed (Fig. 2e).

3.5. Straight Alley Test (Minutes 9–11)

By the closing of two doors (60 cm distant from each other), the runway is then converted to a straight alley in which the subject was constrained. The rat is introduced in one end of the straight alley. Session is initiated when (1) the subject faces the rat; (2) both animals are 40 cm distant from each other. During 30 s, following measures are taken: immobility time, closest distance between the subject and the rat, and the number of approaches/withdrawals (subject must move more than 20 cm forward from the closed door, then return to it). The hand-held rat remains at the place it is introduced during the full 30 s. After this session, it is removed from the straight alley area.

3.6. Forced Contact Test (Minutes 12–13)

Bring the rat up to contact the subject in the straight alley. Direct approaches quickly (within 1 s) to the subject's head. For each such contact, note bites, vocalizations, upright postures, and jump attacks by the subjects. Remove the rat from the apparatus if no defensive threat and/or attack responses are elicited within 15 s. Repeat this test three times. The time interval between each trial is approximately 5 ± 1 s (Fig. 2d).

3.7. Posttest (Minutes 14–16)

Remove the rat immediately after the forced contact test and open the doors to convert straight alley back to an oval runway. Record line crossings, wall rears, wall climbs, and jump escapes (Fig. 2c) during a 3-min session.

3.8. End of Testing

After removal of each animal, the runway field must be carefully mopped using hot soapy water to remove any residual odor due to urine, faces, or to the rat stimulus.

4. Data Analysis and Anticipated Results

Data are analyzed by a one-way analysis of variance (ANOVA) (avoidance distance, flight speeds, immobility time, and closest distance between animals) or the nonparametric Kruskal–Wallis test for some infrequently occurring or highly variable behaviors (avoidance frequencies, reversals, head orientations, approaches/withdrawals, bites, vocalizations, upright postures, and jump attacks). Subsequent comparisons between treatment groups and control is carried out using Newman–Keuls procedures or the non-parametric Mann–Whitney U test. Pre- vs. posttest differences are evaluated by a combined repeated measures ANOVA followed by a Newman–Keuls posthoc comparison (line crossings) or by the Mann–Whitney U test and Wilcoxon matched pair test if the behavior occurs infrequently (wall climbings and jump escapes).

Strain differences are particularly pronounced in the MDTB. When subjects are chased by the rat, C57BL/6 mice use flight as the dominant defense strategy, while the defensive responses of BALB/c, C57BL/6, and DBA/2 mice consist of flight reactions and RA activities. However, when flight or escape is not possible, RA becomes the predominant feature of the defense repertoire in the C57BL/6 mice. When defensive threat/attack behaviors are required, Swiss, BALB/c, DBA/2, and C57BL/6 mice show very similar reactions in terms of the magnitude of the responses observed. CBA mice are poorly defensive in all these test situations. Finally, after the rat is removed from the test apparatus, Swiss, DBA/2, and C57BL/6 mice generally display more vertical activities than BALB/c mice (12). Although in a few experiments female mice have been used, little has been published by way of validation and the influence of the estrus cycle has not been investigated.

Anxiolytic compounds should decrease defensive behaviors, whereas anxiogenic drugs should show opposite effects. However, some responses may be specifically or mainly affected by certain drug classes. Figure 3 shows a graphical representation on how the different classes of drugs affect defensive responses elicited in the MDTB. For example, benzodiazepines (BZs) decrease RA activities of animals chased by the rat and defensive threat and attack responses, while $5\text{-}HT_{1A}$ agents mainly affect contextual defense and defensive threat and attack behaviors. In addition, high-potency BZs, such as alprazolam and clonazepam, as well as antidepressants belonging to the tricyclics, monoamine-oxydase inhibitors, and selective 5-HT reuptake inhibitors have a clearer impact on flight responses than on other defensive reactions. Moreover, drugs that act on the hypothalamic–pituitary–adrenocortical (HPA) axis, including CRF_1 and V_{1b} receptor antagonists, have been shown to

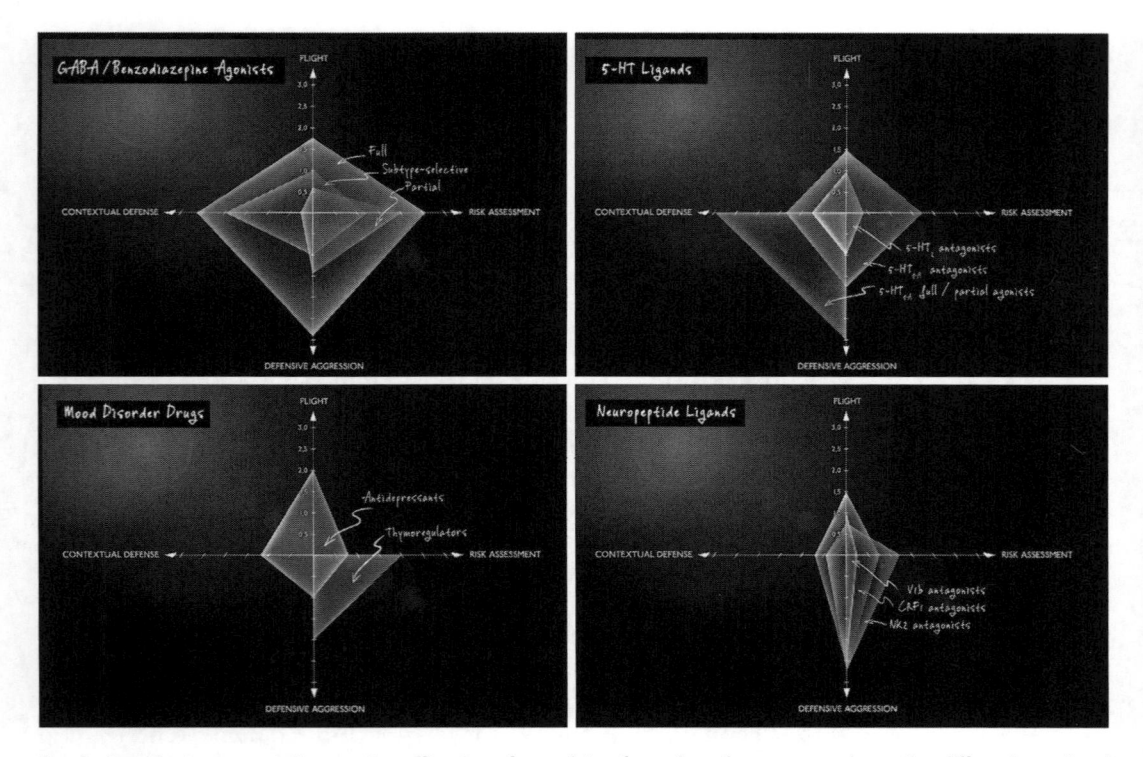

Fig. 3. Graphical representation on the efficacies of a variety of psychoactive compounds on the different emotional behaviors in the mouse defense test battery: flight, risk assessment, defensive aggression, contextual defense. Scores range from 0 (no efficacy) to 3 (highly effective in reducing defensive responses). For more details, see (13).

attenuate the defensive aggression parameters, while leaving the other behaviors unchanged (for a review, see (13)). Taken together, these observations suggest that RA, flight, defensive threat/attack, and escape attempts probably reflect different aspects of anxiety-related reactions, thereby confirming the findings from the factor analysis.

5. Experimental Variables and Troubleshooting

5.1. Training

The MDTB requires about 1 week training for the pre- and post-test measures to be scored reliably, but longer for the defensive measures. A skilled experimenter may be able to perform the test in a satisfactory manner after 2 weeks of practice. A well-trained experimenter is necessary, who is capable to quickly evaluate multiple defensive behaviors, some of which may not be obvious at first glance. Scoring can be live or from tape and must be performed by an observer blind to the drug treatment and test condition.

5.2. Inadequate Levels of Defensive Behaviors

The main problem seems to be the use of batches of animals or strains that give rise to low levels of defensiveness. If this occurs, increase the period of individual housing from 7 to 14 days. Low levels of defensiveness may also occur if the experimental room temperature exceeds 24°C.

5.3. Sedative Drug Effects

The pretest horizontal and vertical activities provide information on sedative or stimulant effects of a drug. These measures can be used to determine the specificity of any changes in defensiveness. If a compound has marked sedative effects, it is likely that all aspects of defensive behaviors will be reduced.

5.4. Reuse of Animals

Use animals only once. Repeated testing may change the nature of the anxiety and hence will also change pharmacological responses. Generally, reexposure to the runway apparatus and the stimulus rat tends to reduce defensiveness.

5.5. Fatigue of the Experimenter

Performing the MDTB requires to be in good shape since the experimenter moves constantly around the runway apparatus and from the holding area to the experimental room. To avoid fatigue and decrease in concentration, limit the testing to about 12–16 animals per day, depending on the experimenter.

6. Concluding Remarks

The MDTB was developed from tests of defensive behaviors in rats, reflecting earlier studies of responses of laboratory and wild rodents to threatening stimuli and situations. It was designed to examine anxiogenic- or anxiolytic-like properties of psychoactive drugs through effects on specific defensive behaviors. Principal component analysis has suggested that the behaviors scored in this procedure may relate to different aspects of anxiety, which relate either to the process of acquiring and analyzing information in the presence of threatening stimuli or to more affective-orientated defense reactions. The MDTB represents a significant improvement over other animal models for evaluating drugs active against emotional disorders since it is capable of responding to and differentiating anxiolytic drugs of different classes through specific profiles of effect on different measures.

References

1. Blanchard RJ, Flannelly KJ, Blanchard DC (1986) Defensive behavior of laboratory and wild *Rattus norvegicus*. J Comp Psychol 100:101–107.

2. Blanchard RJ, Parmigiani S, Agullana R, Weiss SM, Blanchard DC (1995) Behaviors of Swiss-Webster and C57/BL/6 N sin mice in a fear/defense test battery. Aggress Behav 21:21–28.

3. Griebel G, Sanger DJ, Perrault G (1996) The Mouse Defense Test Battery: Evaluation of the effects of non-selective and BZ-1 (ω1) selective, benzodiazepine receptor ligands. Behav Pharmacol 7:560–572.

4. Blanchard DC, Blanchard RJ, Rodgers RJ (1991) Risk assessment and animal models of anxiety. In: Olivier B, Mos J, Slangen JL (eds) Animal Models in Psychopharmacology. Birkhauser Verlag AG, Basel.

5. Griebel G, Blanchard DC, Jung A, Blanchard RJ (1995) A model of 'antipredator' defense in Swiss-Webster mice: Effects of benzodiazepine receptor ligands with different intrinsic activities. Behav Pharmacol 6:732–745.

6. Griebel G, Blanchard DC, Blanchard RJ (1996) Evidence that the behaviors in the Mouse Defense Test Battery relate to different emotional states: A factor analytic study. Physiol Behav 60:1255–1260.

7. Blanchard RJ, Griebel G, Henrie JA, Blanchard DC (1997) Differentiation of anxiolytic and panicolytic drugs by effects on rat and mouse defense test batteries. Neurosci Biobehav Rev 21:783–789.

8. DSM-IV (1994) Diagnostic and Statistical Manual of Mental Disorders, Fourth Edition. American Psychiatric Association, Washington DC.

9. Deakin JFW, Graeff FG (1991) 5-HT and mechanisms of defense. J Psychopharmacol 5:305–315.

10. Graeff FG (1990) Brain defense system and anxiety. In: Burrows GD, Roth M, Noyes R, Jr. (eds) Handbook of Anxiety. vol. 3. The Neurobiology of Anxiety. Elsevier, Amsterdam.

11. Griebel G, Blanchard DC, Blanchard RJ (1996) Predator-elicited flight responses in Swiss-Webster mice: an experimental model of panic attacks. Prog Neuro-Psych Biol Psych 20:185–205.

12. Griebel G, Sanger DJ, Perrault G (1997) Genetic differences in the mouse defense test battery. Aggress Behav 23:19–31.

13. Blanchard DC, Griebel G, Blanchard RJ (2003) The Mouse Defense Test Battery: pharmacological and behavioral assays for anxiety and panic. Eur J Pharmacol 463:97–116.

Chapter 7

Novelty-Suppressed Feeding in the Mouse

Benjamin Adam Samuels and René Hen

Abstract

The use of hyponeophagia, in which exposure to a novel environment suppresses feeding behavior, has been used to assess anxiety-related behavior in animals for over seven decades. More recent work has shown that variations of hyponeophagia, such as the novelty-suppressed feeding test, have become effective paradigms for testing treatment with drugs such as anxiolytics and antidepressants. Most interestingly, unlike many other behavioral paradigms, novelty-suppressed feeding is sensitive to chronic, but not acute, antidepressant treatment, which mirrors the effects of antidepressant treatment in human patients. Here we provide a brief historical overview of novelty-suppressed feeding and provide a protocol for running the test with mice.

Key words: Anxiety, Depression, Mood, Mouse, Antidepressant, Neurogenesis

1. Background and Historical Overview

Novelty-suppressed feeding (hyponeophagia) is a conflict-based test in which an animal that has been deprived of food for a full day faces a choice of approaching and consuming a piece of food in the center of a brightly lit, novel open arena or staying to the side and avoiding the center of this anxiogenic environment. Subjects participating in this test do not require any previous complex training, are not exposed to painful stimuli, and are usually deprived of food for 24 h, less than an animal would face in a normal foraging situation. The main measure of the test is latency to eat (defined as the amount of time it takes for the animal to enter the center of the arena and bite the food pellet with use of forepaws while sitting on its haunches). This is a test of anxiety-related behavior, and therefore, experimental animals with a significantly longer latency to eat than control animals are usually described as more anxious.

Todd D. Gould (ed.), *Mood and Anxiety Related Phenotypes in Mice: Characterization Using Behavioral Tests, Volume II*, Neuromethods, vol. 63, DOI 10.1007/978-1-61779-313-4_7, © Springer Science+Business Media, LLC 2011

Many genetic manipulations resulting in anxiety-related phenotypes in mice, such as the 5-HT1A receptor knockout (1), glucocorticoid receptor knockdown by transgenic antisense (2), and substance P receptor knockout (3), exhibit increased hyponeophagia (4). Furthermore, many drugs with anxiolytic properties, such as benzodiazepines, barbituates, azapirones, and β-adrenergic antagonists, decrease hyponeophagia in rodents (4–6). For a detailed review of pharmacological effects on hyponeophagia, please see Dulawa and Hen (4).

Historically, various precursors of novelty-suppressed feeding have long been used to assess anxiety behavior. Hall first observed a negative correlation between feeding and defecation when animals were exposed to a novel environment in 1934 (7). Classic studies in the 1970s tested the effects of many different anxiolytics in both rats and mice (4, 8). What has become clear is that there are a few essential components of the test that, when altered, can lead to very distinct results. Both hyponeophagia and defecation attenuate with repeated exposures to the environment, suggesting that novelty is a key component of the anxiogenic paradigm. In addition, the lighting used to illuminate the arena can have large effects on anxiety. The greater the intensity of the lighting, the more anxiogenic the environment will be. Finally, when novelty-suppressed feeding is performed in mice, different inbred strains have large baseline differences in hyponeophagia (9), making strain choice critical.

In addition to usefulness in testing anxiety and the effects of anxiolytics, novelty-suppressed feeding has become a popular test for assessing antidepressant efficacy. Commonly used tests for assessing predictive validity for antidepressants, such as forced swim test or tail suspension test, require only a single acute treatment for positive effects. Therefore, from a behavioral perspective, these tests have always lacked face validity as most antidepressants require chronic treatment to yield beneficial effects in humans (10). Importantly, positive effects of antidepressants in novelty-suppressed feeding are only seen after chronic, but not acute or subchronic, treatment in rats (11) and mice (12). Chronic treatment with various antidepressants have been shown to decrease latency to eat, including selective serotonin reuptake inhibitors (SSRIs) such as fluoxetine (prozac) (12–14) and tricyclics (TCAs) such as imipramine (12–14) and desipramine (15).

Anxiety and depression have generally been conceived of as distinct psychiatric disorders, believed to be caused by alterations of different brain circuits (10). However, in reality, anxiety and depression have a high comorbidity with cooccurrence rates up to 60% (16, 17). Therefore, the use of novelty-suppressed feeding to study the effects of chronic antidepressant treatment in various mouse models and the various mechanisms mediating the antidepressant response will shed light on the neurobiology of both anxiety and depression.

One critical mechanism mediating the antidepressant response has already been uncovered using novelty-suppressed feeding as the readout. In addition to decreasing latency to eat in novelty-suppressed feeding, chronic, but not acute or subchronic, treatment of rats and most mouse strains with antidepressants results in an increase in proliferation of adult neural progenitor cells in the dentate gyrus of the hippocampus (12, 18). Interestingly, ablation of this adult neurogenesis niche with a focal radiologic procedure eliminates the antidepressant-induced decrease in latency to eat in novelty-suppressed feeding (12–14), suggesting a requirement of adult neurogenesis to mediate the beneficial effects of antidepressants. Furthermore, multiple stress-induced animal models of depression, such as unpredictable chronic mild stress (UCMS) and chronic corticosterone treatment, yield an increase in latency to eat in novelty-suppressed feeding that can be reversed by chronic antidepressant treatment (13, 14).

Another potentially interesting aspect of novelty-suppressed feeding that has not yet been fully investigated is that animals subjected to chronic antidepressant treatment tend to show a bimodal distribution where some animals show a clear decrease in latency to eat and others do not (Fig. 1). In this example, mice were given fluoxetine (18 mg/kg/day) or saline by oral gavage for 25 days. One of the major drawbacks of antidepressants in psychiatry is that many patients do not respond to treatment. As an example, only 47% of patients respond and only 33% of patients achieve remission

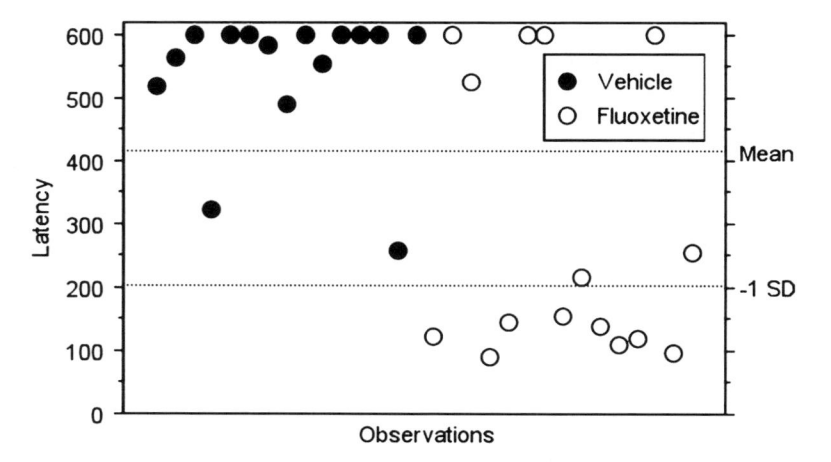

Fig. 1. A univariate scattergram showing response times of individual mouse subjects in novelty-suppressed feeding. In this test, 15 animals were chronically treated (25 days) with the antidepressant fluoxetine (18 mg/kg/day; white circles) and another 15 were treated with saline (*black circles*) by oral gavage. The test ran for 600 s (10 min). Each *circle* represents one animal, with latency to eat plotted along the y axis. The mean and standard deviation of all points are denoted by dashed lines. Fluoxetine significantly decreases the latency to eat in most, but not all, subjects. Note the bimodal distribution within the fluoxetine group in which five subjects are at or near the ceiling of the test. It is possible that these five subjects are models of nonresponders to antidepressant treatment.

in the first line of treatment with a commonly used SSRI (19). Therefore, it is possible that the bimodal distribution of animals in novelty-suppressed feeding following chronic antidepressants may present a model of responders and nonresponders to treatment, and animals could theoretically be separated based on their behavior to study potential mechanisms underlying this divide in responsiveness to treatment.

Finally, when performing novelty-suppressed feeding, it is critical to control for any potential effects of the independent variable of appetite on feeding behavior (4). Many classic anxiolytics, such as benzodiazepines, can stimulate appetite (4). The two most commonly used controls are therefore home cage feeding and percentage weight lost during the deprivation. Ideally, latency to eat in the home cage can also be assessed.

2. Equipment, Materials, and Setup

1. Fifteen to 20 adult mice per group. Mice can be group-housed prior to experiment. All mice should be the same gender and approximately the same age.

2. Experimental arena: Any large container can be used. We use standard mouse shipping containers, available from Taconic Farms. Containers should be cleaned prior to experiment (see Fig. 2 for a picture of animals in the testing arena).

Fig. 2. Animals participating in the novelty-suppressed feeding test. Here, the testing apparatus consists of food pellets attached to a food platform and placed in the center of a large arena (here it is a mouse shipping container).

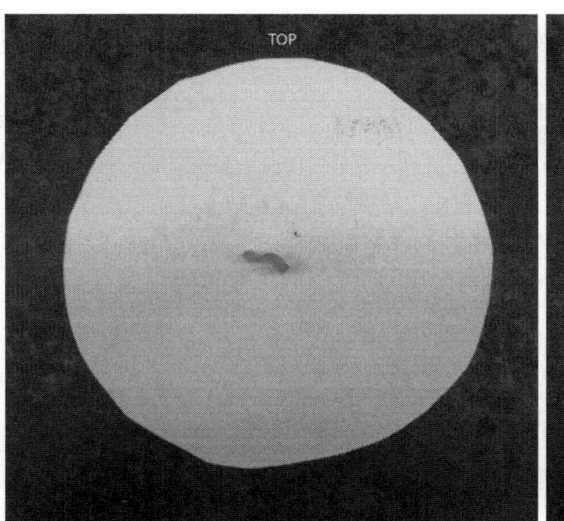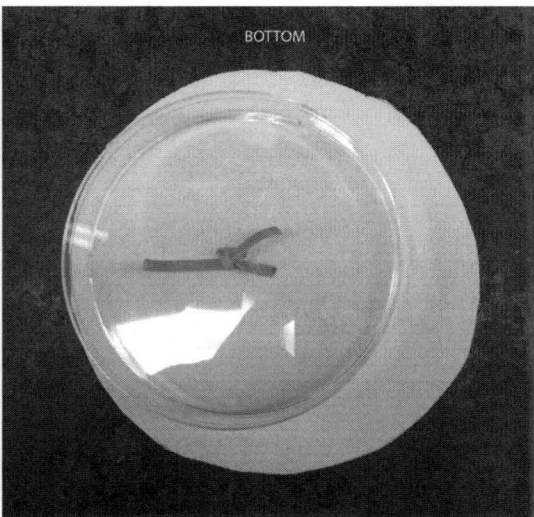

Fig. 3. The food platform. Construction of the platform is simple and requires only a few commonly found lab items. As described in the text, small holes are made in a petri dish and a circle cut from a fresh piece of white Whatman paper. A rubber band is cut, fed through the holes and tied at the bottom. This provides a small loop on the top of the platform to hold the food pellet in place. The walls of the petri dish are then used to hold the platform in place by digging into the bedding of the testing arena.

3. Traceable Daul-Range Light Meter (Luxmeter). Available from VWR International.

4. Standard bedding used in housing cages.

5. Two balances: one with minimum accuracy of 0.5 g for weighing animals, another with minimum accuracy of 0.01 g for weighing food pellets.

6. Platform for food: A petri dish with white circle cut from Whatman paper, tied together with rubber band (Fig. 3). To make the platform: Cut two small holes near the center of a petri dish bottom, approximately 0.5 in. apart. Cut a circle out of a white piece of Whatman paper, diameter approximately 4.5 in. Cut a rubber band to straighten out, then feed through both holes of Whatman paper and both holes of petri dish bottom. Whatman paper should be flush with bottom of petri dish, with dish walls facing away from Whatman paper. Insert a pellet of food into the rubber band loop on top of the Whatman paper. Tighten rubber band and tie on petri dish bottom. These platforms are reusable, but Whatman paper should be replaced regularly. These platforms are necessary to ensure animal does not transport food pellet to the side of arena.

7. Four lamps with 75–100 W bulbs to increase lighting of arena as necessary.

8. A second experimenter to record latencies that is preferably blind to treatment and/or genotype of animals.

9. Two timers and two stopwatches.

10. Clean new cages. Will need 2× new cages for the number that are being run. Therefore, if running eight cages, need 16 clean new cages and 16 lids. Will also need water bottles, food trays, and food for half of the new cages (8 in this example).

11. Plan ahead for order of animal testing and prepare sheets to record latencies, pellet weights, and animal weights (see Fig. 4 and Tables 1 and 2). Can run up to four animals at a time, so plan that groups are distributed evenly across testing. So, for example, if running 40 animals total (8 cages of 5 animals each, one vehicle group of 20 and one treatment group of 20 – see Fig. 4), each run will consist of two animals from the vehicle group and two from the treatment group. This is critical so that across groups animals have been deprived of food for equal

VEHICLE GROUP	TREATMENT GROUP
Cage 1	Cage 5
1	21
2	22
3	23
4	24
5	25
Cage 2	Cage 6
6	26
7	27
8	28
9	29
10	30
Cage 3	Cage 7
11	31
12	32
13	33
14	34
15	35
Cage 4	Cage 8
16	36
17	37
18	38
19	39
20	40

Fig. 4. Designing a new experiment. Here we are outlining the mice, cages, and treatment groups for a new experiment. In this experiment, there are five mice housed per cage and eight cages total. Four cages (cages #1–4; mice #1–20) are the vehicle group, while the other four cages (cages #5–8; mice #21–40) were given a treatment.

Table 1
The log sheet for home cage feeding for the new experiment outlined
in the protocol using the mouse groups in Fig. 4

Home cage feeding

Cage	Animal	Pellet weight start	Pellet weight end	Animal weight predep	Animal weight postdep	Home cage latency
1	1					
1	2					
1	3					
1	4					
1	5					
2	6					
2	7					
2	8					
2	9					
2	10					
3	11					
3	12					
3	13					
3	14					
3	15					
4	16					
4	17					
4	8					
4	19					
4	20					
5	21					
5	22					
5	23					
5	24					
5	25					
6	26					
6	27					
6	28					

(continued)

Table 1
(continued)

Home cage feeding

Cage	Animal	Pellet weight start	Pellet weight end	Animal weight predep	Animal weight postdep	Home cage latency
6	29					
6	30					
7	31					
7	32					
7	33					
7	34					
7	35					
8	36					
8	37					
8	38					
8	39					
8	40					

periods of time. Also, it is critical that only one animal per home cage is run at a time. This is so that at the end of the novelty-suppressed feeding test, individual animals can be placed back in their home cage and food consumption over a set time period can be measured. A log sheet for this example experiment is shown in Tables 1 and 2.

12. Statistical analysis software capable of generating Kaplan–Meier curves (usually considered a type of survival analysis).

3. Procedure

Day 1:

Weigh all animals. Record weight in home cage feeding log sheet (animal weight predep). A sample log sheet is shown in Table 1.

Mark tails with sharpie for quick and easy identification on testing day. This is to avoid stressing animals during behavioral testing by checking toes or ears.

Table 2
The log sheet for novelty-suppressed feeding for the new experiment outlined in the protocol using the mouse groups in Fig. 4

Run #	Cage #	Animal #	Latency	Notes
1	1	1		
	2	6		
	5	21		
	6	26		
2	1	2		
	2	7		
	5	22		
	6	27		
3	1	3		
	2	8		
	5	23		
	6	28		
4	1	4		
	2	9		
	5	24		
	6	29		
5	1	5		
	2	10		
	5	25		
	6	30		
Change bedding in testing arenas				
6	3	11		
	4	16		
	7	31		
	8	36		
7	3	12		
	4	17		
	7	32		
	8	37		
8	3	13		
	4	18		
	7	33		
	8	38		
9	3	14		
	4	19		
	7	34		
	8	39		
10	3	15		
	4	20		
	7	35		
	8	40		

Transfer animals to new, clean home cage. If group-housed, retain all cage mates.

Add new water bottle and food tray, but do not add food. If same water bottles must be used, wipe clean to remove any food particles.

If possible, transfer to testing room for holding overnight during deprivation to acclimatize.

Determine order of testing and prepare log sheets to record latency and home cage feeding. Sample log sheets are shown in Tables 1 and 2.

Day 2:

Testing should begin approximately 24 h after start of food deprivation.

If animals were not held overnight in testing room, transfer to testing room at least 1 h prior to start of testing.

Setup for testing:

Cover bottom of arena with fresh bedding.
Attach 1–2 food pellets to food platform using the rubber band.

Place food platform in center of arena using the walls of the petri dish as a foundation to stabilize location in the bedding.

Ensure lighting is accurate for strain and experimental aim by measuring with luxmeter. To increase lighting, can place lamps with 75–100 W bulbs above arena.

Can run up to four mice at a time, all in different arenas. If running multiple animals at once, ensure even lighting across all arenas by measuring with luxmeter.

If running four mice at once, place all animals from those four home cages into new, clean holding cages with lids only to clear the home cages.

Prepare area for home cage feeding experiment by placing the original four home cages onto a table in a dimly lit area of the room. Weigh a pellet of food for each cage, record the weight (pellet weight start in Table 1), and place it on food tray for animal to have access when returned to home cage. Remove water bottles.

Test can run for 5, 8 or 10 min. In this example, we will assume 10 min.

When ready to begin testing, simultaneously place all subjects in a corner of the arena and then immediately start timer (10 min) and stopwatch simultaneously.

Ensure that the testing room is quiet during the testing period.

Animals will likely first approach the food pellet and sniff without biting. This does not count as eating.

When animal enters center, grasps food pellet with forepaws, and bites, use lap feature to pause the time on the stopwatch.

Immediately remove food platform from testing arena.

Record latency to eat in seconds (latency).

Press lap on stopwatch, which should then revert to the time from when the experiment began.

Repeat until all four animals have eaten or the timer indicates 10 min have passed.

Assign a latency of 600 s to animals that did not eat in the testing period.

Place animals into their original home cage with a single food pellet of known weight in the food tray.

Run timer #2 for 5 min.

If possible, also use stopwatch #2 to also record latency for animals to eat in the home cage. This is usually very quick, within the first 30 s. Record as home cage latency on home cage feeding log sheet (Table 1).

At the end of 5 min, weigh each animal. Record all weights (animal weight postdep in Table 1).

Weigh the food pellet from the home cage and record (pellet weight end in Table 1).

Place each animal into a new home cage with free access to food and water.

Return food pellet that had been weighed to original home cage for next run of testing (pellet weight end of cage 1 animal 1 will be the same as pellet weight start for cage 1 animal 2).

Remove any defecation from the testing arena, and shake the bedding.

When starting second run, ensure original cage mates are placed in the same testing arena.

After five runs when all animals from first set of original home cages have been run through the test, replace bedding and food pellet in testing arena. Also prepare next set of home cages for home cage feeding by transferring animals to a holding cage.

Tip:

If there are two experimenters in the room, it is possible to start the second run of testing while animals from the first run are feeding in the home cage. One experimenter will record latencies in the novelty-suppressed feeding, while the other records animal and pellet weights from the home cage feeding.

4. Data Analysis and Anticipated Results

Data Analysis for novelty-suppressed feeding can be somewhat tricky for two reasons: (1) there will usually be animals that do not eat during the test and (2) the latencies usually do not adhere to a normal distribution. Therefore, standard statistical tests such as ANOVA are inappropriate for analyzing novelty-suppressed feeding latencies. Animals that are assigned 600 s (or the ceiling of the test) are not actually eating at 600 s and thus need to be censored during the statistical analysis. Nonparametric statistics are therefore required. The Kaplan–Meier estimator is particularly useful, because it allows for censoring animals that do not eat during the test. Standard statistical programs such as StatView, SAS, and SPSS are capable of this analysis. It is often found under survival analysis in these programs. Using the same data points shown in the scattergram in Fig. 1, an example of Kaplan–Meier curves is shown in Fig. 5.

To perform statistical analysis, use only total seconds for latencies when performing the Kaplan–Meier estimator to generate a curve. Censor the animals that did not eat in the allotted time. When dealing with multiple variables (e.g., genotype and

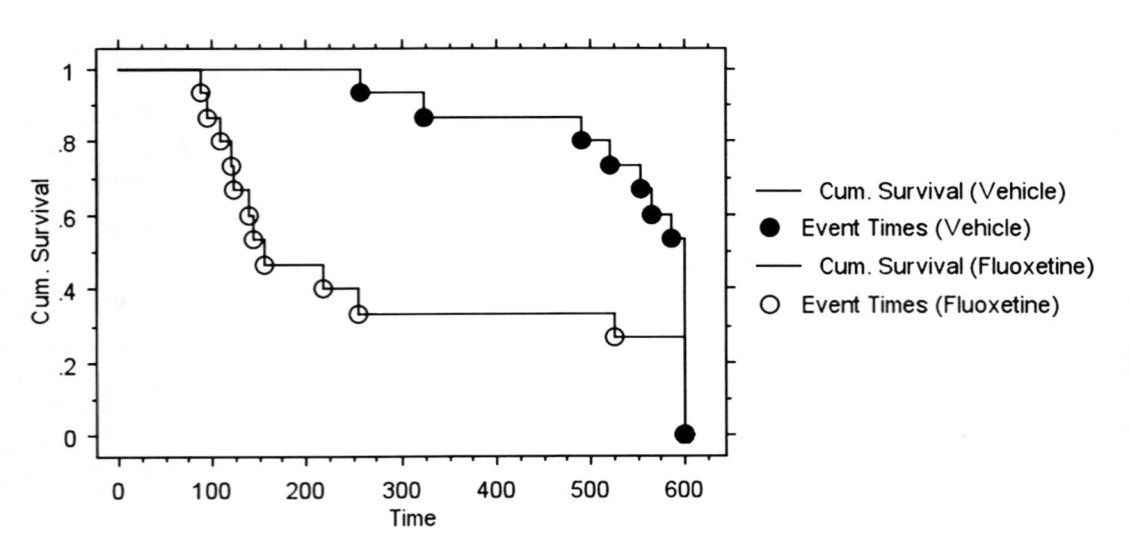

Fig. 5. Data analysis. These are the Kaplan–Meier curves for the animals given fluoxetine and for which the scattergram was shown in Fig. 1. The fluoxetine-treated animals are again denoted by *white circles* while the vehicle-treated are denoted by *black circles*. As an example of how to read this graph, looking at the responders to fluoxetine, approximately 70% ate within the first 250 s of the test (cumulative survival is at approximately 0.3 at 250 s).

Rank Tests for Latency
Censor Variable: Censor
Grouping Variable: Treatment

	Chi-Square	DF	P-Value
Logrank (Mantel-Cox)	5.554	1	.0184
Breslow-Gehan-Wilcoxon	8.238	1	.0041
Tarone-Ware	7.014	1	.0081
Peto-Peto-Wilcoxon	8.238	1	.0041
Harrington-Fleming (rho = .5)	7.014	1	.0081

Fig. 6. Statistical analysis. Multiple tests can be used to compare Kaplan–Meier curves. The most common is the logrank test (Mantel–Cox). For the data shown in Figs. 1 and 5, fluoxetine significantly decreases latency to eat in the novelty-suppressed feeding test ($p=0.0184$).

drug treatment), group the animals with genotype as the strata and the drug treatment as the treatment (sample curves are shown in Fig. 5). The logrank test (often called Mantel–Cox) is used to compare two Kaplan–Meier curves (log rank results for the example experiment are shown in Fig. 6). For home cage feeding, subtract pellet weight at the end of home cage feeding from pellet weight at the start of home cage feeding. Divide the amount eaten (in mg) by the animal's weight. Plot the results and perform analysis of variance (ANOVA) to statistically compare different genotypes and/or treatments. For animal weight lost, subtract weight prior to deprivation from weight following the novelty-suppressed feeding test. Calculate the percentage of weight lost from the weight prior to deprivation. Most mouse strains lose about 10% of their body weight with a 24-h deprivation.

5. Experimental Variables and Troubleshooting

Several variables can lead to dramatically different results in the NSF test. Mouse strain and lighting are the most obvious variables. For mouse strains that are generally anxious, such as 129/SvEv or BALB/c, it is recommended to use relatively lower lighting (approximately 800 lux) or most animals will be on the ceiling of the test (they will not eat). Vice versa, for less anxious strains, such as C57BL/6 or CD-1, it is recommended to use relatively higher lighting (approximately 1,000 lux). For strains of mixed backgrounds, it is highly recommended that extra animals are produced to pilot the lighting conditions. The key is to adjust the anxiogenic properties of the test so that there is a distribution of latencies. If absolutely necessary, it is possible to rerun a cohort of animals through the test, but it is highly recommended that there be a minimum of 10 days between experiments.

Potential confounds of the test are manipulations that will affect feeding behavior. This should be controlled for by comparing both the home cage feeding and weight lost across genotype and/or treatment groups, and if possible, the latency to eat in the home cage.

Though less likely to affect results than alterations in lighting, we have also had some success altering the deprivation period. This only really works if all animals are eating (on the floor of the test) after 24 h. Variations on the deprivation period to 12 h have worked. It is not recommended to increase the deprivation period. However, there are other variations on hyponeophagia that test satiated animals. One example is the novelty-induced hypophagia test (4), in which mice are trained over 3 days to drink sweetened condensed milk. On the fourth day, mice are then presented the sweet milk in a highly anxiogenic environment.

Additionally, an important aspect of the test is the addition of bedding to the arena. The bedding in the testing arena should be the same type of bedding as in the animal's home cage. The bedding is important as it makes the testing arena less aversive around the outside and enhances the contrast with the platform holding the food pellet. Using bedding that is different from the animal's home cage will increase the anxiogenic properties of the test (it will increase latency to eat). Though less common than altering the lighting, using different bedding is also another way to affect results and can be useful if all animals are eating (on the floor of the test) after 24 h.

It is also commonly asked whether the test can be automated. Unfortunately, this has not been sufficiently worked out yet. It is certainly possible to videotape animals participating in the test and measure latency to eat by watching the videos at a later time, but standard video tracking systems have not proved sufficient because they only track an animal's location, not the actual behavior. Animals participating in this test will often enter the center of the arena and sniff the food pellet and then quickly run away, but will not actually take a bite of the food until several minutes later. Currently, standard video tracking software cannot differentiate the sniffing of the food from the actual biting, as in both instances the animal enters the center of the arena and remains for a brief period of time. One way to increase throughput may be to videotape animals in conjunction with video tracking software, and then watch the timepoints when the animal has approached the pellet.

References

1. Gross C, Zhuang X, Stark K et al. (2002) Serotonin1A receptor acts during development to establish normal anxiety-like behaviour in the adult. Nature 416:396–400

2. Rochford J, Beaulieu S, Rousse I et al. (1997) Behavioral reactivity to aversive stimuli in a transgenic mouse model of impaired glucocorticoid (type II) receptor function: effects of diazepam and FG-7142. Psychopharmacology (Berl) 132:145–152

3. Santarelli L, Gobbi G, Blier P et al. (2002) Behavioral and physiologic effects of genetic or pharmacologic inactivation of the substance P receptor (NK1). J Clin Psychiatry 63 Suppl 11:11–17

4. Dulawa SC, Hen R (2005) Recent advances in animal models of chronic antidepressant effects: the novelty-induced hypophagia test. Neuroscience and biobehavioral reviews 29:771–783

5. Shephard RA, Broadhurst PL (1982) Hyponeophagia and arousal in rats: effects of diazepam, 5-methoxy-N,N-dimethyltryptamine, d-amphetamine and food deprivation. Psychopharmacology (Berl) 78:368–372

6. Britton DR, Britton KT (1981) A sensitive open field measure of anxiolytic drug activity. Pharmacol Biochem Behav 15:577–582

7. Hall CS (1934) Emotional behavior in the rat. I. Defecation and urination as measures of individual differences in emotionality. J Comp Psychol 18:385–403

8. Soubrie P, Kulkarni S, Simon P et al. (1975) [Effects of antianxiety drugs on the food intake in trained and untrained rats and mice (author's transl)]. Psychopharmacologia 45:203–210

9. Trullas R, Skolnick P (1993) Differences in fear motivated behaviors among inbred mouse strains. Psychopharmacology (Berl) 111:323–331

10. Nestler EJ, Hyman SE (2010) Animal models of neuropsychiatric disorders. Nat Neurosci 13:1161–1169

11. Bodnoff SR, Suranyi-Cadotte B, Aitken DH et al. (1988) The effects of chronic antidepressant treatment in an animal model of anxiety. Psychopharmacology (Berl) 95:298–302

12. Santarelli L, Saxe M, Gross C et al. (2003) Requirement of hippocampal neurogenesis for the behavioral effects of antidepressants. Science 301:805–809

13. Surget A, Saxe M, Leman S et al. (2008) Drug-dependent requirement of hippocampal neurogenesis in a model of depression and of antidepressant reversal. Biological Psychiatry 64:293–301

14. David DJ, Samuels BA, Rainer Q et al. (2009) Neurogenesis-dependent and -independent effects of fluoxetine in an animal model of anxiety/depression. Neuron 62:479–493

15. Merali Z, Levac C, Anisman H (2003) Validation of a simple, ethologically relevant paradigm for assessing anxiety in mice. Biol Psychiatry 54:552–565

16. Gorman JM (1996) Comorbid depression and anxiety spectrum disorders. Depress Anxiety 4:160–168

17. Leonardo ED, Hen R (2006) Genetics of affective and anxiety disorders. Annu. Rev. Psychol. 57:117–137

18. Malberg JE, Eisch AJ, Nestler EJ et al. (2000) Chronic antidepressant treatment increases neurogenesis in adult rat hippocampus. J Neurosci 20:9104–9110

19. Trivedi MH, Rush AJ, Wisniewski SR et al. (2006) Evaluation of outcomes with citalopram for depression using measurement-based care in STAR*D: implications for clinical practice. Am J Psychiatry 163:28–40

Chapter 8

The Four-Plate Test in Mice

Martine Hascoët and Michel Bourin

Abstract

The four-plate test (FPT) is an animal model of anxiety based on spontaneous response. Animals are exposed to a novel environment. The exploration of this novel surrounding is suppressed by the delivery of mild electric foot shock contingent to quadrant crossing. Animal can only escape from this aversive situation by remaining motionless (passive avoidance). This model of conditioned fear presents several advantages. It is a simple and quick procedure and there is no need for prior training of animals. In this test, benzodiazepines (BZDs) induce a strong antipunishment effect, which has been proposed to be a reflection of their anxiolytic activity. The FPT also allows the detection of anxiolytic effects of other non-BZD anxiolytic compounds such as selective serotonin (5-HT) reuptake inhibitors (SSRI) or mixed serotonin and noradrenaline (NA) reuptake inhibitors (SNRI).

Key words: Anxiolytics, Four-plate test, Mice, Animal model, Receptor ligands

1. Introduction

The use of animal models is well established in the study of biological basis in psychiatric disorders, although there is no evidence for concluding that what occurs in the brain of the animal is equivalent to what occurs in the human brain. Despite traditional difficulties in accepting animal models for psychopathology, these models have become an invaluable tool in the analysis of the multitude of causes, genetic, environmental, or pharmacological that can bring about symptoms homologous to those of patients with a specific disorder. From the onset, it is important to realize that no animal model, natural or induced, can ever fully mirror the human situation which its modeling. Since the last decade, punishment-based assays have fallen out of favor with the growing popularity of more ethologic approaches. On the other hand,

Todd D. Gould (ed.), *Mood and Anxiety Related Phenotypes in Mice: Characterization Using Behavioral Tests, Volume II*, Neuromethods, vol. 63, DOI 10.1007/978-1-61779-313-4_8, © Springer Science+Business Media, LLC 2011

although the majority of the tests are excellent tool for detecting benzodiazepines/GABA compounds, inconsistent results have been reported for drugs that modulate the 5-HT system.

The four-plate test (FPT) introduced by Boissier et al. (1) is based on the suppression of simple innate ongoing behavior, i.e., the exploration of novel surroundings of the mouse. Few true conditioned models of anxiety are employed in mice, e.g., the Vogel test (2, 3), thus the FPT would be probably best classified as a "punishment model." Both freezing behavior and escape or "flight reactions" may be observed in the FPT after the mouse receives a shock and as such the exact fear generated and brain areas associated with this "punishment model" has not yet been totally explored.

The apparatus consists of a floor made of four identical rectangular metal plates. This exploration behavior is suppressed by the delivery of mild electric foot shock contingent on quadrant crossings. Every time the mouse crosses from one plate to another, the experimenter electrifies the whole floor evoking a clear flight-reaction of the animal. Classical anxiolytics, like BDZs, increase the number of punished crossings accepted by the animal (4).

The goal of this chapter is to give a review of the main data obtained with the four-plate test.

2. General Procedure: Four Plate Test

The test is performed in a quiet, darkened room. The mice are placed in holding cages in this room at least 1 h before the test in order to reduce any neophobic response to the test-room environment. After injection (vehicle or treatment), mice are placed in their holding cage. Mice are only used once and were not handled during the housing period. Mice were always tested in a soiled apparatus and no cleaning occurred between trials, but rather at the end of each daily test. For all procedures, observers are unaware of the treatments and all experiments are performed in a randomized manner.

Our apparatus (Bioseb, Chaville, France) consists of a cage ($18 \times 25 \times 16$ cm) floored by four identical rectangular metal plates (8×11 cm) separated from one another by a gap of 4 mm (Fig. 1). The plates are connected to a device that can generate electric foot-shocks (0.6 mA; 0.5 s). Following a 15-s habituation period, the animal is subjected to an electric shock when crossing (transition) from one plate to another, i.e., two legs on one plate and two legs on another (1). The number of punished crossings is calculated for a period of 60 s. An anxiolytic substance is capable of augmenting the number of punished passages.

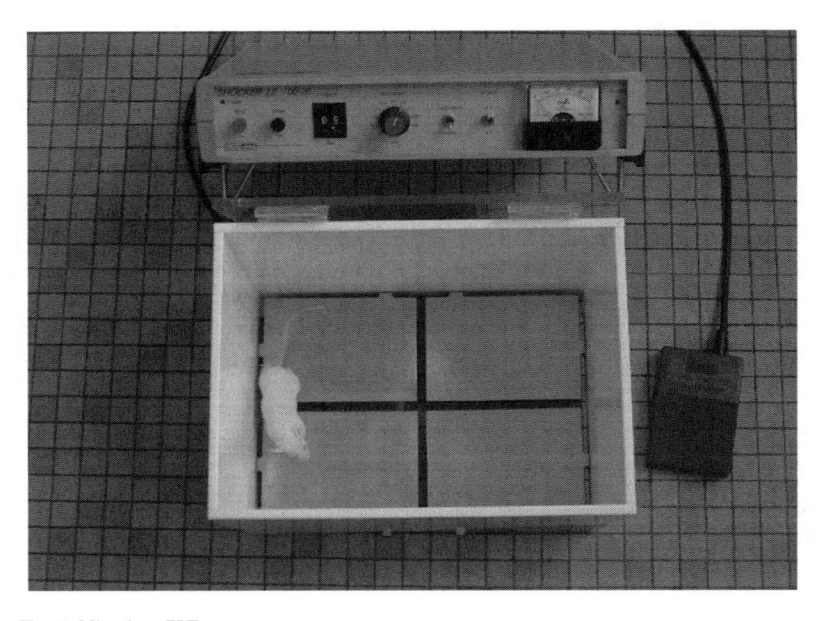

Fig. 1. Mice in a FPT.

3. Drugs

In the present section, the mentioned results from our laboratory were obtained after systemic administration of the drugs intraperitoneally or subcutaneous route.

Experiments were conducted with male mice Swiss weighing 20 ± 2 g.

3.1. FPT and GABAergic System

3.1.1. Benzodiazepines

Diazepam and alprazolam demonstrated a strong antipunishment effect in the FPT when administered 30 min before testing (Fig. 2). The decrease of punished crossings observed for the higher doses was due to the sedative properties of benzodiazepines, but the number of punished crossings was still higher than for saline control animals. Bromazepam demonstrated weak activity and lorazepam failed to show any antipunishment effect in the FPT. Single administration of alprazolam, in comparison with the other four benzodiazepines, produced the highest anxiolytic and lowest sedative effect.

This test, based on unconditioned, punished responses, has been clearly validated for classical anxiolytic compounds (1, 4). However, as with many experimental protocols, drugs that effect general motor function will affect FPT performance. Preliminary screening of locomotor activity appears to be necessary and sufficient to discount stimulant or sedative doses and so for eliminating false positives or negatives drugs. For this purpose, an open field or an actimeter test would be used.

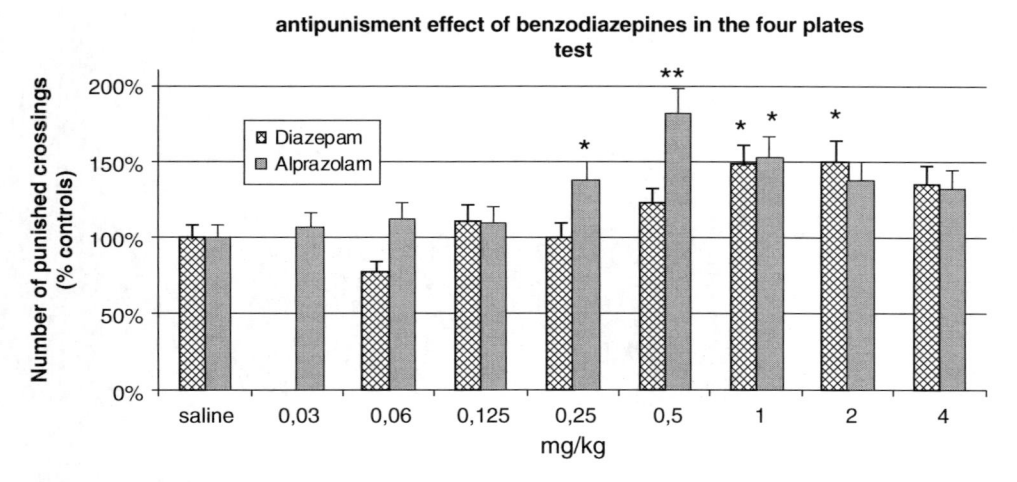

antipunisment effect of benzodiazepines in the four plates test

Fig. 2. Effect of diazepam and alprazolam in the FPT.

3.1.2. Others GABAergic Ligands

Agonists

Alpidem is an imidazopyridine derivative classified as a partial agonist for benzodiazepine receptors. Its effect in the FPT was weak, with only the dose of 4 mg/kg significantly increasing punished crossings (5), upper doses being sedative. Bretazenil another partial agonist for benzodiazepine receptors possessed selective effect on releasing exploratory locomotor activity suppressed by footshock (6). The beta carboline abecarnil demonstrated antipunished behavior in the FPT (6).

The $GABA_A$ receptor agonist muscimol failed to show any effect in the FPT; this might be due to high sedative effect observed in the actimeter test from the dose of 0.25 mg/kg (unpublished data).

Gabapentine is an antiepileptic drug structurally related to GABA (7), but with a mechanism of action still unclear. Gabapentine increases the concentration and synthesis of GABA in the brain (8) and increases the release of GABA from rat striatal brain slice in vitro (9). A recent study has demonstrated that the mechanism of action of gabapentine might be related with an action on $GABA_B$ receptors (10). Gabapentine appears to behave like the benzodiazepine with the same magnitude of effect in animal models of anxiety such as the light dark box in mice, the elevated plus-maze in rat, and an operant conflict paradigm in rat. In the FPT, gabapentine 32 and 64 mg/kg dramatically increased the number of punished passages of mice.

Baclofen, a $GABA_B$ agonist from 0.06 to 2 mg/kg, did not increase punished responding in the FPT.

Antagonists

The $GABA_B$ antagonist CGP 36742 significantly increased the number of punished crossing in the FPT (11).

FG 7142, the benzodiazepine inverse agonist (5), and flumazenil, the benzodiazepine receptor antagonist, do not modify mice behavior in the FPT. Bicuculline (0.25–8 mg/kg) and picrotoxin (0.015–1 mg/kg), two $GABA_A$ receptor antagonists, failed to altered punished behavior in the FPT (unpublished results).

3.2. FPT and Antidepressants

Antidepressants are now recommended as the first-line treatment for many anxiety disorders (12–14).When treating anxiety disorders, the use of antidepressants (ADs) may be an effective alternative to benzodiazepines, which are known to have a high potential for dependency and sedative side effects. The selective serotonin reuptake inhibitors (SSRIs) have been reported to be effective treatments for generalized anxiety disorder (GAD) (15). Although clinical research has revealed promising results concerning the effects of ADs on anxiety, results obtained from animal models of anxiety have remained variable and thus controversial. This has questioned the ability of animal models, validated with BDZs, to demonstrate the anxiolytic-like effects of ADs. Acute administration of established ADs in animals has been reported to produce anxiogenic-like effects in some studies (16, 17) and no specific effects in others (18). However, various authors have shown that ADs may even elicit anxiolytic-like responses (19–21).

In the mice FPT, following acute administration, two ADs, desipramine, maprotiline, two specific inhibitors of norepinephrine reuptake, and trazodone, an atypical ADs, demonstrated anxiogenic-like effects. These results are in accordance with data generally found in the literature. Doses of trazodone and maprotiline, which induced a decrease in punished crossing, were close to sedative doses. However, desipramine did not reduce spontaneous locomotor activity for the doses that induced anxiogenic-like effects; thus it may be concluded that the decrease in number of punished crossings was not due to sedation, but was indeed an anxiogenic-like effect (Table 1).

The SNRIs milnacipran and venlafaxine, together with the SSRIs, paroxetine, fluvoxamine, sertraline, citalopram, but not fluoxetine, dramatically increased the number of punished crossings (Fig. 3). One conclusion could be that the ADs that induce anti-punishment effects seem to act preferentially on the serotoninergic system. More precisely, the selectivity on the uptake of the biogenic amines serotonin (5-HT) and norepinephrine (NE) seems to play a role in the activity of the ADs. The selectivity ratio (r) of the SSRIs as calculated by the IC_{50} NE/IC_{50} 5-HT, (the higher the values, the more selective for 5-HT uptake) decreased from citalopram, ($r = 3400$) through sertraline ($r = 840$), paroxetine ($r = 280$), fluvoxamine ($r = 160$) (all active in the FPT) to fluoxetine ($r = 54$) inactive in the FPT (data from (22)). This preference in serotoninergic selectivity was also corroborated by the fact that at low doses (4, 8, 16 mg/kg) venlafaxine acts preferentially on the 5-HT system, while at higher doses it inhibits both 5-HT and NE reuptake (23). Milnacipran displays the same mechanism of action.

Table 1
Effects of antidepressants in the FPT in mice (from Hascoët et al. (21))

	Vehicle	1 mg/kg	2 mg/kg	4 mg/kg	8 mg/kg	16 mg/kg	32 mg/kg
Nonactive drugs							
Moclobemide	6.0 ± 0.26	5.7 ± 0.47	5.8 ± 0.61	6.2 ± 0.61	5.2 ± 0.32	6.4 ± 0.45	5.7 ± 0.54
Viloxazine	7.8 ± 0.42	–	7.4 ± 0.37	7.6 ± 0.63	7.5 ± 0.67	6.4 ± 0.37	–
Mianserin	6.6 ± 0.40	–	5.9 ± 0.76	6.6 ± 0.67	6.2 ± 0.62	6.2 ± 0.46	4.3 ± 0.45
Imipramine	6.7 ± 0.37	6.8 ± 0.42	6.8 ± 0.57	7.7 ± 0.59	7.4 ± 0.45	6.4 ± 0.52	4.8 ± 0.39
Fluoxetine	$6. \pm 0.401$	6.4 ± 0.65	6.2 ± 0.59	6.8 ± 0.44	6.4 ± 0.60	7.0 ± 0.75	5.0 ± 0.4
Dothiepin	5.1 ± 0.48	5.0 ± 0.44	6.7 ± 0.51	5 ± 0.60	6.6 ± 0.52	6.3 ± 0.45	7.2 ± 0.74
Anxiogenic drugs							
Desipramine	5.3 ± 0.22	–	4.3 ± 0.41	$3.7 \pm 0.26*$	$3.8 \pm 0.25*$	$3.8 \pm 0.35*$	$4 \pm 0.30*$
Trazodone	7.4 ± 0.42	7.5 ± 0.54	7.5 ± 0.70	7.4 ± 0.40	6.7 ± 0.73	8.4 ± 0.45	$3.7 \pm 0.36**$
Maprotiline	7.2 ± 0.92	–	8.3 ± 0.68	5.8 ± 0.70	6.1 ± 0.43	$4.6 \pm 0.47*$	–

*p<0.05
**p<0.01

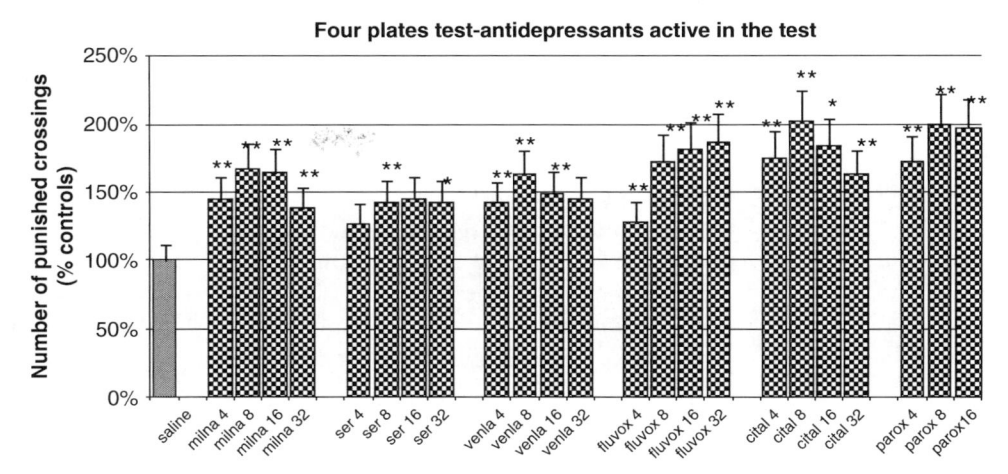

Fig. 3. Effect of various antidepressants in the FPT (from Hascoët et al. (21)).

This also explains the bell-shaped curve with venlafaxine and milnacipran induced by an increased activity of the drugs on NE system.

On the other hand, the decrease of activity observed with citalopram at a dose of 32 mg/kg was probably due to the serotonin syndrome.

The lack of activity in the FPT observed with fluoxetine and imipramine was surprising. But an effect of active metabolites cannot be ruled out.

In conclusion, in the FPT, the kind of behavior observed might be related to anticipatory anxiety with controllable aversive events (24), opposite to depressive anxiety linked with uncontrollable stress. In this case, ADs acting preferentially on serotonin transmission possessed clear anxiolytic-like effects. The exception was the SSRI fluoxetine. The balance between the two transmitters, serotonin and norepinephrine, seemed to be a crucial factor to obtain anxiolytic-like activity in the FPT.

3.3. FPT and Serotonin

As we saw previously, the FPT seems to be sensitive to compounds with high serotonin affinity. 5-HT has been implicated in the neurochemistry of anxiety for a long time, but interest in it has increased dramatically in the past 10 years for a number of reasons: the introduction in clinical practice of the nonbenzodiazepine anxiolytic buspirone (25), and chiefly, the widespread use of antidepressants, in particular the SSRIs for treating anxiety disorders (26–29). The development of specific ligands allows the visualization and the study of the multiplicity of 5-HT receptors and their functional role in health and disease (30, 31). Further, the use of the tryptophan depletion technique in humans (32) and molecular manipulation studies in animals (33, 34) have allowed for a more rigorous assessment of the integrity of the 5-HT system in the CNS.

3.3.1. 5-HT₁ Receptors

In the FPT, only the 5-HT$_{1A}$ receptors subtypes have been studied after acute administration in mice. In animal model of anxiety, these receptors have demonstrated weak or variable effects (35, 36).

Only doses that did not induced stimulant or sedative effect in actimeter test were used in the following study. In the FPT, inconsistent antipunishment effects have been observed with the 5HT$_{1A}$ agonists, and results have not always been replicate across time. 8-OH-DPAT, a specific 5HT$_{1A}$ agonist, failed to induce any antipunishment effect in one study (37) and only has a very weak effect with only two significant doses (1 and 4 mg/kg) in others (unpublished results). Buspirone, a partial 5HT$_{1A}$ agonist, demonstrated an increase in punished crossing for the dose of 0.5 mg/kg and no effect in other studies ((38), unpublished results). WAY 100635, a 5-HT$_{1A}$ antagonist, failed to show any effect in the FPT.

3.3.2. 5-HT₂ Receptors

The identification of the various 5-HT$_2$ receptor subtypes (5-HT$_{2A}$, 5-HT$_{2B}$, and 5-HT$_{2C}$) and their respective localization in areas associated with anxiety in conjunction with the considerable success of SSRIs possessing affinities for various 5-HT$_2$ receptors in affective disorders, have renewed interest in these receptors in anxiety. Early studies implicating the 5-HT$_2$ receptors in anxiety employed nonselective 5-HT$_2$ ligands, suggesting that 5-HT$_2$ antagonism led to a decrease in anxiety, whereas 5-HT$_2$ receptor agonism induced anxiety (39, 40). However, studies in animal models were not as clear cut, with anxiolytic, anxiogenic, or no effects reported depending on the model, species, compound, or experimental conditions utilized (35, 41).

5-HT$_2$ agonists: In the FPT, DOI, a 5-HT$_{2A/2C}$ receptor agonist, produced a consistent anxiolytic-like response from the dose of 0.5–4 mg/kg, with amplitude of response similar to diazepam at doses that did not altered spontaneous locomotor activity (42, 43). BW 723C86, a 5-HT$_{2B}$ agonist, induced antipunishment effect for the doses of 8 mg/kg. RO 60–0175 and mCPP, two 5-HT$_{2C}$ agonists, failed to induce any effect in The FPT (Table 2).

5-HT$_2$ antagonists: All antagonists tested (SR 46349B, a 5-HT$_{2A}$ receptor antagonist; SB 206553 which exhibits high affinity for the 5-HT$_{2B}$ and 5-HT$_{2C}$ receptor and RS 10–2221, a 5-HT$_{2C}$ subtype antagonist) failed to alter mouse behavior in the FPT model of anxiety in Swiss mice. The results were quite surprising, as all 5-HT$_2$ receptor antagonists studied were ineffective in the FPT indicating a lack of anxiolytic effect of these compounds. In contrast, agonism of the 5-HT$_{2A}$ and 5-HT$_{2B}$ receptors seems more implicated in the anxiety generated by the mouse FPT.

3.3.3. 5-HT₃ Receptors

Results obtained with 5-HT$_3$ serotonergic drugs have been less revealing. The 5-HT$_3$ receptor antagonist DAU 6215 was found to have no effect in the FPT (44) and findings with odansetron, also a 5-HT3 antagonist, suggested a weak anxiolytic-like effect in the FPT for the doses of 0.1 and 1 mg/kg (45).

Table 2
Summary of the effects of 5-HT$_2$ ligands in the FPT model of anxiety

	Compound	RS	FPT
5-HT$_2$ agonists	DOI	5-HT2A	+
	BW 723C86	5-HT2B	+
	RO 60-0175	5-HT2B/2C	0
	mCPP	5-HT2B/2C	0
	SR 46349B	5-HT2A	0
	Ketanserin	5-HT2A/2C	0
5-HT$_2$ antagonists	RS 10-2221	5-HT2C	0
	SDZ SER082	5-HT2C/2B	0
	SB 206553	5-HT2B/2C	0
	Deramciclane	5-HT2A; δ	−
	Pirenperone	5-HT2A/2C; α2	0
	Clozapine	5-HT2A/2C; 5-HT3	0
	Cyamemazine	5-HT2A/2C; 5-HT3	0

+ = Anxiolytic-like effect; – = anxiogenic-like effect; 0 no effect. RS = profile of receptor subtype affinity

3.3.4. Other Drugs (from Literature)

See Table 3.

4. The FPT and Analgesia

We have already pointed out the importance of independent spontaneous locomotor activity test such as actimeter or open field in order to conclude to true anxiolytic like effect in the FPT. Another main question has concerned a hypothetic analgesic effect in the FPT.

Indeed, serotonergic, noradrenergic, GABAergic, and dopaminergic systems are implicated in nociception and in anxiety states as well as substance P and opioid systems. The antinociceptive system can also be activated by a variety of stressful stimuli such as footshock and social defeat (64) and stimulation of the periaqueductal gray (PAG) in rat (65). Since mice receive electric foot shock, it is possible that an analgesic action could account for the effects observed in the FPT.

The hot-plate test (66) evaluates the analgesic potential of molecules (selective suppression of pain). The animal is placed on the metal plate heated to 55°C surrounded by a glass cylinder (13–17 cm). The latency (in seconds) of the first jumping is measured. A cut-off time of 2 min is applied. Morphine (4 mg/kg) is used as an internal standard. The hot-plate test is a behavioral model of nociception where organized behaviors such as hind paw-licking

Table 3
Effect of various ligands in the FPT

Drugs		mGLU1	Antagonist	0.5–8 mg/kg	+	Kłodzińska et al. (46)	Mice
AIDA	Glutamatergic ligands	mGLU1	Antagonist	0.5–8 mg/kg	+	Kłodzińska et al. (46)	Mice
LY 354740	Glutamatergic ligands	mGLU2	Agonist	0.5–1 mg/kg	+	Kłodzińska et al. (47)	Mice
MTEP	Glutamatergic ligands	mGLU5	Antagonist	20 mg/kg	+	Rajarao et al. (48)	Mice
MPEP	Glutamatergic ligands	mGLU5	Antagonist	2 mg/kg i.p.	+	Kłodzinska et al. (49)	Mice
AMN082	Glutamatergic ligands	mGLU7	Agonist	3–6 mg/kg i.p.	+	Tatarczyńska et al. (50)	Mice
AP7	NMDA		Antagonist		+	Stachowicz et al. (51)	Mice
SSR125543A	Neuropeptide	CRF1	Antagonist	1–10 mg/kg p.o.	+	Sanger et al. (52)	Mice
Neuropeptide S		NPS		0.2 µg i.c.v.	+	Griebel et al. (53)	Mice
Antalarmin		CRF1	Antagonist	30 mg/kg p.o.	+	Stachowicz et al. (52)	Mice
SSR149415	Vasopressin	AVP	Antagonist	3–10 mg/kg p.o.	+	Leonard et al. (54)	Mice
L-701,324	Glycine B receptor		Antagonist	2 mg/kg	+	Serradeil-Le et al. (55)	Mice
Oxytocin				3–30 mg/kg	+	Przegaliński et al. (56)	Mice
Galnon	Galanin receptor		Agonist	0.1–1 mg/kg i.p.	+	Ring et al. (57)	Mice
IGF-1	Insulin-like growth factor			0.1 µg ICV	+	Malberg et al. (58)	Mice
WAY-200070	Estrogen receptor beta		Agonist	30 mg/kg s.c.	+	Hughes et al. (59)	Mice
JZP-4	Anticonvulsivant			30 mg/kg i.p.	+	Foreman et al. (60)	Mice
Lamotrigine	Anticonvulsivant			10 mg/kg i.p.	+	Foreman et al. (60)	Mice
SR58611A	Beta3 adrenoceptor		Agonist	3–10 mg/kg i.p.	+	Stemmelin et al. (61)	Gerbil
SB 269970	Serotoninergic	5-HT7	Antagonist	1 mg/kg	+	Wesołowska et al. (62)	Mice
SB 399885	Serotoninergic	5-HT6	Antagonist	3–20 mg/kg	+	Wesołowska et al. (63)	Mice

(Column headers: Drugs | | | | Tested doses | Effect | Refs. |)

and jumping are elicited following noxious thermal stimulus. These reactions are controlled by supraspinal mechanisms. Licking is a rapid response elicited by painful thermal stimuli that is a direct indicator of nociceptive threshold (67). In contrast, jumping represents a more elaborated response with a longer latency and encompasses an emotional component of escaping (68).

In a previous study (69), we have compared drug effects in the FPT and in the hot-plate test, in order to disambiguate drug-induced antipunishment from alterations in pain sensitivity.

4.1. Effect of Morphine

Morphine induced strong antinociceptive effects in the hot-plate test for doses that did not induce any effect in the spontaneous locomotor activity test, the actimeter. At doses active in alleviating pain in various tests, morphine did not increase the number of shocks received in the FPT (Table 4) (1).

4.2. Effects of Benzodiazepines

Our results show an antinociceptive effect for alprazolam at the highest dose tested (1 and 2 mg/kg) and for diazepam at doses of 0.25 and 2 mg/kg. Previous works have shown that BZDs significantly inhibit antinociceptive effects of a range of environmental stimuli such as footshock (70). However, there are conflicting reports on the ability of BZDs to induce antinociceptive effects (71, 72). Diazepam induces a dose-dependent attenuation of the effect of opiates in various animal models of pain, such as tail-flick and hot-plate tests (73, 74). However, the data of interaction studies between BZDs and opioids are contradictory. With the exception of the 0.25 mg/kg dose for diazepam, alprazolam and diazepam have both been found previously to decrease spontaneous locomotor activity (Table 4) and to induce a strong dose-dependent myorelaxation at the doses effective in the hot-plate test (4). These effects may account for the results obtained in the hot-plate test (Table 3). Furthermore, in the FPT, alprazolam was found to be anxiolytic from 0.25 to 1 mg/kg and diazepam at 1 and 2 mg/kg (4) (Table 4). In contrast to the hot-plate test, anxiolytic-like effects of these BZDs observed in the FPT were not altered by

Table 4
Comparison of the effects of antidepressants in the hot-plate test, FPT, and locomotor activity

	Hot-plate	FPT	Locomotor activity test
Morphine	+(4–8 mg/kg)	0	0
Diazepam	+(0.25 and 2 mg/kg)	+(1 and 2 mg/kg)	–(1 and 4 mg/kg)
Alprazolam	+(1 and 2 mg/kg)	+(4–16 mg/kg)	–(0.5–4 mg/kg)

+Indicates an analgesic or anxiolytic-like effect or/and locomotor stimulant; –indicates a hyperalgic or anxiogenic-like effect or sedative effect; 0 indicates no effect their sedative and myorelaxation properties. The antinociceptive effects of these BZDs are therefore questionable. Only the effect induced by diazepam 0.25 mg/kg may thus represent a pure antinociceptive effect in the hot-plate test. In addition, this dose has not been found to induce anxiolytic-like effect in the FPT (4). This suggests that GABAergic anxiolytic and antinociceptive processes are triggered differently, each being dependent on different doses and/or brain regions. The antipunishment effect of the two BZDs studied can thus not be attributed to an antinociceptive effect. This is consistent with other reports of the independence of antinociception and behavioral defensive mechanisms (75).

4.3. Effects of Antidepressants

Only the NA reuptake inhibitors (desipramine and maprotiline) and the 5-HT$_{2A}$ antagonist/5-HT reuptake inhibitor trazodone induced analgesia in the hot-plate test. The SSRIs and SNRIs were found to be without effect. Trazodone induced a particularly strong dose-dependent antinociceptive effect. These results are in agreement with earlier studies showing antinociceptive activity for desipramine (76, 77), maprotiline (78), and trazodone (79, 80) in thermal, mechanical, and/or chemical pain tests. However, discrepancies found among articles might be attributable to different experimental conditions and variables measured, noxious stimuli, doses administered, species, and strain used.

Interestingly, the compounds found to be analgesic in the hot-plate test (desipramine, maprotiline, and trazodone) were also found to be either anxiogenic or without any effect (trazodone) in the FPT (21) at the same doses (Table 5). On the contrary, ADs reported to be without effect in the hot-plate test have been previously found to be anxiolytic in the FPT (21) (Table 5). In view of the absence of a relation between antinociceptive and antipunishment effects and inconsistent results on the ability of ADs to induce analgesia in various pain models, we suggest that analgesia and antiaversive effects are completely independent. The antipunishment effects obtained in the FPT following acute administration of ADs cannot be attributed to analgesia or increased locomotor activities.

In summary, if the neural substrates for aversion and pain were the same, then one would expect that anxiolytic and anxiogenic compounds would, respectively, produce increase or decrease of pain threshold. But our results show no relation between antinociceptive and antiunishment effects. It can thus be inferred that, although the neurotransmitters are implicated not only in pain but also in aversive mechanisms, they act through different pathways, in different brain areas and/or receptors according to the situation. Finally, taking into account the inconsistencies of data among the various pain models, the ability of various drugs to modify pain threshold independently of behavioral defensive mechanisms (75, 83) leads us to suggest that the reported drug-induced antipunishment effect in the FPT cannot be attributed to modification of the pain threshold, but to a pure anxiolytic-like effect.

Table 5
Comparison of the effects of antidepressants in the hot-plate test, FPT, and locomotor activity

	Hot-plate	FPT	Locomotor activity test
Fluoxetine	0	0	0 (19)
Milnacipran	0	+(4–16 mg/kg) (21)	0 (81)
Venlafaxine	0	+(4–16 mg/kg) (21)	+(16 mg/kg) (73)
Paroxetine	0	+(4–16 mg/kg) (21)	0 (23)
Imipramine	0	0 (21)	0 (82)
Desipramine	+(16 mg/kg)	−(2–16 mg/kg) (21)	0 (82)
Maprotiline	+(8–6 mg/kg)	−(2–16 mg/kg) (21)	0 (19)
Trazodone	+(2–16 mg/kg)	0 (21)	−(16 mg/kg)
Nomifensine	0	0	+(8 and 16 mg/kg)

+ Indicates an analgesic or anxiolytic-like effect or/and locomotor stimulant; − indicates a hyperalgic or anxiogenic-like effect or sedative effect; 0 indicates no effect

5. Discussion

This paradigm has not been commonly used in other laboratories, making it difficult to formulate interlaboratory comparisons. As such, the various factors potentially influencing the behavioral response of mice have not been profoundly studied. However, results obtained in the FPT have been highly reproducible across time (more than 20 years). The test can be easily used due to its simplicity and superfluous training of animals.

Its success in our laboratory and the demonstration of an anxiolytic-like effect of ADs in this model (in comparison to many of the traditional paradigms employed) emphasizes the validity of this model (21) as AD drugs have a wider spectrum of action than BDZs in the treatment of anxiety disorders. More robust and reproducible results were observed in the FPT in comparison to the EPM and L/D paradigms.

The FPT is now being increasingly used (49, 52, 62, 84, 85). However, in others laboratories, it is mainly used only for the screening of potentially new anxiolytics.

In our laboratory, this test has allowed to highlight neurobiological mechanisms (21, 42, 43, 86–89). Results presented in the present chapter have also demonstrated the robustness of the FPT. This paradigm has demonstrated potent and reproducible properties in predicting the anxiolytic-like profiles of 5-HT_{2A}, 5-HT_{2B} receptor agonists, and ADs. It was used to investigate the influence of the serotonergic system in the anxiolytic activity of ADs.

It has also permitted the investigation of the role of the each 5-HT_2 receptor subtype in their mechanisms of action and indicates a central position for the 5-HT_2 receptor and, in particular, the 5-HT_{2A} and 5-HT_{2B} subtypes in anxiety.

A preliminary study (unpublished results) has demonstrated neurobiological modifications following FPT exposition. Among them, a decrease in 5-HT and 5-HIAA concentrations was found in the hippocampus. In addition, in order to further understand mechanisms implicated in the FPT, we have used local injections paradigm to localize structures implied in anxiolytic-like effects of the benzodiazepine, diazepam, and the $5\text{-HT}_{2A/2C}$ agonist, DOI (90, 91). The periacqueductal grey substance, three sub-regions of hippocampus (CA1–CA3), and two nuclei of amygdala (BLA and LA) have been studied. Diazepam had an anxiolytic-like effect in naïve mice, only when injected into lateral nucleus of amygdala. DOI injections elicit anxiolytic-like effects only when injected into CA2, comparable to the anxiolytic-like effect of diazepam 1 mg/kg injected intraperitoneally. In amygdala and PAG, DOI exerted an anxiogenic-like effect. These results have helped us to understand mechanisms of action of ligands in the FPT, probably through an interaction with several neurotransmitter systems, including serotonergic and GABAergic systems.

Our laboratory equally reported that a single prior undrugged exposure to the FPT reduces punished responding on retest at intervals ranging from 24 h to 42 days (37). Furthermore, prior experience attenuates the anxiolytic response to the benzodiazepines diazepam and lorazepam, similar to results observed in the EPM and L/D (89, 92). Previously described by File and Lister, the abolishment of anxiolytic-like effect of benzodiazepines in experienced mice has been widely studied, especially in elevated plus-maze (93). The loss of effect observed with diazepam in the retest session with FPT can be considered as a one-trial tolerance phenomenon. Punishment is not the triggering factor, while knowledge of the environment seems to be the main reason in the appearance of one-trial tolerance to benzodiazepines. FPT may be a good candidate to study one-trial tolerance because punishments act as a potentiator in this model (94).

6. Conclusion

The FPT creates now renewed interest due to its simplicity with reproducible results. The FPT does not seem to be influenced by variability in test condition that contributes to discrepancies among results, including a wide range of experimental animals used (age, gender, strain) and procedures adopted (housing conditions, handling, time of testing, illumination, method of scoring, routes of drug administration). There is no need of training compared with other conflict tests. For screening purpose, the FPT might be the first choice test. However, before any conclusion can be drawn, it is necessary to verify that drug has no analgesic effects and does not modify spontaneous locomotor activity. This is easily verified utilizing a hot-plate apparatus, employing morphine as the control compound and an actimeter test or open field test. Finally, the FPT allows the investigation of more complex mechanisms' implication neurotransmitter systems and/or receptors' localization in discrete regions of the brain.

References

1. Boissier JR, Simon P, Aron C (1968) A new method for rapid screening of minor tranquilizers in mice. Eur J Pharmacol 4:145–151

2. Liao JF, Hung WY, Chen CF (2003) Anxiolytic-like effects of baicalein and baicalin in the vogel conflict test in mice. Eur J Pharmacol 464:141–146

3. Umezu T (1999) Effects of psychoactive drugs in the vogel test in mice. Jpn J Pharmacol 80:111–118

4. Bourin M, Hascoët M, Mansouri B, Colombel MC, Bradwejn J (1992) Comparison of behavioral effects after single and repeated administrations of four benzodiazepines in three mice behavioral models. J Psychiatry Neurosci 17:72–77

5. Hascoët M and Bourin M (1997) Anticonflict effect of alpidem as compared with the benzodiazepine alprazolam in Rats. Pharmacol Biochem Behav 2:317–324

6. Jones GH, Schneider C, Schneider HH, Seidler J, Cole BJ and Stephens DN (1994) Comparison of several benzodiazepine receptor ligands in two models of anxiolytic activity in the mouse: an analysis based on fractional receptor occupancies. Psychopharmacol (Ber) 114:191–199

7. Sills GJ (2006) The mechanisms of action of gabapentine and pregabalin. Curr Opin Pharmacol. 6:108–113

8. Taylor CP, Gee NS, Su TZ, Kocsis JD, Welty DF, Brown JP, Dooley DJ, Boden P, Singh L (1998) A summary of mechanistic hypotheses of gabapentine pharmacology. Epilepsy Res 29:233–249

9. Götz E, Feuerstein TJ, Lais A, Meyer DK (1993) Effects of gabapentine on release of gamma-aminobutyric acid from slices of rat neostriatum. Arzneimittelforschung 43:636–638

10. Roberto M, Gilpin NW, O'Dell LE, Cruz MT, Morse AC, Siggins GR, Koob GF (2008) Cellular and behavioral interactions of gabapentine with alcohol dependence. J Neurosci 28:5762–5771

11. Partyka A, Kłodzińska A, Szewczyk B, Wierońska JM, Chojnacka-Wójcik E, Librowski T, Filipek B, Nowak G, Pilc A (2007) Effects of GABAB receptor ligands in rodent tests of anxiety-like behavior. Pharmacol Rep 59:757–762

12. Feighner JP (1999) Overview of antidepressants currently used to treat anxiety disorders. J Clin Psychiatry 60:18–22

13. Gorman JM, Kent JM (1999) SSRIs and SNRIs: broad spectrum of efficacy beyond major depression. J Clin Psychiatry 60:33–38

14. Zohar J, Westenberg HG (20004) Anxiety disorders: a review of tricyclic antidepressants and selective serotonin reuptake inhibitors. Acta Psychiatr Scand Suppl 03:39–49

15. Rocca P, Fonzo V, Scotta M, Zanalda E, Ravizza L (1997) Paroxetine efficacy in the treatment of generalised anxiety disorder. Acta Psychiatr Scand 95: 444–450

16. File SE (1985). Animal models for predicting clinical efficacy of anxiolytic drugs: social behaviour. Neuropsychobiology 13:55–62

17. Linnoila M, Eckhardt M, Durcan M, Lister R, Martin P (1987) Interactions of serotonin with ethanol: clinical and animal studies. Psychopharmacology Bull 23:452–457

18. Chopin P, Briley M (1987) Animal models of anxiety: the effects of compounds that modify 5-HT neurotransmission. TIPS 8:383–389

19. Bourin M, Redrobe JP, Hascoet M, Colombel MC, Baker GB (1996) A schematic representation of the psychopharmacological profile of

antidepressants. Prog Neuro-Psychopharmacol Biol Psychiat 20:1389–1402

20. Handley SL, McBlane JW (1992) Opposite effects of fluoxetine in two animal models of anxiety. Br J Pharmacol 107:446P (suppl)

21. Hascoët M, Bourin M, Colombel MC, Fiocco AJ, Baker GB (2000) Anxiolytic-like effects of antidepressants after acute administration in a four-plate test in mice. Pharmacol Biochem Behav 65:339–344

22. Hyttel J (1996) Pharmacological characterisation of selective serotonin reuptake inhibitors (SSRIs). Int Clin Psychopharmacol 9:19–26

23. Redrobe JP, Bourin M, Colombel MC, Baker GB(1998) Dose-dependent noradrenergic and serotonergic properties of venlafaxine in animal models indicative of antidepressant activity. Psychopharmacology 138:1–8

24. Griebel G (1996) Variability in the effect of 5-HT related compounds in experimental models of anxiety : evidence for multiple mechanism of 5-HT in anxiety or never-ending story?. Polish J Pharmacol 48: 129–136

25. Eison MS (1989) The new generation of serotonergic anxiolytics: possible clinical roles. Psychopathology 22:13–20

26. Ables AZ, Baughman OL 3rd (2003) Antidepressants: update on new agents and indications. Am Fam Physician 67:547–554

27. Bourin M, Lambert O (2002) Pharmacotherapy of anxious disorders. Hum Psychopharmacol 17:383–400

28. Nemeroff CB (2003) Anxiolytics: past, present, and future agents. J Clin Psychiatry 64:3–6

29. Vaswani M, Linda FK, Ramesh S (2003) Role of selective serotonin reuptake inhibitors in sychiatric disorders: a comprehensive review. Prog Neuropsychopharmacol Biol Psychiatry 27:85–102

30. Lucki I (1996) Serotonin receptor specificity in anxiety disorders. J Clin Psychiatry 7:5–10

31. Passchier J, van Waarde A (2001) Visualisation of serotonin-1A (5-HT1A) receptors in the central nervous system. Eur J Nucl Med 28:113–129

32. Bell C, Abrams J, Nutt D (2001) Tryptophan depletion and its implications for psychiatry. Br J Psychiatry 178:399–405

33. Murphy DL, Wichems C, Li Q, Heils A (1999) Molecular manipulations as tools for enhancing our understanding of 5-HT neurotransmission. Trends Pharmacol Sci 20:246–252

34. Zhuang X, Gross C, Santarelli L, Compan V, Trillat AC, Hen R (1999) Altered emotional states in knockout mice lacking 5-HT1A or 5-HT1B receptors. Neuropsychopharmacology 21:52 S-60 S

35. Griebel G (1995) 5-Hydroxytryptamine-interacting drugs in animal models of anxiety disorders: more than 30 years of research. Pharmacol Ther 65:319–395

36. Millan MJ (2003) The neurobiology and control of anxious states. Prog Neurobiol 70:83–244.

37. Hascoët M, Bourin M, Couetoux du Tertre A (1997) Influence of prior experience on mice behavior using the four-plate test. Pharmacol Biochem Behav 58:1131–1138

38. Hascoët M, Bourin M, Nic Dhonnchadha BA (2000) The influence of buspirone, and its metabolite 1-PP, on the activity of paroxetine in the mouse light/dark paradigm and four plates test. Pharmacol Biochem Behav 67:45–53

39. Charney DS, Woods SW, Goodman WK, Heninger GR (1987) Serotonin function in anxiety. II. Effects of the serotonin agonist MCPP in panic disorder patients and healthy subjects. Psychopharmacology (Berl) 92:14–24

40. Hensman R, Guimarães FS, Wang M, Deakin JF (1991) Effects of ritanserin on aversive classical conditioning in humans. Psychopharmacology (Berl) 104:220–224

41. Griebel G, Perrault G, Sanger DJ (1997) A comparative study of the effects of selective and non-selective 5-HT2 receptor subtype antagonists in rat and mouse models of anxiety. Neuropharmacology 36:793–802

42. Nic Dhonnchadha BA, Bourin M, Hascoët M (2003) Anxiolytic-like effects of 5-HT$_2$ ligands on three mouse models of anxiety. Behav Brain Res 140:203–214

43. Nic Dhonnchadha BA, Hascoët M, Jolliet P, Bourin M (2003) Evidence for a 5-HT$_{2A}$ receptor mode of action in the anxiolytic-like properties of DOI in mice. Behav Brain Res 147:175–184

44. Borsini F, Brambilla A, Cesana R, Donetti A (1993) The effect of DAU 6215, a novel 5HT-3 antagonist, in animal models of anxiety. Pharmacol Res 27:151–164

45. Dooley DJ, Klamt I (1993) Differential profile of the CCK$_B$ receptor antagonist CI-988 and diazepam in the four-plate test. Psychopharmacology (Berl) 112:452–454

46. Kłodzińska A, Tatarczyńska E, Stachowicz K, Chojnacka-Wójcik E (2004) The anxiolytic-like activity of AIDA (1-aminoindan-1,5-dicarboxylic acid), an mGlu 1 receptor antagonist. J Physiol Pharmacol 55:113–126

47. Kłodzińska A, Chojnacka-Wójcik E, Pałucha A, Brański P, Popik P, Pilc A (1999) Potential anti-anxiety, anti-addictive effects of LY 354740, a selective group II glutamate metabotropic receptors agonist in animal models. Neuropharmacology 38:1831–1839

48. Rajarao SJ, Platt B, Sukoff SJ, Lin Q, Bender CN, Nieuwenhuijsen BW, Ring RH, Schechter LE, Rosenzweig-Lipson S, Beyer CE (2007) Anxiolytic-like activity of the non-selective galanin receptor agonist, galnon. Neuropeptides 41:307–320

49. Klodzinska A, Tatarczyńska E, Chojnacka-Wójcik E, Nowak G, Cosford ND, Pilc A (2004) Anxiolytic-like effects of MTEP, a potent and selective mGlu5 receptor agonist does not involve GABA(A) signaling. Neuropharmacology 47:342–350

50. Tatarczyńska E, Klodzińska A, Chojnacka-Wójcik E, Palucha A, Gasparini F, Kuhn R, Pilc A (2001) Potential anxiolytic- and antidepressant-like effects of MPEP, a potent, selective and systemically active mGlu5 receptor antagonist. Br J Pharmacol 132:1423–1430

51. Stachowicz K, Brañski P, Kłak K, van der Putten H, Cryan JF, Flor PJ, Andrzej P (2008) Selective activation of metabotropic G-protein-coupled glutamate 7 receptor elicits anxiolytic-like effects in mice by modulating GABAergic neurotransmission. Behav Pharmacol 19:597–603

52. Sanger DJ, Joly D (1991) The effects of NMDA antagonists on punished exploration in mice. Behav Pharmacol 2:57–63

53. Griebel G, Simiand J, Steinberg R, Jung M, Gully D, Roger P, Geslin M, Scatton B, Maffrand JP, Soubrié P (2002) 4-(2-Chloro-4-methoxy-5-methylphenyl)-N-[(1 S)-2-cyclopropyl-1-(3-fluoro-4-methylphenyl)ethyl]5-methyl-N-(2-propynyl)-1, 3-thiazol-2-amine hydrochloride (SSR125543A), a potent and selective corticotrophin-releasing factor(1) receptor antagonist. II. Characterization in rodent models of stress-related disorders. J Pharmacol Exp Ther 301:333–345

54. Leonard SK, Dwyer JM, Sukoff Rizzo SJ, Platt B, Logue SF, Neal SJ, Malberg JE, Beyer CE, Schechter LE, Rosenzweig-Lipson S, Ring RH (2008) Pharmacology of neuropeptide S in mice: therapeutic relevance to anxiety disorders. Psychopharmacology (Berl) 197: 601–611

55. Serradeil-Le Gal C, Wagnon J 3 rd, Tonnerre B, Roux R, Garcia G, Griebel G, Aulombard A (2005) An overview of SSR149415, a selective nonpeptide vasopressin V(1b) receptor antagonist for the treatment of stress-related disorders. CNS Drug Rev 11:53–68

56. Przegaliński E, Tatarczyńska E, Chojnacka-Wójcik E (1998) Anxiolytic- and antidepressant-like effects of an antagonist at glycineB receptors. Pol J Pharmacol 50:349–354

57. Ring RH, Malberg JE, Potestio L, Ping J, Boikess S, Luo B, Schechter LE, Rizzo S, Rahman Z, Rosenzweig-Lipson S (2006) Anxiolytic-like activity of oxytocin in male mice: behavioral and autonomic evidence, therapeutic implications. Psychopharmacology (Berl) 185:218–225

58. Malberg JE, Platt B, Rizzo SJ, Ring RH, Lucki I, Schechter LE, Rosenzweig-Lipson S (2007) Increasing the levels of insulin-like growth factor-I by an IGF binding protein inhibitor produces anxiolytic and antidepressant-like effects. Neuropsychopharmacology 32:2360–2368

59. Hughes ZA, Liu F, Platt BJ, Dwyer JM, Pulicicchio CM, Zhang G, Schechter LE, Rosenzweig-Lipson S, Day M (2008) WAY-200070, a selective agonist of estrogen receptor beta as a potential novel anxiolytic/antidepressant agent. Neuropharmacology 54:1136–1142

60. Foreman MM, Hanania T, Eller M (2009) Anxiolytic effects of lamotrigine and JZP-4 in the elevated plus maze and in the four plate conflict test. Eur J Pharmacol 14:602:316–320

61. Stemmelin J, Cohen C, Terranova JP, Lopez-Grancha M, Pichat P, Bergis O, Decobert M, Santucci V, Françon D, Alonso R, Stahl SM, Keane P, Avenet P, Scatton B, le Fur G, Griebel G (2008) Stimulation of the beta3-Adrenoceptor as a novel treatment strategy for anxiety and depressive disorders. Neuropsychopharmacology 33:574–587

62. Wesołowska A, Nikiforuk A, Stachowicz K, Tatarczyńska E (2006) Effect of the selective 5-HT7 receptor antagonist SB 269970 in animal models of anxiety and depression. Neuropharmacology 51:578–586

63. Wesołowska A, Nikiforuk A (2007) Effects of the brain-penetrant and selective 5-HT6 receptor antagonist SB-399885 in animal models of anxiety and depression. Neuropharmacology 52:1274–1283

64. Grisel JE, Fleshner M, Watkins LR, Maier SF (1993) Opioid and nonopioid interactions in two forms of stress-induced analgesia. Pharmacol Biochem Behav 45:161–172

65. Fardin V, Oliveras JL, Besson JM (1984) A reinvestigation of the analgesic effects induced by stimulation of the periaqueductal gray matter in the rat: IR The production of behavioral side effects together with analgesia. Brain Res 306:105–123

66. Jacob JJ, Tremblay EC, Colombel MC (1994) Enhancement of nociceptive reactions by naloxone in mice and rats. Psychopharmacologia 37:217–223

67. Espejo EF, Mir D (1993) Structure of the rat's behaviour in the hot plate test. Behav Brain Res 56:171–176

68. Espejo EF, Stinus L, Cador M, Mir D (1994) Effects of morphine and naloxone on behaviour in the hot plate test: an ethopharmacological study in the rat. Psychopharmacology 113:500–510

69. Ripoll N, Hascoët M, Bourin M (2006) The four-plates test: anxiolytic or analgesic paradigm? Prog Neuropsychopharmacol Biol Psychiatry 30:873–880

70. Drugan RC, Ryan SM, Minor TR, Maier SF (1984) Librium prevents the analgesia and shuttlebox escape deficit typically observed following inescapable shock. Pharmacol Biochem Behav 21:749–754

71. Gatch MB (1999) Effects of benzodiazepines on acute and chronic ethanol-induced nociception in rats. Alcohol Clin Exp Res 23:1736–1743

72. Nadeson R, Guo Z, Porter V, Gent JP, Goodchild CS (1996) Gamma-aminobutyric acidA receptors and spinally mediated antinociception in rats. J Pharmacol Exp Ther 278:620–626

73. Pakulska W, Czarnecka E (2001) Effect of diazepam and midazolam on the antinociceptive effect of morphine, metamizol and indomethacin in mice. Pharmazie 56:89–91

74. Rosland JH, Hole K (1990) Benzodiazepine-induced antagonism of opioid antinociception may be abolished by spinalization or blockade of the benzodiazepine receptor. Pharmacol Biochem Behav 37:505–509

75. Borges PC, Coimbra NC, Brandao ML (1988) Independence of aversive and pain mechanisms in the dorsal periaqueductal gray matter of the rat. Braz J Med Biol Res 21:1027–1031

76. Fasmer OB, Hunskaar S, Hole K (1989) Antinociceptive effects of serotonergic reuptake inhibitors in mice. Neuropharmacology 28:1363–1366

77. Otsuka N, Kiuchi Y, Yokogawa F, Masuda Y, Oguchi K, Hosoyamada A (2001) Antinociceptive efficacy of antidepressants: assessment of five antidepressants and four monoamine receptors in rats. J Anesth 15:154–158

78. Yokogawa F, Kiuchi Y, Ishikawa Y, Otsuka N, Masuda Y, Oguchi K, et al.(2002) An investigation of monoamine receptors involved in antinociceptive effects of antidepressants. Anesth Analg 95:163–168

79. Okuda K, Takanishi T, Yoshimoto K, Ueda S (2003) Trazodone hydrochloride attenuates thermal hyperalgesia in a chronic constriction injury rat model. Eur J Anaesthesiol 20:409–415

80. Schreiber S, Backer MM, Herman I, Shamir D, Boniel T, Pick CG (2000) The antinociceptive effect of trazodone in mice is mediated through both mu-opioid and serotonergic mechanisms. Behav Brain Res 114:51–56

81. Bourin M, Masse F, Dailly E, Hascoët M (2005) Anxiolytic-like effect of milnacipran in the four-plate test in mice: mechanism of action. Pharmacol Biochem Behav 81:645–656

82. David DJ, Renard CE, Jolliet P, Hascoet M, Bourin M (2003) Antidepressant-like effects in various mice strains in the forced swimming test. Psychopharmacology 166:373–382

83. Prado WA, Roberts MH (1985) An assessment of the antinociceptive and aversive effects of stimulating identified sites in the rat brain. Brain Res 340:219–228

84. Beyer CE, Dwyer JM, Platt BJ, Neal S, Luo B, Ling HP, Lin Q, Mark RJ, Rosenzweig-Lipson S, Schechter LE (2010) Angiotensin IV elevates oxytocin levels in the rat amygdala and produces anxiolytic-like activity through subsequent oxytocin receptor activation. Psychopharmacology (Berl) 209:303–311

85. Czopek A, Byrtus H, Kołaczkowski M, Pawłowski M, Dybała M, Nowak G, Tatarczyńska E, Wesołowska A, Chojnacka-Wójcik E (2010) Synthesis and pharmacologicalevaluationofnew5-(cyclo)alkyl-5-phenyl-and 5-spiroimidazolidine-2,4-dione derivatives. Novel 5-HT1A receptor agonist with potential antidepressant and anxiolytic activity. Eur J Med Chem 45:1295–1303

86. Massé F, Hascoët M, Bourin M (2005) alpha2-Adrenergic agonists antagonise the anxiolytic-like effect of antidepressants in the four-plate test in mice. Behav Brain Res 164:17–28

87. Massé F, Hascoët M, Dailly E, Bourin M (2006) Effect of noradrenergic system on the anxiolytic-like effect of DOI (5-HT2A/2 C agonists) in the four-plate test. Psychopharmacology (Berl) 83:471–481

88. Nic Dhonnchadha BA, Ripoll N, Clenet F, Hascoët M, Bourin M (2005) Implication of 5-HT2 receptor subtypes in the mechanism of action of antidepressants in the four plates test. Psychopharmacology (Berl) 179:418–429

89. Ripoll N, Hascoët M, Bourin M (2006) Implication of 5-HT(2A) subtype receptor in DOI activity in the four-plates test-retest paradigm in mice. Behav Brain Res 166:131–139

90. Massé F, Petit-Démoulière B, Dubois I, Hascoët M, Bourin M (2008) Anxiolytic-like effect of DOI microinjections into the hippocampus (but not the amygdala nor the PAG) in the mice test. Behav. Brain Res 188: 291–297

91. Petit-Demoulière B, Massé F, Cogrel N, Hascoët M, Bourin M (2009) Brain structures

implicated in the four-plate test in naïve and experienced Swiss mice using injection of diazepam and the 5-HT$_{2A}$ agonist DOI. Behav Brain Res 204:200–205

92. Ripoll N, Nic Dhonnchadha BA, Sébille V, Bourin M, Hascoët M (2005) The four-plates test-retest paradigm to discriminate anxiolytic effects. Psychopharmacology (Berl) 180: 73–83

93. File SE (1990) "One-trial tolerance to the anxiolytic effects of chlordiazepoxide in the plus-maze." Psychopharmacology (Berl) 100:281–282

94. Petit-Demoulière B, Hascoët M, Bourin M (2008) Factors triggering abolishment of benzodiazepines effects in the Four-Plate Test–retest in mice. Eur Neuropsychopharmacol 18:41–47

Chapter 9

The Vogel Punished Drinking Task as a Bioassay of Anxiety-Like Behavior of Mice

Alicia A. Walf and Cheryl A. Frye

Abstract

There are a plethora of whole animal models that are utilized to assess the neurobiological substrates involved in complex constructs affecting the human condition, such as anxiety. One such behavioral measure that utilizes a conditioned response that can be utilized to assess the neurobiological underpinnings of anxiety is the Vogel punished drinking task. The Vogel punished drinking task produces reliable and valid results in rats and mice in a single 3-min testing session, which does not require prior training. Briefly, mice are water-deprived (typically 18–24 h) in their home cages. Mice are then placed in the testing chamber which contains a drinking bottle with an electrified sipper that delivers a mild shock to the subject after a predetermined number of licks are made. The dependent measures in this task are the latency for subjects to engage in drinking from the bottle, the number of punished licks made, the number of shocks delivered (which always covaries with the number of licks made), and the number of drinking bouts. The most commonly reported measure is the number of punished licks made by subjects, with an increase in the number of licks reflecting antianxiety-like responding. This paper will detail the methods for conducting this task in mice and discuss data from our laboratory and others related to subjects' variables to consider (e.g., hormone status, age, strain), with a focus on assessing the role of steroids for anxiety in murine models. Indeed, the Vogel task is an indispensable whole animal model for investigating the neurobiological underpinnings and potential therapeutic targets of anxiety.

Key words: Anxiety, Affect, Mood, Murine, Knockout, Steroid, Conflict

1. Background and Historical Overview

1.1. Background

Fundamental questions about the neurobiological underpinnings of complex emotional and neuropsychiatric conditions, such as anxiety, and development of therapeutics targeting these substrates, can be effectively addressed using a whole animal model. Anxiety may simply be defined as a maladaptive fear response. Whereas fear is an adaptive stress response initiated following the perception of encountering threatening internal or external stimuli to the

Todd D. Gould (ed.), *Mood and Anxiety Related Phenotypes in Mice: Characterization Using Behavioral Tests, Volume II*,
Neuromethods, vol. 63, DOI 10.1007/978-1-61779-313-4_9, © Springer Science+Business Media, LLC 2011

individual to maintain or reinstate homeostasis, anxiety occurs in the absence of discrete threatening stimuli and/or genuine risk and can be a chronic or irreversible state. Despite this relatively simple definition of what anxiety may constitute, there are clear complexities in the behavioral, physiological, and endocrinological profile of anxiety (ranging from an acute state to a neuropsychiatric condition) among people. As such, it is fruitful to use whole animal models in which potential brain targets and mechanisms of anxiety can be systematically manipulated, in concert with findings obtained in clinical settings in which such manipulations are generally limited. Indeed, involvement of neurotransmitters, such as GABA and serotonin, as well as brain targets, such as the amygdala, hippocampus, and other limbic regions with anxiety, are well-founded in basic and clinical research. Despite what is known about these most well-studied targets, there is great heterogeneity in the etiology of anxiety, as well as its manifestation and efficacy of treatments, that point to the importance of investigating other targets. As such, using an animal model is important to elucidate potential targets in the context of other processes that can be manipulated and/or measured, which may better mimic the clinical situation. Generally, anxiety as a neuropsychiatric condition is considered to involve both state and trait dimensions. State anxiety refers to current interoceptive status and can be modified by the stimuli being encountered by the individual at that moment. Conversely, trait anxiety refers to a long-lasting, or permanent, predisposition to anxiety feelings in an individual. In most animal models, including the Vogel punished drinking task, state anxiety is being assessed. However, we must consider that one strength of using murine models of anxiety is that mice of different strains or genetic mutations can be utilized to investigate trait anxiety.

There are a number of reliable and valid tasks utilized in animals to determine deviations from the normal anxiety/fear response as a way to further elucidate the brain targets and mechanisms of anxiety, with the aim to improve diagnoses and treatment of anxiety. This paper will focus on the utility of the Vogel punished drinking task (referred to as the "Vogel task" from this point forward). The topics covered in this paper will be the historical overview of the use of the Vogel task, reliability and validity of using this task, a description of typical methodology for this task in mice and anticipated results (independent, dependent, and subjects' variables), and troubleshooting issues in using this task. A major focus will be on the role of sex steroids for behavior in this task, given the paucity of these results reported in the literature, particularly in mice. For the purpose of this paper, to the extent possible, a focus will be on the use of the Vogel task in mice, but critical findings in rats (which allude to cross-species similarities) will also be included as appropriate and necessary.

1.2. Historical Overview

There is a long and rich history of using tasks that employ conflict to assess anxiety in rodent models. The Vogel task is a well-established conflict task used to assess anxiety states in rodents, and the mechanisms which may underlie these effects. The element of conflict occurs when rodents are given a choice between stimuli that are desired (e.g., a reward stimuli) and undesired (e.g., an aversive stimuli). These conflicts can be trained (conditioned) or natural (unconditioned, and not subject to extensive training). One well-known and often-used unconditioned conflict task is the elevated plus maze, in which there is a conflict for rodents based upon their proclivity to explore and avoid open areas (i.e., the elevated open arms of the maze (1, 2)). The Vogel task was proposed about 40 years ago by Vogel and is one in which an unconditioned stimuli is used and rodents are not trained to avoid or respond to the novel stimuli utilized in the task (3). The Vogel task produces a conflict in water-deprived rodents for the desired stimuli (water) with an opposing and concurrent undesired stimuli (intermittent shock from the water bottle) (3–8). The Vogel task can be compared to the Geller–Seifter (8) conflict task in which rodents are initially trained to respond for a food reward and then have this food reward coupled with an aversive stimuli such as shock (9). In this task, rodents are therefore producing a conditioned and punished response. Conversely, rodents in the Vogel task do not have to employ a learned response to complete the task. Thus, the development of the Vogel task has provided a conflict task in which rodents do not have to go through extensive training for an anxiety response to be assessed.

1.3. Reliability and Validity of the Vogel task

An important consideration to make regarding measures of psychological function, such as the Vogel task, is whether the behavioral assays utilized are reliable (i.e., repeatable and consistent). Of particular interest is whether the measure has interrater reliability. Interrater reliability for the Vogel task is high in our laboratory and is presumed high in others given that there is an automatic measurement component to this task. Indeed, it is not tenable to manually count the number of licks made by rodents, so this is accomplished by using a "lickometer," typically supplied with the apparatus when purchased. In our laboratory, all experimenters are trained by others who have expertise in using the Vogel task until there is little variability in the scores obtained by each experimenter under the same experimental conditions.

The Vogel task is considered to have face, construct, and predictive validity. In terms of face validity, or whether the task measures what it is supposed to measure, the Vogel task is a valid measure of anxiety-like behavior of rodents. It measures the propensity for rodents to avoid aversive, or threatening, stimuli. Furthermore, the Vogel task has construct validity, or the ability of an observable measure (i.e., number of punished licks made)

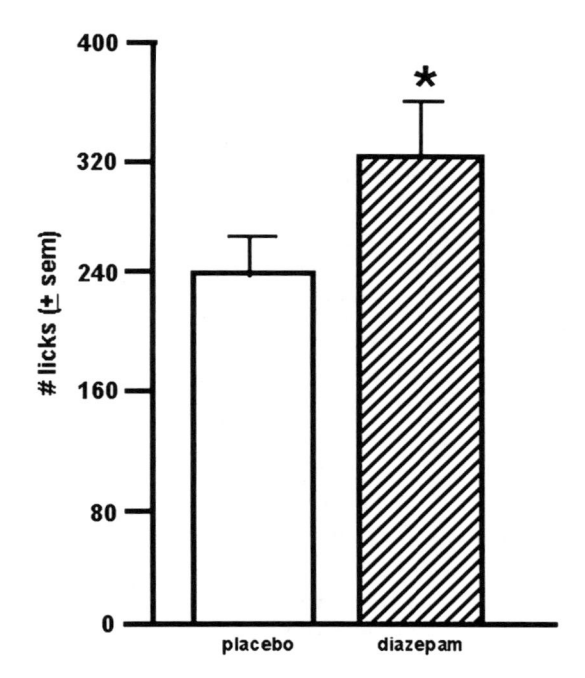

Fig. 1. Validation of the Vogel task. The number of punished licks made by rodents was increased in those treated with a positive control (diazepam, 10 mg/kg IP) compared to placebo vehicle. *Asterisk* indicates significantly higher than vehicle administration.

reflects a state that is not observable (i.e., anxiety). Indeed, in the Vogel task anxiogenic (i.e., negative control) and anxiolytic (i.e., positive control) drugs, respectively, decrease and increase the number of punished licks made (4, 5). As an example in our laboratory, administration of the anxiolyic, diazepam, significantly increases the number of punished licks made compared to placebo vehicle (Fig. 1). Furthermore, the Vogel task also has predictive validity, or the ability of a dependent variable to measure one construct to predict behavior on a related measure. For example, mice that make more punished licks during the Vogel task also have increased antianxiety-like behavior in other anxiety measures (e.g., open field, light-dark, elevated plus maze (9–12)).

2. Equipment, Materials, and Setup

2.1. Vogel Task: Materials

In our laboratory, the following items are needed to test mice in the Vogel task. Before testing, temporary housing cages are needed. To collect data, a Vogel task (e.g., Anxio-meter Columbus Instruments, Columbus, OH; Fig. 2), a video-tracking system with a web cam or video camera, computer with Microsoft Excel, data sheets, data binder, pen, and stopwatch are needed. To analyze data, a statistical program is needed.

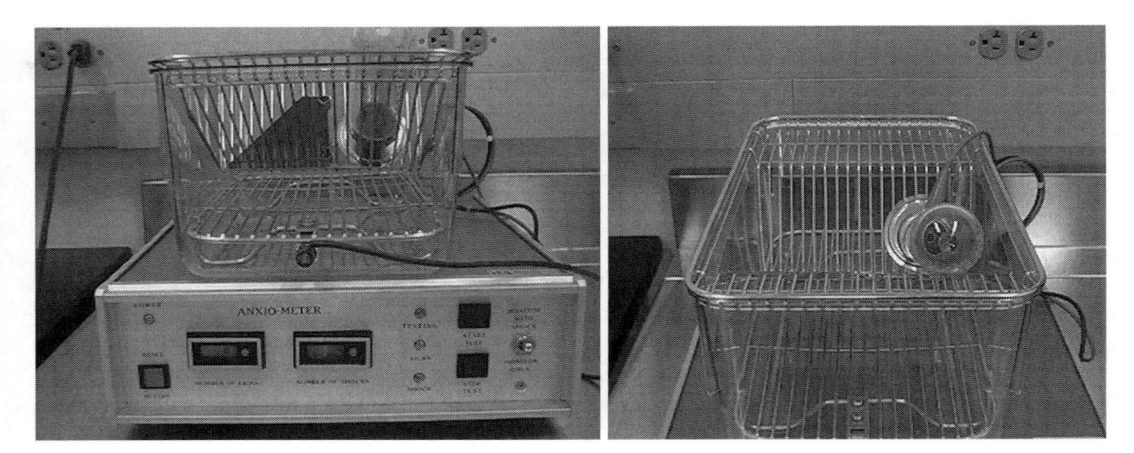

Fig. 2. The Vogel apparatus used in the studies described in mice in this Chapter. A view from the front is on the left and the view from the top is on the right. Anxio-meter Vogel Apparatus was purchased from Columbus Instruments.

2.2. Vogel Task: Setting Up

The Vogel task is set up in a brightly lit room with six, 32-W fluorescent overhead lights that produce consistent illumination of ~2,800 lumens each. The Vogel task is set up on a counter in the room.

2.3. Vogel Task: Data Collection

Data of rodents in the Vogel task are collected by a trained experimenter. The latency for the task to begin is recorded. The task begins after the rodent licks from the bottle 20 times. The number of drinking bouts is defined by the rodent drinking from the bottle and then walking away from the bottle. Hatchmarks are made on the data sheet to record drinking bouts. After 3 min, when the task is complete, the experimenter records the read-out in the Vogel task for the number of licks and shocks received. Additionally, data are typically recorded by a video camera or a video-tracking system (e.g., Any-Maze, Stoelting). As such, there are then multiple ways that the data are stored, and videos can be reviewed as necessary. The trained experimenter collects data on data sheets that are securely stored in the laboratory. As well, by having the experimenter hand-collect data in the Vogel task, notations can be made about the rodents' behavior (e.g., freezing), which typically cannot be automatically tracked by the video-tracking system, and can be collected simultaneously (see Sect. 9.3, *dependent variables*, below).

2.4. Elevated Plus Maze: Calibration/ Validation

The Vogel task should be calibrated and validated before use. For example, the experimenter should perform a pilot experiment with mice that have been administered positive and negative control compounds, e.g., anxiolytics or anxiogenics, respectively. These data can be compared to results reported in the literature on these anxiolytics' and anxiogenics' effects (5).

3. Procedure

Step 1: The Vogel apparatus is set up as per the manufacturer's instructions. The water bottle is filled and attached to the "lickometer." The current is checked by making sure there are readings when the experimenter touches the tip of the water bottle. Different intensities of shock are utilized based upon species and length of water deprivation. We have found it to be advantageous to use the lowest possible intensity and shortest length of water deprivation (e.g., 0.25 mA and 18 h of water deprivation) to obtain valid and normative anxiety responses of mice in this task. See Millan and Brocoo (5) for a detailed description of this point (5).

Step 2: Experimental mice are transported in their home cages from their housing room. Mice are brought on a cart in their home cages to the hallway outside the behavioral testing room in which they will be assessed in the Vogel task.

Step 3: In the hallway, mice are singly caged in opaque temporary housing cages.

Temporary housing: To reduce the potential confounds of different experiences between animals before behavioral testing, all mice are briefly held in temporary housing cages immediately before testing. It is critical that all experimental mice consistently: (1) spend a brief time in temporary housing before testing (as is standard in our laboratory), or (2) remain in their home cages before testing.

Step 4: Fill out data sheets. Data sheets have the subject numbers, date, coded conditions, and experimenter's initials. The video-tracking system is turned on and is made ready for use.

Step 5: Carry each mouse to be tested in its temporary cage into the behavioral testing room.

Step 6: The experimenter then removes the mouse from temporary caging by picking it up by its tail. The mouse is placed in the Vogel chamber at the distal end from where the bottle is, facing towards the bottle. It is critical that each experimental mouse be placed in the chamber in this same way.

Step 7: Start the stopwatch and video-tracking system. The behavior of mice must be consistently recorded throughout the task, so it is important that data collection commences immediately from when the mouse is placed in the chamber. The stopwatch needs to be started immediately as mice are given a 15-min latency to begin drinking from the bottle.

Step 8: Data are collected manually by the experimenter and automatically by the apparatus for 3 min following the first twenty licks made by the mouse.

Dependent variables: The typical measures of antianxiety-like behavior of mice in the Vogel task are the number of punished licks that are made. Punished licks occur with every 20 unpunished licks made by mice. The number of shocks received (1 for every 20 licks) and total licks can also be recorded. The number of drinking bouts, which are operationally defined by the mouse drinking from the water bottle and then moving away from the bottle and ceasing to drink from it, can be recorded as a proxy measures for ambulation in this task. The pattern of effects observed for number of punished licks made and these other measures are the same and it is typical to only report the number of punished licks made. Additionally, experimenters can record other behaviors of the mice in this task. Importantly, we record "flinch-jump" ratings in this task (0, no response; 1, flinch; 2, jump; 3, jump and squeak) as an index of responses to the shock stimulus. Mice and rats typically show a range of 0–1 for flinch-jump responses with the intensity of shock that is utilized in our laboratory. Furthermore, any time spent freezing by rodents (defined as cessation of movement beyond breathing and blinking) is recorded as another measure of stress/anxiety.

Step 9: The mouse is removed from the chamber by its tail and placed inside its temporary housing cage, which is then carried to the hallway. The mouse is then returned to its home cage. When all mice are tested, they are immediately returned to their housing room and given access to water ad libitum in their home cages.

Step 10: The Vogel task is cleaned with Quatricide and thoroughly dried with paper towels between each mouse is tested.

Step 11: Behavioral data are analyzed.

Data analysis: The effects of independent variables for behavior in the Vogel task are typically determined by calculating differences in means of dependent variables using analyses of variance. In our laboratory, a commercially available statistical analyses program, such as Statview or SPSS, is utilized, but freeware statistics programs are also available to analyze results. An appropriate p value to use in experiments using the Vogel task is less than 0.05 to demonstrate statistically significant effects of independent variables on the dependent variables. When significant main effects are found, *post hoc* tests, such as Tukey's HSD, which corrects for multiple comparisons, or Fisher's LSD tests, are utilized to determine differences between groups.

4. Anticipated Results

In our laboratory, we are focused on understanding the effects and mechanisms of steroids for anxiety-like responding. The following results exemplify data we have obtained using the Vogel task protocol, as described in this report, to investigate the antianxiety effects of estrogens (e.g., 17β-estradiol; E_2) (13–15) and progestogens (11),

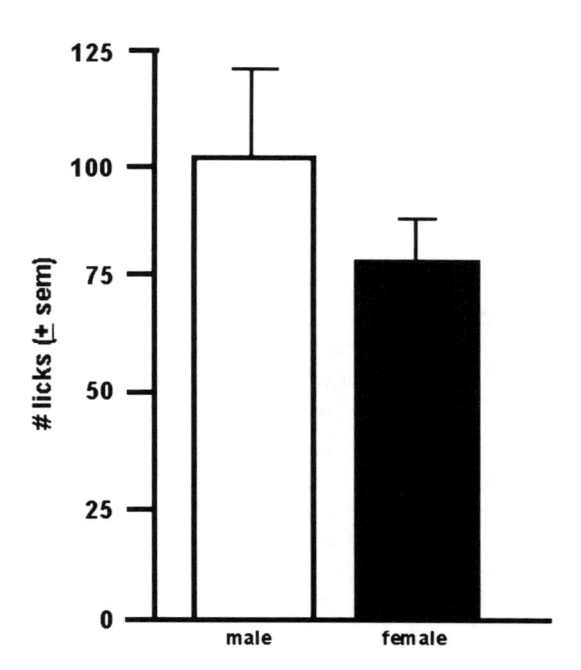

Fig. 3. Sex differences – gonadally intact, adult male and female mice were tested in the Vogel task. Male mice make more punished licks than do female mice.

which are hormones secreted by the ovaries, and androgens secreted by the testes (10). To date, the results that we have obtained using the Vogel task have been robust, reliable, and valid (see Fig. 1). In support, there are sex differences in the response in the Vogel task similar to what has been observed in other tasks with C57Bl/6J female mice showing greater anxiety-like behavior than male mice (Fig. 3). As such, it is critical to decide upon the sex and reproductive status of mice to be utilized before initiating a study using the Vogel. Historically, many studies have used male rodents in the Vogel and other anxiety tasks to reduce potential confounds from the influence of circulating hormones across the estrous cycle. It is important to embrace and consider these sex and hormonal effects as more women than men are diagnosed with general anxiety disorder. Of note, these sex differences among rodents are less robust in anxiety tasks when the endogenous hormonal states of females are taken into account, suggesting a role of ovarian steroids among females. Indeed, ovariectomized rats subcutaneously administered E_2 (0.09 mg/kg, which produces physiological, proestrous-like E_2 levels) make more punished licks than rats administered a lower dosage of E_2 or placebo vehicle (Fig. 4). Furthermore, young adult female C57Bl/6J mice administered physiological dosing of E_2 make more punished licks than aged mice (18–24 months old) of the same strain (Fig. 5). Interestingly, aged mice are responsive to coadministration of E_2 and the clinically used hormone replacement therapy, raloxifene, for increasing punished licks in the Vogel

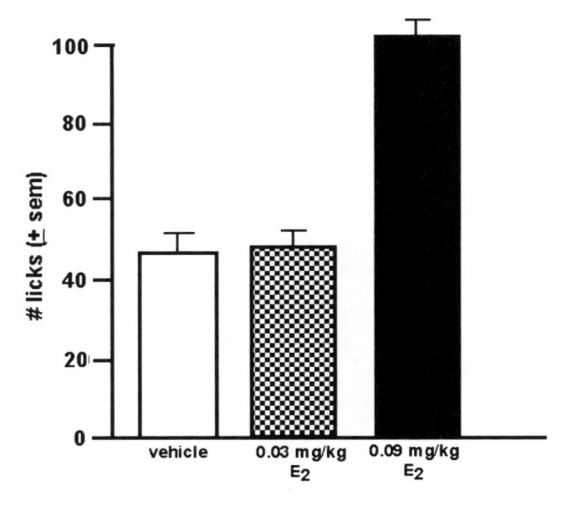

Fig. 4. Effects of physiological estrogen dosing – adult female rats were ovariectomized and administered placebo vehicle, a low (0.03 mg/kg) or moderate (0.09 mg/kg) dosage of 17β-estradiol (E_2) subcutaneously 44–48 h before behavioral testing. Rats administered 0.09 mg/kg E_2 (which produces proestrous-like levels of E_2 in circulation and the brain) made significantly more punished licks than the other groups.

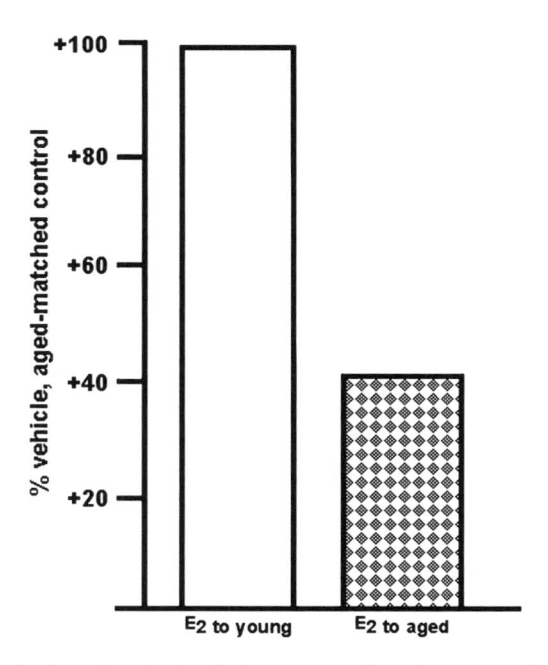

Fig. 5. Effects of aging – female mice that were young (6 months) or aged (18–24 months) were administered physiological, subcutaneous dosing with 17β-estradiol (E_2; 0.1 mg/kg) and compared to young, ovariectomized control mice. Young mice administered E_2 had a greater response than aged mice for the number of punished licks made as percentage of the licks made by the control mice.

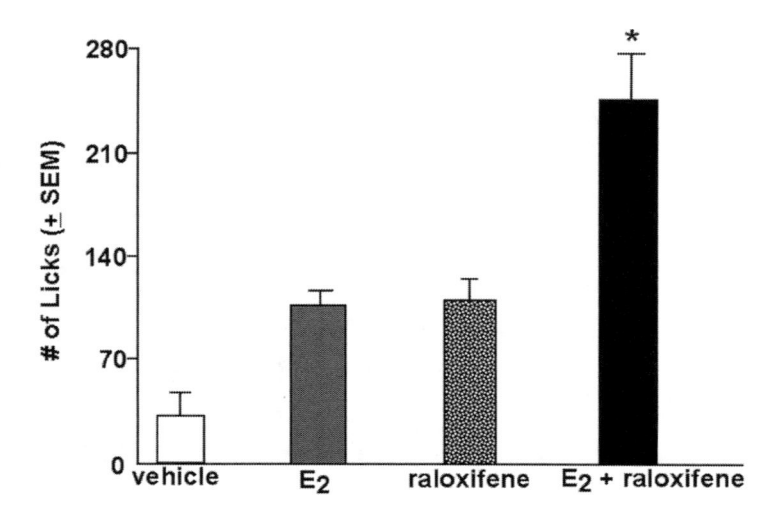

Fig. 6. Effects of hormone replacement therapies – ovariectomized female rats were administered physiological, subcutaneous dosing with placebo vehicle, 17β-estradiol (E_2; 0.1 mg/kg), raloxifene (a hormone replacement therapy; 10 mg/kg), or both E_2 and raloxifene. Coadministration of E_2 and raloxifene had the clearest effects to increase the punished licks made. A similar effect has been reported among aged (18–24 months) mice (14). *Asterisk* indicates significantly higher than vehicle administration.

task compared to vehicle or E_2 or raloxifene alone (14). We have observed the same effect among ovariectomized rats (Fig. 6). Thus, these data demonstrate that the Vogel task can be utilized to investigate the effects of steroids for anxiety-like states of mice.

Another question of interest is the mechanisms for these effects in the Vogel task. We have used two general approaches to investigate this question: (1) pharmacology, and (2) mutant mice. First, comparisons were made between administration of selective estrogen receptor modulators (SERMs) that act at ERα (17α-E_2), ERβ (coumestrol), or both (E_2). When administered to female rats (13) and male mice (Fig. 7), SERMs with activity at ERβ increase punished licks compared to vehicle. A question is the role of progestogens in these effects, given that E_2 fluctuates with progestogens naturally among females and E_2 can increase progestogen production. It has previously been demonstrated that progesterone or its neuroactive metabolite increased punished licks in the Vogel task of young and aged mice (9, 11). Second, we utilized mutant mice models to investigate the mechanisms for these effects of steroids. Mice that were wildtype or heterozygous or homozygous for the progestin receptor knockout (PRKO) were administered vehicle or progesterone. Administration of progesterone increased punished licks, irrespective of genotype, suggesting that PRs are not required for these effects (Fig. 8). Additionally, we compared effects of E_2 administration to wildtype mice or those that lack the capacity to form progesterone metabolites (5α-reductase knockout mice).

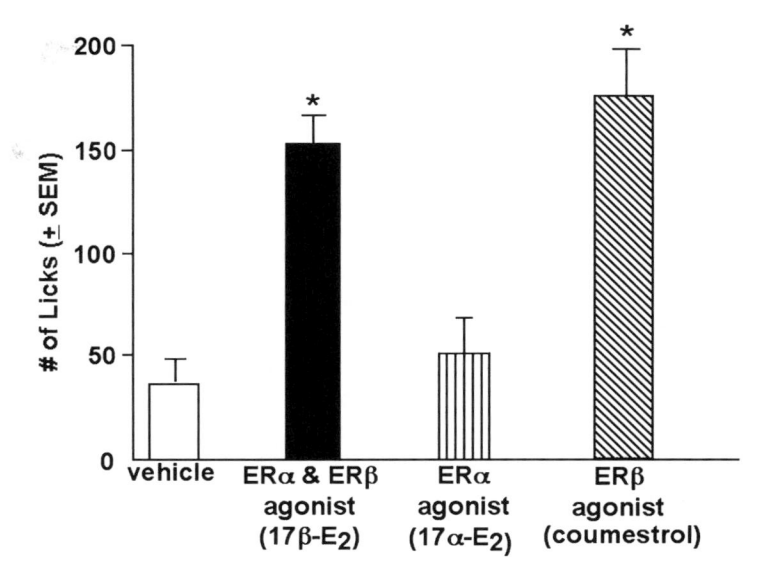

Fig. 7. Effects at estrogen receptor (ER) subtypes – male mice that were aged (18–24 months) and administered placebo vehicle, 17β-estradiol (E$_2$; 0.1 mg/kg; which has actions at ERα and ERβ), 17α-E$_2$ (0.1 mg/kg; which has greater affinity for ERα than ERβ), and coumestrol (0.01 mg/kg which has greater affinity for ERβ than ERα), 44–48 h before testing – were compared. Mice administered compounds that have actions at ERβ had an increased number of punished licks than did mice administered vehicle. *Asterisk* indicates significantly higher than vehicle administration.

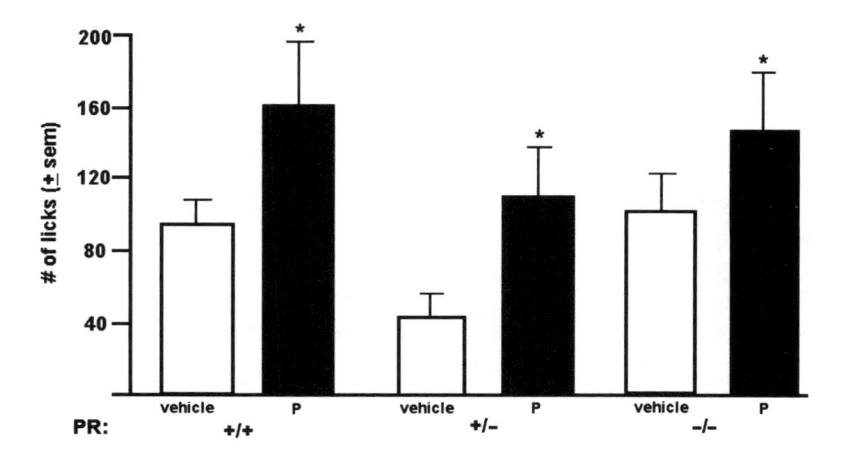

Fig. 8. Effects in progestin receptor (PR) knockout mice – male wildtype (+/+) mice or those heterozygous (+/−) or homozygous (−/−) for the progestin receptor knockout were administered vehicle or progesterone (P$_4$; 10 mg/kg; 1 h before testing). P$_4$ increased punished licking compared to vehicle administration, irrespective of genotype. *Asterisk* indicates significantly higher than vehicle administration.

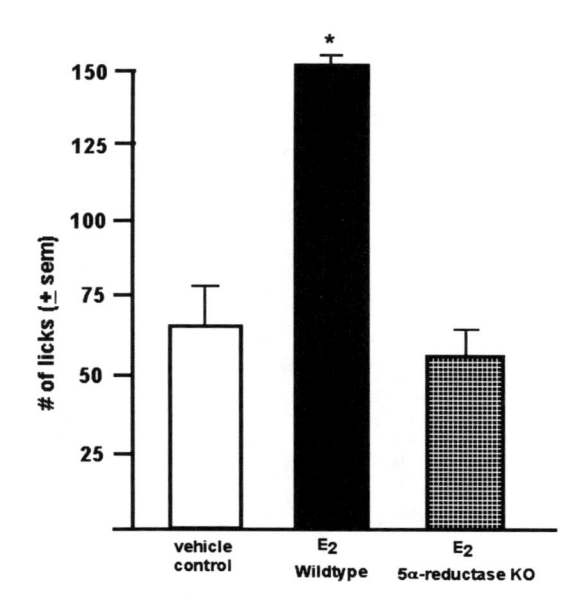

Fig. 9. Effects in 5α-reductase knockout mice – female, ovariectomized wildtype (+/+) mice or those homozygous (–/–) for the 5α-reductase knockout were administered 17β-estradiol (E₂; 0.1 mg/kg), 44–48 h before testing, and behavior was compared to that in mice administered placebo vehicle. E₂ increased punished licking of wildtype, but not 5α-reductase knockout mice, compared to vehicle administration. *Asterisk* indicates significantly higher than vehicle administration.

E₂ was only effective among wildtype mice (Fig. 9), suggesting metabolism is important for these effects in the Vogel task. Thus, these data briefly describe experiments that can be performed with pharmacological techniques and mutant mice to investigate mechanisms for anxiety-like responding using the Vogel task.

5. Experimental Variables

5.1. Experimental Subjects

All experiments involving animal subjects must be performed with approval of a governing body and in accordance with relevant guidelines and regulations regarding animal usage in experimental procedures. Laboratory-bred mice that are typically tested for anxiety-related behaviors in the Vogel task are adults (~55 days of age or older). Several animal subjects' variables need to be considered when designing experiments with the Vogel task, such as strain (12), sex, estrous cycle, and age (see above for description of data investigating sex, hormonal, and age effects). C57BL/6 mice are the most commonly used inbred strain of mice, and this strain is utilized in behavioral studies in our laboratory. Indeed, knockout mouse models in our laboratory are back-crossed onto a C57Bl/6 background. We typically obtain all experimental mice from the breeding colonies that we have and maintain in the Life Sciences Research Buildings in the Laboratory Animal Care Facility at The University at Albany, SUNY (original stock from Jackson Laboratory, Bar Harbor, ME). Of note, in the studies described in this report, rodents were bred and raised in a brand-new facility that was undergoing construction and similar intermittent and unpredictable disturbances that can be characteristic of new facilities. Thus, given the potential for differences based upon these subjects' variables, it is critical that these variables and the following procedural variables be seriously considered when initiating studies using the Vogel task to assess anxiety-like responding.

5.2. General Health and Normative Response Screening Procedure

All experimental mice are checked for normative responses before initiation in behavioral studies as per modified methods (16). General health status is determined by daily observations of their general appearance (normal whisker movements, clean and unruffled fur, good posture, normal gait, good muscle tone) and normative behavior (no evidence of fighting with cagemates, presence of fur grooming, building nests in home cages with Nestlets, huddling/sleeping with other mice in their home cages, ability to cage climb, ability to withdraw paw when gently pulled). Normative sensory responses are determined by the presence of a blink reflex to a cotton swab placed close to the eyes and an ear twitch reflex when a swab was lightly placed against ears. Sensitivity to heat stimulus can be determined with the paw lick or tailflick tasks (17, 18). Evaluations were done by observers blind to the conditions of mice so that potential differences could be used to justify inclusion of mice in behavioral studies. Current studies in the laboratory assessing functional effects of steroids among mice show that the majority of mice ($95 + 5\%$) are generally healthy and have normative responses before behavioral testing.

5.3. Handling/ Habituation Procedure

Prior experience with stressors from the environment or handling by experimenters may alter Vogel task behavior, especially among mice as they may be more sensitive to arousal than rats (19–21). To minimize these potential confounding effects, mice typically receive 5 days of handling and habituation prior to behavioral testing (22). Day 1: Mice are picked up from their homecage, handled for 15 s, and returned to their homecage. Day 2: Mice are picked up from their home cage and transferred to a novel cage. Day 3: Mice are picked up from their home cage and transferred to a novel cage and weighed and then replaced to their home cage. Day 4: Mice are transferred to another room via a cart while in their home cage. Day 5: Mice are transferred to another room via a cart while in their home cage, injected with 0.2 mL placebo vegetable oil vehicle subcutaneously, and placed in novel environment for 5 min.

5.4. Housing of Mice

Experimental mice are always group-housed (3–5 per cage) in our laboratory to obviate potential for social isolation stress (a model of depression among mice) to alter anxiety-like responding (23, 24). Mice are housed in standard polycarbonate, shoebox cages, without filter tops, with one Nestlet so that species-typical nest-building activities can occur. Mice are housed in a temperature-controlled room ($21 \pm 1°C$) on a 12/12 h reversed-light cycle (lights off at 8:00 a.m.) with continuous access to typical rodent chow. Tap water is continuously provided to mice in their home cages except for the 18 h immediately preceding behavioral testing in the Vogel task.

5.5. Timing of Testing

Mice are housed on a 12/12 h reversed-light cycle in our laboratory. Mice are always behaviorally tested in the early phase of their active, dark-cycle. This is important to consider so that there is consistency in hormonal milieu and because anxiety, as well as other behaviors that could influence results in the Vogel task (stress, arousal, motor behavior, etc.), is influenced by circadian rhythms/light cycle (25).

5.6. Length of Test

Mice are given 15 min to begin licking from the bottle from when they are placed in the Vogel chamber. Once they have made 20 licks from the bottle and received their initial punished response, mice are assessed for 3 min in the Vogel task.

6. Troubleshooting

The most frequent problems associated with running this task are that mice do not drink from the bottle in the maximum latency that they are given. Factors that commonly can contribute to this are disturbances during testing, ataxia or poor motor behavior of subjects, and the duration of water deprivation. Care must be taken to avoid noise or other disturbances during testing. In our laboratory, we hang signs outside the testing room hallway and close room doors to inform others in the laboratory that Vogel testing is in progress and loud noises and other disturbances should be minimized. In the case of a disturbance, a mouse may freeze in this task. In this event, the experimenter should record the duration of freezing. Another possibility for mice not drinking from the water bottle may be related to potential neurological effects of experimental manipulations (i.e., those that may cause sedation, dizziness, hypoactivity, etc.), which may be pharmacological or due to genetic mutation. In this case, the importance of screening mice for normative behavior is clear (see above procedure). As well, experimenters may want to consider adding a habituation phase to the task, where mice are allowed to explore the chamber, with or without the bottle. During habituation, the bottle should not be electrified so as not to produce test-decay effects. If these factors have been accounted for, it may be of interest to modestly lengthen the duration of water deprivation (e.g., from 18 h to 24 h). The task should be validated based upon the length of water deprivation before experimental subjects are run (see above).

7. Summary

The Vogel task is one of many whole animal models that are utilized to assess the neurobiological substrates involved in complex constructs affecting the human condition, such as anxiety. The Vogel task is a widely-used, reliable, and valid measure of anxiety-related behavior of mice that can be easily implemented in the laboratory. The Vogel task utilizes a conditioned response in a single 3-min testing session, which does not require prior training. The Vogel task has great value as a model because it allows for systematic investigations in an in vivo system on the role of genes, neuromodulators, and hormones in several potential brain targets important for anxiety. Thus, a clear strength of utilizing animal models, such as the Vogel task, is the ability to probe constructs, such as anxiety, and gain insight about their neurobiological underpinnings and potential therapeutic targets, which can be applied to the human condition.

Acknowledgments

The research described in this paper from our laboratory was supported in part by grants from the National Institute of Mental Health (MH06769801), National Science Foundation (IBN03-16083), and U.S. Army Department of Defense (BC051001). Technical assistance provided by Dr. Madeline Rhodes, Kassandra Edinger, Carolyn Koonce, Danielle Llaneza, and Kanako Sumida is greatly appreciated.

References

1. Walf AA, Frye CA. The use of the elevated plus maze as an assay of anxiety-related behavior in rodents. Nat Protoc. 2007;2:322–8.

2. Walf AA, Frye CA. Using the elevated plus maze as a bioassay to assess the effects of naturally-occurring, and exogenously-administered compounds, to influence anxiety-related

behaviors of mice. In: Gould, T, ed. Mood and Anxiety Related Phenotypes in Mice: Characterization Using Behavioral Tests, Volume 1. Springer Protocols, 2009:225–246.

3. Vogel JR, Beer B, Clody DE. A simple and reliable conflict procedure for testing anti-anxiety agents. Psychopharmacologia. 1971;21:1–7.

4. Millan MJ. The neurobiology and control of anxious states. Prog Neurobiol. 2003;70:83–244.

5. Millan MJ, Brocco M. The Vogel conflict test: procedural aspects, gamma-aminobutyric acid, glutamate and monoamines. Eur J Pharmacol. 2003;463:67–96.

6. Treit D. Animal models for the study of antianxiety agents: a review. Neurosci Biobehav Rev. 1985;9:203–22.

7. Pollard GT, Howard JCL. Effects of drugs on punished behaviour: preclinical test for anxiolytics. Pharmacol. Ther. 1989;45: 403–424.

8. Geller I, Seifter J, 1960. The effects of meprobamate, barbiturate, D-amphetamine and promazine on experimentally induced conflict in the rat. Psychopharmacologia 1960;1:482–492.

9. Carboni E, Wieland S, Lan NC, Gee KW. Anxiolytic properties of endogenously occurring pregnanediols in two rodent models of anxiety. Psychopharmacology. 1996;126:173–8.

10. Frye CA, Edinger K, Sumida K. Androgen administration to aged male mice increases antianxiety behavior and enhances cognitive performance. Neuropsychopharmacology. 2008;33:1049–61.

11. Frye CA, Sumida K, Dudek BC, Harney JP, Lydon JP, O'Malley BW, Pfaff DW, Rhodes ME. Progesterone's effects to reduce anxiety behavior of aged mice do not require actions via intracellular progestin receptors. Psychopharmacology. 2006;186:312–22.

12. van Gaalen MM, Steckler T. Behavioural analysis of four mouse strains in an anxiety test battery. Behav Brain Res. 2000;115:95–106.

13. Walf AA, Frye CA. ERbeta-selective estrogen receptor modulators produce antianxiety behavior when administered systemically to ovariectomized rats. Neuropsychopharmacology. 2005;30:1598–609.

14. Walf AA, Frye CA. A review and update of mechanisms of estrogen in the hippocampus and amygdala for anxiety and depression behavior. Neuropsychopharmacology. 2006;31:1097–111.

15. Walf AA, Frye CA. Estradiol reduces anxiety- and depression-like behavior of aged female mice. Physiol Behav. 2010;99:169–74.

16. Crawley JN, Chen T, Puri A, Washburn R, Sullivan TL, Hill JM, Young NB, Nadler JJ, Moy SS, Young LJ, Caldwell HK, Young WS. Social approach behaviors in oxytocin knockout mice: comparison of two independent lines tested in different laboratory environments. Neuropeptides. 2007;41:145–63.

17. Frye CA, Walf AA, Rhodes ME, Harney JP. Progesterone enhances motor, anxiolytic, analgesic, and antidepressive behavior of wild-type mice, but not those deficient in type 1 5α-reductase. Brain Res. 2004;1004: 116–24.

18. Walf AA, Koonce CJ, Frye CA. Estradiol or diarylpropionitrile decrease anxiety-like behavior of wildtype, but not estrogen receptor beta knockout, mice. Behav Neurosci. 2008;122: 974–81.

19. Mong JA, Pfaff DW. Hormonal and genetic influences underlying arousal as it drives sex and aggression in animal and human brains. Neurobiol Aging. 2003;24:S83–8.

20. Morgan MA, Pfaff DW. Estrogen's effects on activity, anxiety, and fear in two mouse strains. Behav Brain Res. 2002;132:85–93.

21. Morgan MA, Pfaff DW. Effects of estrogen on activity and fear-related behaviors in mice. Horm Behav. 2001;40:472–82.

22. Frye CA, Sumida K, Dudek BC, Harney JP, Lydon JP, O'Malley BW, Pfaff DW, Rhodes ME. Progesterone's effects of aged mice do not require actions via intracellular progestin receptors. Psychopharmacology 2006;186:312–22.

23. Hunt C, Hambly C. Faecal corticosterone concentrations indicate that separately housed male mice are not more stressed than group housed males. Physiol. Behav. 2006;87:519–26.

24. Zhu SW, Yee BK, Nyffeler M, Winblad B, Feldon J, Mohammed AH. Influence of differential housing on emotional behaviour and neurotrophin levels in mice. Behav. Brain Res. 2006;169:10–20.

25. Jones N, King SM. Influence of circadian phase and test illumination on pre-clinical models of anxiety. Physiol. Behav. 2001;72:99–106.

Chapter 10

A Vogel Conflict Test Using Food Reinforcement in Mice

Abstract

Effects of compounds on suppressed or punished responding are often used to predict anxiolytic efficacy in humans. The use of mice in these tests has many advantages including the ability to evaluate transgenic animals. In contrast to steady state baselines that can take months to establish, the present method can deliver dose-response data within a week. Mice are food-deprived and placed in an experimental chamber with two nose poke holes. Every nose poke (FR1) produces a 20 mg food pellet. After 2 days of training, a drug or drug vehicle is administered and the mice are exposed to a mixed FR1 (food), FR1 (food + shock) schedule in alternating, unsignalled periods of 4 and 10 min for three cycles. In the 10-min period, nose pokes produce both food plus brief electrification of the grid floor (0.5 mA for 100 ms); during the 4-min period, only food is delivered with each nose poke. Under these conditions, the introduction of shock can substantially decrease nose pokes during the 10-min punishment period without significantly affecting responding during the non-punishment periods. Results are shown that document the comparability of the drug effects in this procedure to that of other punishment procedures requiring protracted training. This was exemplified with the increases in suppressed responding by the anxiolytic chlordiazepoxide, but not by either *d*-amphetamine or morphine.

Key words: Vogel conflict test, Mouse, Anxiety, Anxiolytic, Geller–Seifter conflict test, Operant response

1. Background and Historical Significance

Responding suppressed by punishment is widely used as a method for detecting anxiolytic activity of compounds since increases in punished responding by drugs have been used to predict clinical efficacy for the management of generalized anxiety disorder and acute anxiety states (1, 2). The ability to genetically alter receptor proteins and other potential drug targets in mice has put an increased demand on behavioral methods in this species.

Although punished responding has been established in mice, there have been very few studies (3, 4). The scarcity of such data is likely related to several factors. Food or water deprivation in mice

Todd D. Gould (ed.), *Mood and Anxiety Related Phenotypes in Mice: Characterization Using Behavioral Tests, Volume II*, Neuromethods, vol. 63, DOI 10.1007/978-1-61779-313-4_10, © Springer Science+Business Media, LLC 2011

must be very precisely controlled and monitored so as to preclude health problems and mortality. The training time required to establish stable baselines of responding under punishment procedures can be several months. Finally, faster procedures to evaluate anti-anxiety agents in mice exist and include the elevated plus maze and a host of other ethologically based methods such as open field, marble-burying, and light/dark transit studies (see present volumes).

Another method that has been used to some extent in mice is the Vogel conflict procedure that is a variant of the Geller–Seifter conflict procedure in which water-deprived animals are exposed to a drinking spout and, after a period of drinking, every lick produces an electric current that suppresses the rate of drinking. Drugs used in the treatment of anxiety prevent the shock-induced suppression (5, 6). A drawback with the Vogel test is that it has not been studied in much detail with mice and the difficulties in establishing a validated model have been acknowledged (7). A method has been described in ICR mice in which diazepam, pentobarbital, but not buspirone prevented the suppression of drinking (4). Second, the traditional Vogel procedure may not be pharmacologically isomorphic with the food-based conflict methods that are more firmly under schedule control. For example, buspirone does not generally increase punished responding under the Geller conflict method, but increases drinking in the Vogel conflict procures with much greater proclivity (8). False positives have also been reported. For example, isoproterenol increased Vogel conflict responding and water consumption, but did not increase Geller conflict responding (9).

The present procedure describes a Vogel-like method in mice utilizing food presentation and an arbitrary operant response (original report in (10)). We used the nose poke response for this purpose since it is readily emitted at a high basal rate so as to reduce the time for training. The nose poke response is also not the consummatory response as is drinking in the Vogel procedure and thus we hoped to minimize the influence of appetite stimulants on this behavior.

2. Equipment, Materials, and Setup

2.1. Animals

Mice of any genotype that can eat food pellets and poke their head into a hole can be used. We have studied male, NIH Swiss mice (Harlan Sprague Dawley, Indianapolis, IN) weighing between 28 and 34 g, and a number of receptor knockout mice on both a C57bl/6 or a CD1 background. The mice can be either housed in groups initially or in standard, individual mouse cages with continuous access to standard rodent chow and water. Our experience with this procedure has utilized a 12-h light dark cycle (lights on: 06:00–18:00 h) and experiments are conducted toward the dark cycle period.

2.2. Test Environment   Experiments are conducted in mouse operant conditioning chambers (MED Associates, Inc., St. Albans, VT) (Fig. 1). These chambers were equipped with two holes (13 mm diameter), 1 cm above a grid floor and spaced 9.7 cm apart. The holes could be illuminated with white light. A water bottle was located on the opposite wall. Nose pokes within the hole (>6 mm) were counted as responses and produced the audible click of a relay. Food pellets (20 mg, BioServe, Frenchtown, NJ) could be delivered to a food trough (centered between the nose poke holes) coincident with an audible tone for 100 ms. The chambers were located within sound-attenuating cubicles supplied with ventilation and white noise to mask extraneous sounds. Scrambled electric shock could be delivered to the grid floor of the chamber by a constant current AC source (MED Associates). Experimental events and data were collected with MED-PC software (MED Associates).

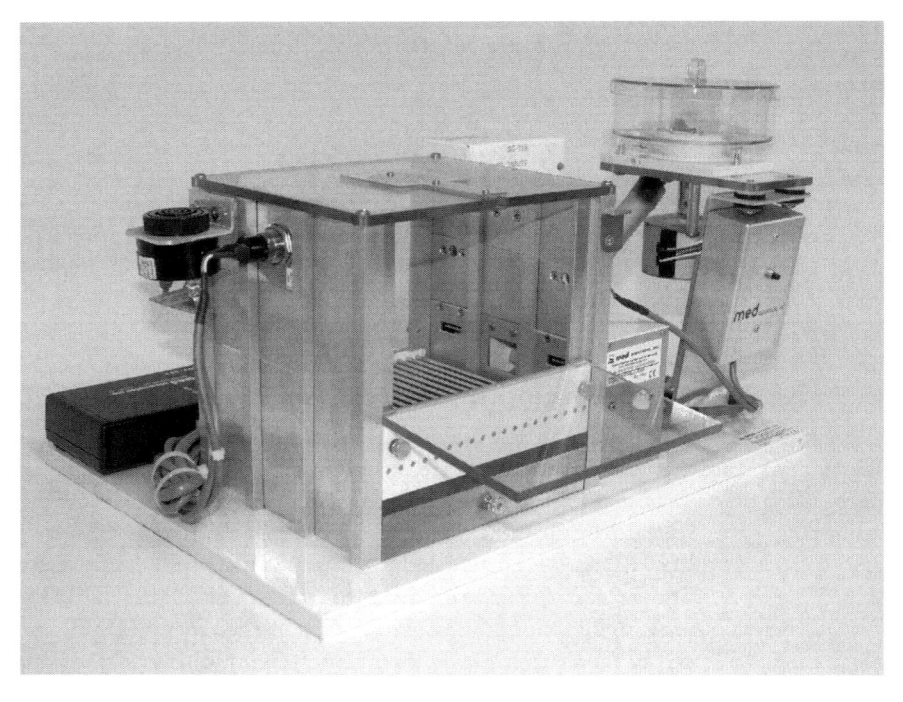

Fig. 1. Mouse operant conditioning chamber. The figure shows on the front panel an opening (center near grid floor) into which food pellets can be delivered by a pellet-dispensing device (behind front panel). On either side of the pellet opening would be the two nose poke holes (not shown in this photograph; instead retractable levers are pictured). A mechanical relay (not shown) is mounted behind the front panel to generate an audible click with each nose poke. Also not shown is a small sound speaker to deliver white masking noise into the chamber. A water bottle would be secured to the back of the chamber with the spout accessible to the mouse. The entire chamber is then placed within a sound-attenuating enclosure and operated and monitored by a remote computer. Photograph courtesy of Steve Walsh, Med Associates, Inc.

2.3. Setup

Mice are initially brought into the vivarium for about 3 days to 1 week prior to set up in order for acclimation to occur. Mice ideally will be individually housed in order to best control food intake in individual animals. However, creative variations with group housing are possible, the key being to control food intake in individual mice. During the acclimation period, mice are given continuous access to standard rodent diet and water.

During the last 3 days of acclimation, food pellets to be used as reinforcers (see Sect. 10.3.1) are added to the food supply in order for the mice to become accustomed to these pellets such that their initial introduction does not induce pellet avoidance.

On the day prior to training, food restriction begins with each mouse being given ~3 g standard rodent chow plus two reinforcer pellets at about 16:00 h. Water remains continuously available.

2.4. Dependent Measures

Rates of responding (responses/min) under each component of the mixed schedule are calculated for the day of pharmacological testing (Sect. 10.4). Dose-effect and/or time-effect curves for a compound are constructed from the data collected with drug vehicle and different drug doses and/or times postdosing.

Food intake measures can be independently determined if desired. A number of methods can be utilized.

3. Behavioral Procedure

Mice are brought into the experiment room with the test chambers and allowed to acclimate in their home cages for approximately 60 min. During the 60 min acclimation period, 5 reinforcer pellets are placed within the food hopper of the operant chambers. The experimenter should also check that pellet delivery works properly and that there are no pellets on the chamber floor or stuck in the pellet delivery tube. After 60 min, the mice are weighed and placed into the operant test chambers.

3.1. Day 1: Training

Initial training occurs nearly 24 h postremoval of food, but variation is likely tolerated. During training, mice remain within the operant chamber overnight with water available. During the evening experimental session beginning at about 17:00 h, the mice are placed within the experimental chamber for the first time. After about 120 s, the left nose hole is illuminated and every nose poke response within the opposite hole produces a single food pellet. The hole was illuminated as a discriminative stimulus associated with the experimental session being in operation. However, nose pokes within the illuminated hole had no scheduled consequences. A total of 200 food pellets can be delivered within this first session that terminates after that point or after 8 h, whichever occurs first. At ~09:00 the next morning (day 2), the mice are returned to their living cages. If the mice received at least 15 food pellets in the training evening, they are not fed. If mice received less than 15 pellets, they are eliminated from the study.

3.2. Day 2: Further Training

At ~15:00 h of day 2, mice are returned to the test room, acclimated for about 60 min and then dosed with physiological saline or relevant drug vehicle by the desired route of administration. The mice are then placed into the operant chambers without any reinforcer pellets having been placed into the food hoppers. Two minutes later, the left nose hole is illuminated and, as before, every response in the right nose hole produces a single food pellet. This component ends after 4 min or, if there are no responses, the component terminates after a total of 15 min. Four minutes after termination of component 1, a 10-min period is introduced without any change in stimulus conditions; every response within the right nose hole continues to produce a food pellet during this period. The 4- and 10-min periods then alternate for three cycles; the component durations are fixed with the exception of the first component as just described. The reason for the potential time extension in component 1 is to maximize the probability of occurrence of at least one reinforced response before the punishment component where shock will be introduced for the first time on experimental day 3. After the experimental session, the mice are returned to their home cages with ~3 g food. Mice that meet a response criterion of at least 18 responses on day 2 are then advanced to the day 3 schedule. Mice that do not can either be eliminated from the study or run the next day as a third training session.

3.3. Day 3: Pharmacological Testing

Mice that have advanced to this stage are brought back into the test room at about 15:00 h and acclimated to the test room for about 60 min. Mice are given a dose of a compound or given the compound vehicle, returned to their home cages for the appropriate pretreatment interval, and then placed within the experimental chamber. The experimental session on this day is comparable to that used on training day 2 with the exception that electrification of the grid floor (0.5 mA for 100 ms) occurs in conjunction with food delivery during each of the three 10-min periods.

3.4. No Punishment Controls

A separate group of mice can be used to ascertain the levels of responding when electric shock is omitted on the day 3 test.

4. Data Analysis

Rates of responding (responses/min) under each component of the mixed schedule are calculated on day 2 and day 3. Dose-effect curves for each compound are constructed from the data collected on day 3 and analyzed by one-way ANOVA for nonpunished and punished responding separately. Given significant dose-effect changes, rates of responding during administration of drug vehicle are compared to the rates occurring during each drug dose by posthoc Dunnett's test (e.g., GraphPad Prism Software). At least 6 mice are studied for each drug dose and a total of from 8 to 12 are evaluated for vehicle controls. Statistical probabilities of less than 0.05 were considered to be significant.

5. Anticipated Results

In NIH Swiss mice, our experience has seen good training results on day 1 with 33 ± 2.7 (mean $\pm$ SEM) and 100 ± 7.8 responses for left and right nose pokes observed, respectively. Performances on day 2 resulted in over 92.6% of mice advancing directly to the day 3 procedure by meeting the 18 right nose poke response criterion on day 2. Introduction of shock on day 3 should result in the rate of unpunished responding being essentially unchanged, whereas the rate of responding in the shock component should be significantly reduced (by about 90% in NIH Swiss mice). Vehicle control responding has averaged 2.1 ± 0.2 and 0.2 ± 0.02 responses/min in components 1 and 2, respectively, with the range of punished responses being between 3 and 12 responses) (10).

Classical anxiolytic agents that act upon $GABA_A$ receptors would all be anticipated to increase suppressed responding as exemplified by the effects of the benzodiazepine anxiolytic chlordiazepoxide (Fig. 2). In contrast, drugs that are not generally used for their antianxiety effects per se are expected to decrease suppressed responding as exemplified by d-amphetamine and morphine (Fig. 2). These compounds are also generally inactive in other more traditional punishment procedures (2).

6. Experimental Variables and Troubleshooting

6.1. Time of Experimental Sessions

Although times of day were provided in the procedure sections, variation is likely without major impact. The first day of training that occurs overnight seems valuable for getting the most rapid training possible. Subsequent training sessions could likely be within variations of ± 6 h without marked alterations in behavioral effect.

6.2. Food Deprivation Levels

The amount of food deprivation needed for acquisition and maintenance of the behaviors described here will vary depending upon a host of factors that include genotype (see below). Again the specific methods described above have worked well with several strains,

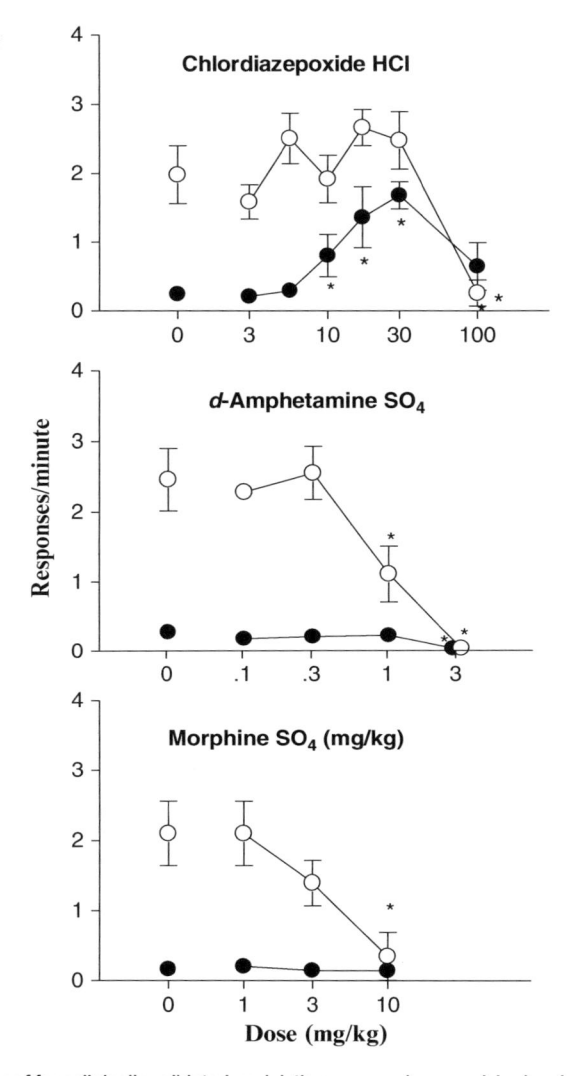

Fig. 2. Effects of four clinically validated anxiolytic compounds on punished and nonpunished responding of mice. *Open circles*: nonpunished responding. *Filled circles*: punished responding. Each drug point represents the mean ± SEM of at least 6 mice; vehicle control values are from 8 to 12 mice. Significant differences ($p < 0.05$, Dunnett's test) from vehicle control values (points above 0) are represented by an *asterisk*. This figure is a modification of that in Witkin et al. (10).

but empirical data for all are not at hand. In all cases, it is critical that mice be carefully observed every day and that body weights be taken. It is recommended that during training days, food be placed on the floor of the cage for easy access as climbing to eat from a cage top might be difficult under conditions of food restriction.

6.3. Training Variations   The training parameters given above have been successfully implemented, but the full range of parametric variation that is acceptable and indeed optimal has not been determined. It is likely that a host of changes could be enabled for specific purposes. For example, mice that do not meet criterion on one training day could be brought in for additional training at other times.

6.4. Test Variations

As with the training variables discussed above, testing variations are likewise possible. For example, mice could be used for drug testing on multiple occasions. Additional variations could include withholding of shock on the test day; this might enable increased sensitivity to drug effects; however, baseline changes over the course of the test session would also be anticipated. Of course, the procedure is also amenable to the detection of time course by appropriate drug pretreatments and to drug combination studies, etc.

6.5. Genotype Comparisons

Different mouse genotypes have inherent differences in their food and water regulation and body weights. Hence, it is critical to know your mice from this vantage point to determine parameters around the duration of food deprivation, the amount of food to deliver postexperimental session, and perhaps the number of days of training needed to reach criterion performances. When conducting experiments with transgenic or receptor-deleted mice, a contemporaneous wild-type control mouse will have to be put through the same procedures. Use of appropriate counter-balancing practices for the experimental day is encouraged. Parametric variation might be needed for different strains. For example, the intensity of the punisher might need adjusting to have the same behavioral effects across strains.

6.6. Experimental Chambers

The experimental chambers used in the current procedure are relatively expensive and space consuming when considering the conduct of experiments in many mice. Given this fact, relatively few chambers can generally be made available for simultaneous mouse training or testing. These drawbacks are counterbalanced by advantages in the multiple and flexible uses of this equipment. Alternatives are possible, but are not commercially available to my knowledge. Thus, a stand-alone apparatus can be envisioned in which the minimal structure required for the conflict procedure is built.

6.7. Statistical Power

The number of mice required for detection of a given level of drug effect has not been explicitly explored. Current use of this method has used six mice per dose group and 8–12 mice for vehicle control with good success.

6.8. Use of No-punishment Control Group

The data from this control group are used to determine the degree to which responding in the punishment group is reduced. Once these data are collected, routine dose-response studies could, if desired, be collected, using the punishment baseline without the use of the no-punishment control group. Since there is an ongoing nonpunishment component built into the procedure, a concurrent no-punishment baseline is essentially always present.

6.9. Use of Food-Intake Controls

A host of ancillary behavioral assessments can be made against which to compare the effects of drugs on punished responding. Food intake is one that can provide information about the relationship of the drug effects on punished responding to food motivation (see Sect. 10.7 below).

7. Concluding Remarks

The Vogel conflict test using water drinking in mice is rarely seen in the experimental literature. In a recent report by Pisu and colleagues (11), 4 days were required for training and dose-effect analysis compared to the three reported here. Very few studies using operant methods in mice have likewise been disclosed (12).

In contrast, methods using ethological methods (e.g., elevated plus maze, dark/light box, marble-burying) have dominated the work with mice and rats in the anxiolytic research area (c.f. see other chapters in this book series). The apparent greater promiscuity of the ethological methods for diverse pharmacological mechanisms along with the ease of testing is a likely contributor to their use in anxiolytic drug assessments. Although comparison of methods is not within the scope of the current discussion, the scientific literature suggests some of the drawbacks of such methods that include a lack of reproducibility of findings across labs that likely results from the use of behaviors that are less well controlled than those described in here.

Under the current method, drug effects on suppressed responding can be evaluated after only 2 days of training in mice. In addition, parameters such as shock intensity and duration do not need to be individually calibrated for each mouse to get a specified degree of suppression or a significant drug effect. Parametric manipulation is time consuming and yet is critical to determining the behavioral effects of drugs (3, 7, 13).

Despite the fast turnaround for drug effects on operant behavior which involved minimal training, only 7.4% of NIH, Swiss mice tested by us (10) did not meet the training criteria on day 2. Moreover, vehicle control rates on test day 3 were generally comparable across many groups of mice. The reproducibility of the drug effects was additionally verified in triplicate experiments with chlordiazepoxide.

Although the methods described permit rapid and reliable detection of drug effects on suppressed responding, no attempt was made to fully optimize the assay or to fine-tune it toward particular pharmacological activities. Parameters that may be worthy of manipulation could include addition of another training day, addition of discriminative stimuli associated with the nonpunishment and punishment components, component response durations, the inclusion of a schedule component under which responding has no scheduled consequences, and the use of intermittent schedules of food or shock delivery.

The possibility of detecting false positives in the mouse punishment procedure due to increases in food consumption per se was also a concern in the present assay (see (9) for a good example). The present procedure was designed to minimize the influence of effects of compounds on feeding on their effects on punished responding by having the operant response separate from the consummatory response. In addition, separate experiments have been conducted to evaluate the effects of the compounds studied on food consumption in mice showing that these behaviors can be dissociated (10).

Finally, it is important to remember that punishment procedures are not fully functional models of anxiety. Instead, as with most well-controlled behaviors, these methods provide stable baselines against which specific classes of pharmacological agents can produce increases in response rates. Although capable of detecting the activity of a host of antianxiety agents used in clinical practice, others mechanisms are not always positive. For example, although selective serotonin-uptake inhibitors are widely used in first-line treatment of some anxiety disorders, they are generally not reported to increase suppressed responding of this type but are active under other conditions in mice (e.g., marble-burying or nestlet shredding (14–16)).

References

1. Geller I, Seifter J (1960) The effects of meprobamate, barbiturates, *d*-amphetamine and promazine on experimentally induced conflict in the rat. Psychopharmacologia 1:482–492.
2. Cook L, Davidson AB (1973) Effects of behaviorally-active drugs in a conflict-punishment procedure in rats. In: Garratini S, Mussini E, Randall LO, eds.), The Benzodiazepines, Raven Press, New York, 327–345.
3. Spealman RD, Katz JL (1980). Some effects of clozapine on punished responding by mice and squirrel monkeys. J Pharmacol Exp Ther 212:435–440.
4. Umezu T (1999) Effects of psychoactive drugs in the Vogel conflict test in mice. Jpn J Pharmacol 80:111–118.
5. Vogel JR, Beer B, Clody DE (1971) A simple and reliable conflict procedure for testing antianxiety agents. Psychopharmacologia 21:1–7.
6. Millan MJ, Brocco M (2003) The Vogel conflict test: procedural aspects, γ-aminobutyric

acid, glutamate and monoamines. Eur. J. Pharmacol. 463:67–96.
7. Umezu T (1995) Assessment of anxiolytics (5) – Vogel-type conflict task in mice. Nihon Shinkei Seishin Yakurigaku Zasshi 15: 305–314.
8. Barrett JE, Witkin JM (1991) Buspirone in animal models of anxiety. In Tunnicliff G. Eison AS, Taylor DP, eds, Buspirone: Mechanisms and Clinical Aspects, Academic Press, New York, 37–79.
9. Patel JB and Malick JB (1980) Effects of isoproterenol and chlordiazepoxide on drinking and conflict behaviors in rats. Pharmacol. Biochem. Behav. 12: 819–821
10. Witkin JM, Morrow D, Li X (2004) A rapid experimental conflict procedure for detection of anxiolytic compounds in mice. Psychopharmacology 172:52–57.
11. Pisu C, Pira L, Pani L (2010) Quetiapine anxiolytic-like effect in the Vogel conflict test is

serotonin dependent. Behav Pharmacol 21: 649–653.

12. Barbano MF, Castañé A, Martín-García E, Maldonado R (2009) Delta-9-tetrahydrocannabinol enhances food reinforcement in a mouse operant conflict test. Psychopharmacology 205:475–487.

13. Witkin JM, Barrett JE (1976) Effects of pentobarbital on punished behavior at different shock intensities. Pharmacol. Biochem. Behav. 5:535–538.

14. Borsini F, Podhorna J, Marazziti, D (2002) Do animal models of anxiety predict anxiolytic effects of antidepressants? Psychopharmacology 163:121–141.

15. Li X, Morrow D, Witkin JM (2006) Decreases in nestlet shredding of mice by serotonin uptake inhibitors: comparison with marble burying. Life Sci. 78:1933–1939.

16. Witkin JM (2008) Animal models of obsessive-compulsive disorder. Current Protocols in Neuroscience, Chapter 9, Unit 9.30.

Chapter 11

Fear Conditioning and Extinction as a Model of PTSD in Mice

Georgette M. Gafford and Kerry J. Ressler

Abstract

A primary symptom in the diagnosis of PTSD is the inability to control fear. Therefore, the study of fear and its inhibition are essential in understanding the disorder. This chapter will provide detailed protocols for fear conditioning and extinction which can be used as a model to further understand and study PTSD. Background, significance, and required materials will also be discussed.

Key words: Posttraumatic stress disorder, Unconditioned stimulus, Unconditioned response, Conditioned stimulus, Conditioned response, Species-specific defense response

1. Background and Significance

One defining characteristic of posttraumatic stress disorder (PTSD) is the inability to adequately suppress fear (1). The protocols of fear conditioning and its inhibition (extinction) outlined in this chapter are relevant to the study of PTSD given that the disorder involves learned fear and the inability to suppress it. Support for the idea that PTSD involves a disrupted ability to extinguish learned fear has been demonstrated in studies showing that fear learning and extinction are deficient in subjects with PTSD compared to trauma-exposed subjects that do not develop PTSD (2). Additionally, subjects with PTSD show impaired recall of fear extinction compared to trauma-exposed controls (3). Results from the study of fear conditioning and extinction have already been applied in the clinical setting in treating fear-related disorders. Extinction-based exposure therapy has been successfully used to treat anxiety disorders (4–7). More recently, clinical extinction paradigms have been combined with pharmacological agents that enhance extinction learning (8–11), and several ongoing studies are currently examining these approaches in the treatment of PTSD (12, 13).

Todd D. Gould (ed.), *Mood and Anxiety Related Phenotypes in Mice: Characterization Using Behavioral Tests, Volume II*, Neuromethods, vol. 63, DOI 10.1007/978-1-61779-313-4_11, © Springer Science+Business Media, LLC 2011

Fear conditioning provides a robust tool to investigate fear learning. In a typical fear-conditioning task, a neutral cue (e.g., a tone) which normally does not elicit a fear response is paired with an aversive cue such as footshock which normally does elicit a fear response. Animals quickly learn to associate the presentation of the tone itself with a fear response and this is called acquisition (14). The neutral cue in the above example is called the conditioned stimulus (CS), the shock is the unconditioned stimulus (US), and the fear response is the unconditioned response (UR). After a delay, animals can be reexposed to the CS alone and they will show a fear response to it without further footshock (i.e., recall or retrieval). Once animals learn the association between the tone and footshock, their fear response is referred to as the conditioned response (CR). Fear conditioning provokes multiple conditioned responses in the rodent including heart rate changes (15, 16), suppression of ongoing appetitive behavior (17), enhancement of avoidance behavior (18), potentiation of the acoustic startle response (19), and the species-specific defensive response (SSDR) called freezing (20). This chapter will limit its focus to discussion of freezing behavior as the CR, as it is the most often used and reported behavioral output in the study of fear conditioning in mice.

The protocols discussed below will primarily focus on mice, as they remain the most widely studied from a genetic perspective, and for this reason, they are increasingly used in the study of learning and memory. The behavioral output of "freezing" is defined as the lack of all movement except that required for respiration (20). A typical fear conditioning training curve is shown in Fig. 1a.

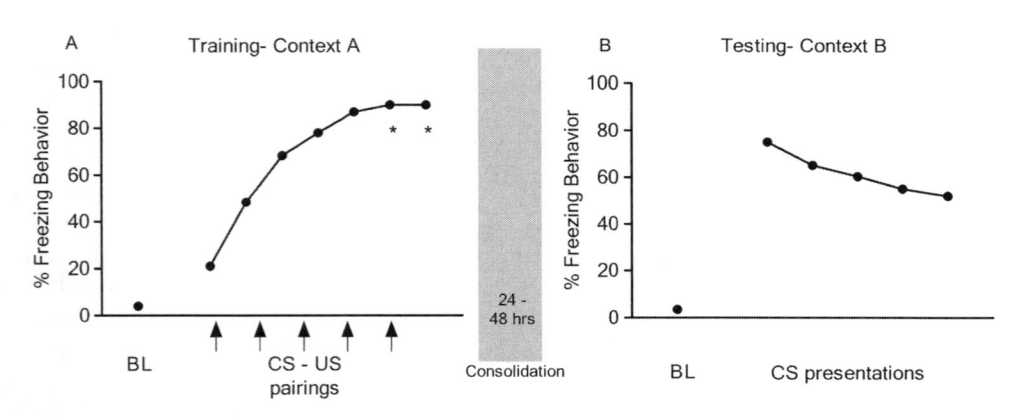

Fig. 1. (a) Typical freezing behavior during fear conditioning training (Context A). Baseline (BL) freezing indicates freezing prior to presentation of any stimuli. Animals show little freezing during BL prior to the presentation of 5 CS (e.g., tone)–US (e.g., shock) pairings. After each CS–US pairing, freezing increases. After the training stimuli, animals show strong post-training freezing indicated by *asterisks*. (b) After 24–48 h, animals are tested in a novel context (Context B) to the same CS given during training without delivery of any US. Typical freezing behavior during the fear conditioning test is shown. Prior to presentations of the CS, baseline freezing should be minimal indicating animals successfully discriminate between the training context (Context A) and the novel context (Context B).

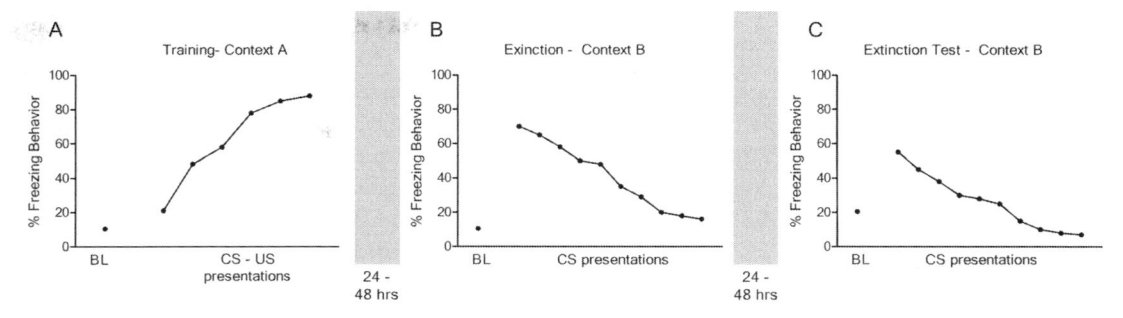

Fig. 2. (**a**) Typical freezing behavior during fear conditioning training. (**b**) 24–48 hrs later baseline freezing during extinction should be low to indicate good discrimination between the training context (Context A) and the novel extinction context (Context B). Freezing behavior decreases over multiple presentations of the CS during extinction training. (**c**) The extinction test shows the same pattern as during extinction training except with higher baseline and a faster decrease in freezing behavior than during extinction training.

Several tone–shock (or alternatively, light-shock, or odor-shock) pairings lead to an increase in freezing behavior. However, animals can learn this association quickly enough to show memory for the tone–shock association after only a single trial (21).

If the experimenter gives only a few exposures to the CS in the absence of the US, this allows for testing to determine the strength of a memory for the CS, without markedly affecting the memory trace (Fig. 1b). In contrast, exposing animal to many CSs without the corresponding US initiates extinction learning (Fig. 2). Note that the baseline level of freezing prior to extinction training is very low (Fig. 2b). This indicates the animals distinguish between the context in which they were trained and this new context in which they will be extinguished. Through extinction acquisition, as shown in Fig. 2b, animals reduce their CR (freezing) to the CS (e.g., tone) because they learn it is no longer followed by the US (e.g., shock). To test memory for extinction, usually after 24–48 h, mice are brought back into the same context where they learned extinction and are again exposed to multiple presentations of the tone (again usually more than 10). This procedure is referred to as the extinction test (i.e., extinction recall). An example of freezing during the extinction test is shown in Fig. 2c. Note that during the extinction test, freezing is initially high, but quickly dissipates as more tones are played. The decrease in freezing during the extinction test should occur more quickly than during extinction training. Multiple days of extinction testing are sometimes run if freezing is particularly high, thereby disrupting the ability to discern differences between the groups. The process of extinction does not permanently erase the memory because animals are able to recall the original CS–US association with the passage of time (spontaneous recovery), with a reminder of the US (e.g., footshock) in a new context (reinstatement), or with CS exposure in a context different from extinction (renewal) (22, 23).

The neural circuitry and cellular mechanisms that mediate fear conditioning have been extensively characterized (24–33). Most of these manipulations are found to disrupt fear conditioning by acting on the amygdala, a brain structure crucial for emotional learning (34). Manipulation of many of the same molecular targets that impair or facilitate fear conditioning similarly effect fear extinction (35–45). However, fear extinction is also dependent on unique neural circuitry such as the infralimbic portion of the prefrontal cortex (46). While other behavioral tasks may be applicable to the study of PTSD (47–59), the well-characterized neural pathway of fear conditioning and extinction as well as the ability of animals to quickly learn the fear conditioning task makes it ideal for use in a controlled laboratory setting in the study of fear-related memory.

2. Equipment Materials and Setup

2.1. Animals

In our group, fear conditioning and extinction are generally conducted with mice between 6 and 8 weeks of age with "wild type" C57BL/6J (Jackson Laboratories) background. Note that many mouse strains begin to lose hearing beyond 12–15 weeks of age. Fear conditioning and extinction are observed in both sexes; however, male mice have been shown to display more robust freezing behavior (60).

The increase in the creation and use of transgenic mouse strains means that genetically modified mice are trained and tested in behavioral procedures established in rats. In many cases, these tests have not been fully validated in mice. Some research that has been done in this area shows that mice often behave differently from rats in significant ways (61–64). Furthermore, as is the case in rats, each mouse strain has a unique behavioral profile (65–74). The behavioral differences between mouse strains are important for assessing fear conditioning and extinction because some strains differ in baseline levels of anxiety which may affect their tendency to freeze (67, 70, 73). Mice are available with many different genetic deletions as well as a variety of inducible knockouts which allow scientists to disrupt a gene at different time points during behavior. Inducible knockouts are particularly useful for studying fear memory because different phases of memory can be manipulated (e.g., acquisition or recall) by inducing the gene knockout at different time points relative to the learning procedure.

2.2. Test Environment

Training and testing should be conducted at about the same time of day within the animals' light cycle. Mice are fear-conditioned and tested in two different environments (for example see Fig. 3a, b). One environment is referred to as the training context (Context A) and the other is the testing or extinction context (Context B).

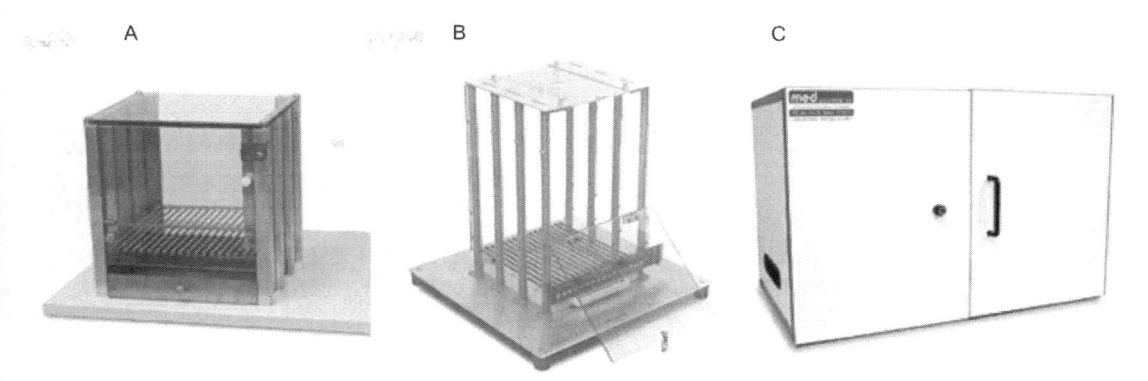

Fig. 3. Examples of (**a**) fear conditioning chamber (Med Associates, ENV-307A) (**b**) Novel Context B differs in shape and size (Coulbourn Instruments, H10-11M-TC) (**c**) sound attenuating box (Med Associates, ENV-018MD).

Context B should differ in its contextual cues (i.e., shape, odor, lighting, and flooring). If a separate type of testing apparatus is not available, combinations of removing or covering the shock grid, changing the lighting, and wiping down the chamber with a solution with a different odor are often sufficient to provide for successful context discrimination. Low baseline levels of freezing during testing or extinction training indicate that animals have discriminated between the training and testing or extinction chambers. All training and testing in mice should be done in an environment protected from noise. Ideally, the chambers should be housed in a sound attenuated outer box (such as Fig. 3c) within a room protected from noise.

2.3. Stimuli

The stimuli used in fear conditioning can affect the strength of fear learning (see Table 1). It is important to ensure that the CS (e.g., tone, odor, taste) is not aversive to the animal, thereby acting as a US on its own. A simple way to examine this is to test several animals only with presentations of the CS and measure freezing. If little to no freezing occurs, then the CS can be used as an appropriately neutral stimulus. Work has shown that the number of CS–US pairings, as well as the duration and intensity of the US can modulate the strength of training (e.g. (75–81)). Other factors that can be manipulated to strengthen the behavioral response to the CS include increasing the number of days of training or extinction and manipulating the interval between trials (e.g. (82)). Increased number of trials, strength, and duration of the US (e.g., shock) generally lead to increased fear (freezing) and delayed extinction (e.g. (80, 83)). It is also important to consider whether the hypothesis of the experiment requires you to look for an enhancement or disruption of fear or extinction. To more easily see an enhancement of extinction (i.e., decreased freezing), it may be beneficial to train with a strong protocol. In contrast, it will be

Table 1
General training guidelines

Strength of conditioning	Weak	Moderate	Strong
Shock intensity (milliamps)	0.2–0.4	0.5–0.8	1.0–2.0
Shock duration (seconds)	0.25–0.4	0.5–0.8	1–2
Number of trials	1–3	4–7	8–10

Table 2
Typical lab parameters

Type of session	Days	Shock intensity (milliamps)	Shock duration (seconds)	Number of trials	Tone or white noise (decibels)	Time between trials (minutes)
Training	1	0.6–1.0	0.5	5	72–78	5
Testing	1	N/A	N/A	5	72–78	1.5
Extinction	1	N/A	N/A	15	72–78	1.5
Extinction test	1–2	N/A	N/A	30	72–78	1.5

easier to see a disruption of extinction (i.e., enhanced freezing) with a weak training protocol. Typical fear conditioning and extinction parameters from our lab are shown in Table 2. These protocols are routinely used in our C57BL/6J and transgenic mice.

2.4. Setup

2.4.1. Group Assignment

Prior to training, the group assignment of animals should be counterbalanced so that time of day of training/testing is matched across groups avoiding circadian rhythm fluctuations that may affect your experiment. Another consideration when assigning animals to groups is to counterbalance across training and testing boxes. All of one group should not be placed in one chamber as there may be some small but measurable effect of placement within that particular box. Baseline levels of freezing or anxiety (e.g., open-field test or elevated plus maze measures) is also useful information for counterbalancing, to ensure equivalent behavior across experimental groups. Dependent upon when your drug manipulation will occur, assignment to groups may be done after training to ensure that your groups show similar levels of freezing prior to any manipulation.

2.4.2. Dependent Measures

As mentioned earlier, freezing is the lack of all movement except for respiration. It is the most commonly used dependent measure in fear conditioning and extinction. Many companies offer automated video-based systems for scoring freezing (e.g., Clever systems Freezescan, Coulbourn Instruments Freezeframe). These systems should be routinely tested by hand scoring a few sessions to be sure they are correctly scoring freezing (84, 85). Hand scoring of freezing should be done from videotape of the behavior session. The amount of time needed to learn how to score freezing is minimal. Two different people should score the freezing behavior of the same animal and compare their results. There should be 90–95% agreement on the freezing score between individuals doing the scoring. To hand score video, assign one number to active behavior and another to freezing behavior and record whether the animal is freezing or active on paper or in a computer program once per second. More than one cage can be scored at a time. For example, if four animals are trained at a time, each animal can be scored once every 4 s. A new column of numbers should be started for each minute so that changes in freezing may be assessed over time. For each animal, the freezing scores for each minute should be added up, then divided by 60 s and multiplied by 100 to get the percentage. For example, if an animal froze 30 times in the first minute of training, they are freezing 50% of the time (e.g., $30/60 \times 100 = 50\%$). To keep track of time while scoring, it is necessary to have a metronome to play an audible tone or click marking every second. Alternatively, one could also use a stopwatch to track the total amount of time the animal spends freezing over the entire session.

On occasion, lack of movement may actually be sleeping. This is easily detected with hand scoring, but automated scoring systems are not designed to differentiate between freezing and sleeping. If the session is so long the animal sleeps, the session may need to be shortened, otherwise it will have to be hand scored. Any video-based scoring system will come with video recording capability because the camera is necessary to score the freezing. These video files are quite large and will require access to a sufficient amount of storage. These files should be stored either as individual experiments on a DVD or on a computer with a large amount of storage space. Another potential dependent measure is to record the number of fecal boli after the behavior session. This is a simple additional indicator of the level of fear during the behavioral session (86).

2.4.3. Time Considerations

Time needed to conduct these experiments depends on the number of training and testing chambers available and the level of extinction you aim to achieve for the experiment. The number of days used for extinction tests may be determined by the level of freezing. If freezing is too high, it may be necessary to run multiple extinction tests to see differences between the groups. Experiments should be scheduled beforehand to assure all training and testing can be done during the animals' light cycle. Basic fear conditioning and testing would require 4 days (2 days of handling followed by training on day 3 and testing on day 4). Handling of the animal is crucial prior to training to decrease the stress response to the noise of transport and to the experimenter. Fear extinction would require

5 days (2 days of handling followed by fear conditioning on day 3, extinction on day 4, and extinction test on day 5).

Procedures for the use of fear conditioning and extinction as a model of PTSD in a between-subjects design are presented below.

1. If animals are ordered from outside the vivarium, allow 1 week for them to habituate to their new environment. Animals should be a minimum of 6–8 weeks old at the beginning of the experiment. A typical experiment should have 8–10 male mice per group.

2. After habituation to the vivarium, mice should be handled for at least 5 min each over 2 days. Animals should also be transported as on the day of the experiment to get them accustomed to the transport procedure. The goal is to habituate them to the handling and transport procedure and decrease anxiety unrelated to fear conditioning.

3. As noted previously, counterbalance the groups for anything that may influence your experiment (chamber, weight, time of day, etc.).

4. Whatever CS you plan to use should be tested immediately before the start of training to assure it is at the desired level. Additionally, quantitative calibration of CS intensity should be performed when possible, such as with a luminometer to measure light intensity (lux) when using a light CS or an audio multimeter or frequency meter to measure sound frequency (Hertz) and intensity (decibel level) when using an audio CS.

5. It is also advisable to assure the shockers are working correctly by verifying the shock (most easily by touching the shockers when shock is being delivered) in all the boxes. Assess that the shock feels similar across boxes and that it is actually being delivered. An oscilloscope or voltmeter is also frequently used to more precisely calibrate shock intensity across shock grids in different training chambers.

6. Mice should be brought into the experimental room one group at a time and placed into the conditioning chamber. Other animals used in the study should be kept in the vivarium or somewhere protected from the noise of the session.

7. If drug injections are being used within the experiment, inject control vehicle or drug prior to fear conditioning or extinction to examine its effects on acquisition learning. The timing of the injection may be manipulated to investigate different aspects of fear memory formation or extinction (e.g., acquisition, recall). In contrast, to examine the consolidation period following fear or extinction training, one would inject posttraining.

8. Begin the testing program that will deliver the stimuli. (Be sure electronic data are backed up on more than one computer!).

9. After training is over, return the animals to their home cage in the vivarium.

3. Data Analysis

Repeated measures ANOVAs are generally all the statistical analysis needed for fear conditioning and extinction. Repeated measures ANOVA will measure within group (over time) and between-group (control vs. experimental) differences in freezing behavior during the tone–shock (during training) or tone presentations (during testing or extinction). Freezing during training should increase and freezing during extinction should decrease significantly over the training session resulting in a significant difference ($p < 0.05$) for the within group measure. Between-group significance indicates differences between the experimental compared to the control condition. Different hypothetical examples of extinction behavior illustrating a variety of different possible statistical effects are shown in Fig. 4a–c. For statistical analysis of testing or extinction, freezing behavior during the baseline as well as during the tones should be analyzed. Freezing during extinction often starts out equivalently, only revealing differences after multiple tone presentations. In this instance, it may be important to also compare early (trials 1–5) to late extinction (trials 6–15).

4. Troubleshooting

Low freezing: Too little freezing (i.e., a floor effect) can make it difficult to determine if the experimental drug is resulting in a behavioral effect. To increase freezing, one can manipulate multiple factors including the level of aversive stimulation (e.g., shock), intervals between trials, number of training trials, or the number of days of training.

High freezing: Too much freezing (i.e., a ceiling effect) can make it difficult to determine if the experimental drug is resulting in a behavioral change. To decrease freezing, one can decrease strength

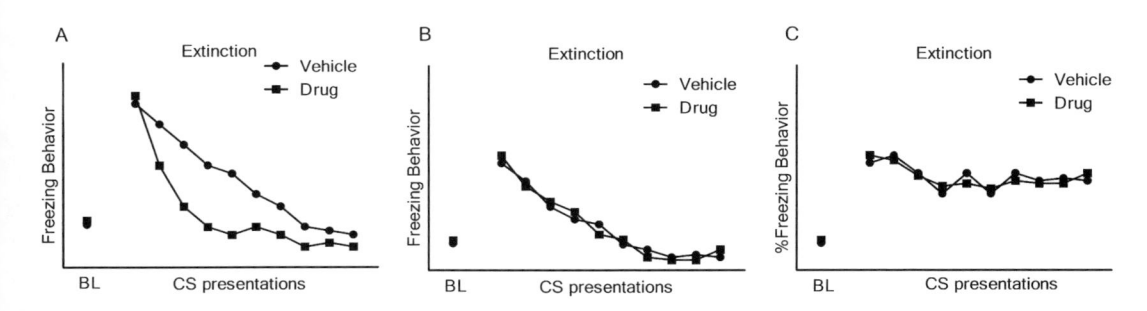

Fig. 4. **(a)** An illustration of a significant within and between subjects effect; **(b)** a significant within subjects effect only; and **(c)** No significant between or within subjects effect.

of aversive stimulation (e.g., shock), decrease the number of US trials, or decrease the number of days of training if using multiple days.

Inability to discriminate between contexts: Fear conditioning and fear extinction rely upon the ability of the animal to distinguish between the training context (A) and the test or extinction context (B). To increase discriminability, many aspects of the context can be manipulated such as flooring (i.e., shock grid vs. Plexiglas or even sandpaper), context shape, lighting (i.e., white vs. red light), and olfactory cues (acetic acid vs. acetophonone). It is a good idea to test the discriminability between the training (Context A) and testing/extinction (Context B) contexts prior to the start of any experiments.

No differences between experimental and control group: Although lack of differences between the experimental group and control group may simply represent the null hypothesis, there are many experimental reasons that may lead to a false negative result. For example, if a drug only affects acquisition of fear learning, but it is being tested in a consolidation paradigm or an extinction paradigm, no effect may be seen. Conversely, if a drug is being tested in a fear recall paradigm but it only affects consolidation learning, a false negative result may be observed. Thus, it is critical to appreciate the multiple emotional learning/expression components that can be manipulated with these relatively simple fear behaviors.

5. Concluding Remarks

Fear conditioning and its extinction have proven to be an excellent model to investigate the neurobiology of disorders of fear dysregulation, including panic disorder, phobia, and PTSD. As a technique, it is quickly learned and simple to measure. Despite the simple technical manipulations, which make it quite attractive, there are a large number of learning and emotional constructs that can be separated depending on the time of experimental manipulation. For example, fear acquisition, fear consolidation, rate of emotional memory forgetting, extinction acquisition, extinction consolidation, spontaneous recovery, renewal, and reinstatement can all be examined with the simple few training and testing manipulations outlined above. Using these models, one can test the usefulness of drugs to facilitate fear, block consolidation, or enhance extinction. These approaches, combined with the power of mouse genetics, allow for a molecular dissection of the differential components of fear and extinction learning and, perhaps one day, will be translated to treat the excessive fear and lack of fear inhibition seen in patients with PTSD and other fear-related disorders.

References

1. Orr SP, Metzger LJ, Lasko NB, Macklin ML, Hu FB, Shalev AY, Pitman RK: Physiologic responses to sudden, loud tones in monozygotic twins discordant for combat exposure: association with posttraumatic stress disorder. Arch Gen Psychiatry 2003;60:283–88.

2. Milad MR, Pitman RK, Ellis CB, Gold AL, Shin LM, Lasko NB, Zeidan MA, Handwerger K, Orr SP, Rauch SL. Neurobiological basis of failure to recall extinction memory in posttraumatic stress disorder. Biol Psychiatry. 2009 Dec 15;66(12):1075–82.

3. Rothbaum BO, Schwartz AC. Exposure therapy for posttraumatic stress disorder Am J Psychother. 2002;56(1):59–75.

4. Wolpe J. Basic principles and practices of behavior therapy of neuroses. Am J Psychiatry. 1969 Mar;125(9):1242–7.

5. Barlow, DH Long-term outcome for patients with panic disorder treated with cognitive-behavioral therapy. J Clin Psychiatry. 1990;51 Suppl A:17–23.

6. Barad M. Fear extinction in rodents: basic insight to clinical promise. Curr Opin Neurobiol. 2005 Dec;15:710–5.

7. Norberg MM, Krystal JH, Tolin DF. A meta-analysis of D-cycloserine and the facilitation of fear extinction and exposure therapy. Biol Psychiatry. 2008 Jun 15;63(12):1118–26.

8. Hofmann SG, Meuret AE, Smits JA, Simon NM, Pollack MH, Eisenmenger K, Shiekh M, Otto MW. Augmentation of exposure therapy with D-cycloserine for social anxiety disorder. Arch Gen Psychiatry. 2006 Mar;63(3):298–304.

9. Ressler KJ, Rothbaum BO, Tannenbaum L, Anderson P, Graap K, Zimand E, Hodges L, Davis M. Cognitive enhancers as adjuncts to psychotherapy: use of D-cycloserine in phobic individuals to facilitate extinction of fear. Arch Gen Psychiatry. 2004;61(11):1136–44.

10. Davis M, Ressler K, Rothbaum BO, Richardson R. Effects of D-cycloserine on extinction: translation from preclinical to clinical work. Biol Psychiatry. 2006;60(4):369–75.

11. Rauch SL, Shin LM, Phelps EA. Neurocircuitry models of posttraumatic stress disorder and extinction: human neuroimaging research--past, present, and future. Biol Psychiatry. 2006 Aug 15;60(4):376–82.

12. Orr SP, Milad MR, Metzger LJ, Lasko NB, Gilbertson MW, Pitman RK. Effects of beta blockade, PTSD diagnosis, and explicit threat on the extinction and retention of an aversively conditioned response. Biol Psychol. 2006 Oct;73(3):262–71.

13. Pavlov, I.P. Conditioned Reflexes (Translated by G.V. Anrep.) London: Oxford University Press, 1927.

14. de Toledo L. Changes in heart rate during conditioned suppression in rats as a function of US intensity and type of CS. J Comp Physiol Psychol. 1971 Dec;77(3):528–38.

15. Sutterer JR, Obrist PA. Heart rate and general activity alterations of dogs during several aversive conditioning procedures. J Comp Physiol Psychol. 1972;80(2):314–26.

16. Annau Z, Kamin LJ. The conditioned emotional response as a function of intensity of the US. J Comp Physiol Psychol. 1961;54:428–32.

17. Rescorla R, Lolordo V. Inhibition of avoidance behavior. J Comp Physiol Psychol. 1965 Jun;59:406–12.

18. Davis M. Pharmacological and anatomical analysis of fear conditioning using the fear-potentiated startle paradigm. Behav Neurosci. 1986 Dec;100(6):814–24.

19. Fanselow MS, Bolles RC. Naloxone and shock-elicited freezing in the rat. J Comp Physiol Psychol. 1979 Aug;93(4):736–44.

20. Fanselow, M.S. Factors governing one-trial contextual conditioning. Animal Learning & Behavior 1990;18(3):264–270.

21. Myers KM, Davis M. Mechanisms of fear extinction. Mol Psychiatry. 2007 Feb;12(2):120–50.

22. Quirk GJ, Mueller D. Neural mechanisms of extinction learning and retrieval. Neuropsychopharmacology. 2008 Jan;33(1):56–72.

23. Han JH, Kushner SA, Yiu AP, Cole CJ, Matynia A, Brown RA, Neve RL, Guzowski JF, Silva AJ, Josselyn SA. Neuronal competition and selection during memory formation. Science. 2007; 316(5823):457–60.

24. Han JH, Kushner SA, Yiu AP, Hsiang HL, Buch T, Waisman A, Bontempi B, Neve RL, Frankland PW, Josselyn SA. Selective erasure of a fear memory. Science. 2009 Mar 13; 323(5920):1492–6.

25. Fanselow MS, Kim JJ. Acquisition of contextual Pavlovian fear conditioning is blocked by application of an NMDA receptor antagonist D,L-2-amino-5-phosphonovaleric acid to the basolateral amygdala. Behav Neurosci. 1994; 108(1):210–2.

26. Schafe GE, LeDoux JE. Memory consolidation of auditory pavlovian fear conditioning requires protein synthesis and protein kinase A in the amygdala. J Neurosci. 2000;20(18):RC96.

27. Rodrigues SM, Schafe GE, LeDoux JE. Intra-amygdala blockade of the NR2B subunit of the

NMDA receptor disrupts the acquisition but not the expression of fear conditioning. J Neurosci. 2001;21(17):6889–96.

28. Rumpel S, LeDoux J, Zador A, Malinow R Postsynaptic receptor trafficking underlying a form of associative learning. Science. 2005; 308(5718):83–8.

29. Davis M, Hitchcock JM, Bowers MB, Berridge CW, Melia KR, Roth RH. Stress-induced activation of prefrontal cortex dopamine turnover: blockade by lesions of the amygdala. Brain Res. 1994;664(1–2):207–10.

30. Paré D, Quirk GJ, Ledoux JE. New vistas on amygdala networks in conditioned fear. J Neurophysiol. 2004 Jul;92(1):1–9.

31. Heldt SA, Ressler KJ. Training-induced changes in the expression of GABAA-associated genes in the amygdala after the acquisition and extinction of Pavlovian fear. Eur J Neurosci. 2007; 26(12):3631–44.

32. Ehrlich I, Humeau Y, Grenier F, Ciocchi S, Herry C, Lüthi A. Amygdala inhibitory circuits and the control of fear memory. Neuron. 2009; 62(6):757–71.

33. LeDoux JE. Emotion circuits in the brain. Annu Rev Neurosci. 2000;23:155–84.

34. Santini E, Muller RU, Quirk GJ. Consolidation of extinction learning involves transfer from NMDA-independent to NMDA-dependent memory. J Neurosci. 2001;21(22):9009–17.

35. Suzuki A, Josselyn SA, Frankland PW, Masushige S, Silva AJ, Kida S. Memory reconsolidation and extinction have distinct temporal and biochemical signatures. J Neurosci. 2004 May 19;24(20):4787–95.

36. Cain CK, Blouin AM, Barad M. L-type voltage-gated calcium channels are required for extinction, but not for acquisition or expression, of conditional fear in mice. J Neurosci. 2002;22:9113–21.

37. Barad M, Blouin AM, Cain CK. Like extinction, latent inhibition of conditioned fear in mice is blocked by systemic inhibition of L-type voltage-gated calcium channels. Learn Mem. 2004;11(5):536–9.

38. Marsicano G, Wotjak CT, Azad SC, Bisogno T, Rammes G, Cascio MG, Hermann H, Tang J, Hofmann C, Zieglgänsberger W, Di Marzo V, Lutz B. The endogenous cannabinoid system controls extinction of aversive memories. Nature. 2002 Aug 1;418(6897):530–4.

39. Varvel SA, Anum EA, Lichtman AH. Disruption of CB(1) receptor signaling impairs extinction of spatial memory in mice. Psychopharmacology (Berl). 2005 Jun;179(4):863–72.

40. Chhatwal JP, Myers KM, Ressler KJ, Davis M. Regulation of gephyrin and GABAA receptor binding within the amygdala after fear acquisition and extinction. J Neurosci.2005;25, 502–6.

41. Pamplona FA, Prediger RD, Pandolfo P, Takahashi RN. The cannabinoid receptor agonist WIN 55,212-2 facilitates the extinction of contextual fear memory and spatial memory in rats. Psychopharmacology (Berl). 2006 Nov;188(4):641–9.

42. Varvel SA, Wise LE, Niyuhire F, Cravatt BF, Lichtman AH. Inhibition of fatty-acid amide hydrolase accelerates acquisition and extinction rates in a spatial memory task. Neuropsychopharmacology. 2007 May;32(5): 1032–41.

43. McNally GP, Westbrook RF. Opioid receptors regulate the extinction of Pavlovian fear conditioning. Behav Neurosci. 2003 Dec;117(6):1292–301.

44. Parsons RG, Gafford GM, Helmstetter FJ. Regulation of extinction-related plasticity by opioid receptors in the ventrolateral periaqueductal gray matter. Front Behav Neurosci. 2010 Aug 3;4. pii: 44.

45. Sotres-Bayon F, Quirk GJ. Prefrontal control of fear: more than just extinction. Curr Opin Neurobiol. 2010;20(2):231–5

46. Liberzon I, Krstov M, Young EA. Stress-restress: effects on ACTH and fast feedback. Psychoneuroendocrinology. 1997 Aug;22(6): 443–53

47. Wakizono T, Sawamura T, Shimizu K, Nibuya M, Suzuki G, Toda H, Hirano J, Kikuchi A, Takahashi Y, Nomura S. Stress vulnerabilities in an animal model of post-traumatic stress disorder. Physiol Behav. 2007 Mar 16;90(4): 687–95.

48. Shimizu K, Sawamura T, Nibuya M, Nakai K, Takahashi Y, Nomura S. [An animal model of posttraumatic stress disorder and its validity: effect of paroxetine on a PTSD model in rats]. Nihon Shinkei Seishin Yakurigaku Zasshi. 2004 Oct;24(5):283–90.

49. Siegmund A, Wotjak CT. Hyperarousal does not depend on trauma-related contextual memory in an animal model of Posttraumatic Stress Disorder. Physiol Behav. 2007 Jan 30;90(1):103–7.

50. Siegmund A, Wotjak CT. A mouse model of posttraumatic stress disorder that distinguishes between conditioned and sensitised fear. J Psychiatr Res. 2007 Nov;41(10):848–60.

51. Rau V, DeCola JP, Fanselow MS. Stress-induced enhancement of fear learning: an animal model of posttraumatic stress disorder. Neurosci Biobehav Rev. 2005;29(8):1207–23.

52. Cohen H, Zohar J. An animal model of post-traumatic stress disorder: the use of cut-off

behavioral criteria. Ann N Y Acad Sci.2004;
1032:167–78.

53. Adamec R, Head D, Blundell J, Burton P, Berton O. Lasting anxiogenic effects of feline predator stress in mice: sex differences in vulnerability to stress and predicting severity of anxiogenic response from the stress experience. Physiol Behav. 2006;88(1–2):12–29.

54. Cohen JA, Mannarino AP, Perel JM, Staron V. A pilot randomized controlled trial of combined trauma-focused CBT and sertraline for childhood PTSD symptoms. J Am Acad Child Adolesc Psychiatry. 2007;46:811–9.

55. Cohen JA, Mannarino AP, Staron VR. A pilot study of modified cognitive-behavioral therapy for childhood traumatic grief (CBT-CTG). J Am Acad Child Adolesc Psychiatry. 2006;45: 1465–73.

56. Cohen H, Jotkowitz A, Buskila D, Pelles-Avraham S, Kaplan Z, Neumann L, Sperber AD. Post-traumatic stress disorder and other co-morbidities in a sample population of patients with irritable bowel syndrome. Eur J Intern Med. 2006;17:567–71.

57. Stam R. PTSD and stress sensitisation: a tale of brain and body Part 2: animal models. Neurosci Biobehav Rev. 2007;31(4):558–84.

58. Shimizu K, Kikuchi A, Wakizono T, Suzuki G, Toda H, Sawamura T, Nibuya M, Takahashi Y, Soichiro N. [An animal model of posttraumatic stress disorder in rats using a shuttle box]. Nihon Shinkei Seishin Yakurigaku Zasshi. 2006 Apr;26(2):93–9.

59. Wiltgen BJ, Sanders MJ, Behne NS, Fanselow MS. Sex differences, context preexposure, and the immediate shock deficit in Pavlovian context conditioning with mice. Behav Neurosci. 2001 Feb;115(1):26–32.

60. Whishaw IQ, Tomie J. Of mice and mazes: similarities between mice and rats on dry land but not water mazes. Physiol Behav. 1996 Nov;60(5):1191–7.

61. Frick KM, Stillner ET, Berger-Sweeney J. Mice are not little rats: species differences in a one-day water maze task. Neuroreport. 2000; 11(16):3461–5.

62. Podhorna J, Didriksen M. Performance of male C57BL/6J mice and Wistar rats in the water maze following various schedules of phencyclidine treatment. Behav Pharmacol. 2005 Feb;16(1):25–34.

63. Cressant A, Besson M, Suarez S, Cormier A, Granon S. Spatial learning in Long-Evans Hooded rats and C57BL/6J mice: different strategies for different performance. Behav Brain Res. 2007 12;177:22–9.

64. Upchurch M, Wehner JM. DBA/2Ibg mice are incapable of cholinergically-based learning in the Morris water task. Pharmacol Biochem Behav. 1988 Feb;29(2):325–9.

65. Crawley JN, Paylor R. A proposed test battery and constellations of specific behavioral paradigms to investigate the behavioral phenotypes of transgenic and knockout mice. Horm Behav. 1997;31:197–211.

66. Montkowski A, Poettig M, Mederer A, Holsboer F. Behavioural performance in three substrains of mouse strain 129. Brain Res. 1997 Jul 11;762(1–2):12–8.

67. Holmes A, Wrenn CC, Harris AP, Thayer KE, Crawley JN. Behavioral profiles of inbred strains on novel olfactory, spatial and emotional tests for reference memory in mice. Genes Brain Behav. 2002 Jan;1(1):55–69.

68. Koopmans G, Blokland A, van Nieuwenhuijzen P, Prickaerts J. Assessment of spatial learning abilities of mice in a new circular maze. Physiol Behav. 2003 Sep;79(4–5):683–93.

69. Bothe GW, Bolivar VJ, Vedder MJ, Geistfeld JG. Genetic and behavioral differences among five inbred mouse strains commonly used in the production of transgenic and knockout mice. Genes Brain Behav. 2004 Jun;3 149–57.

70. Brooks SP, Pask T, Jones L, Dunnett SB. Behavioural profiles of inbred mouse strains used as transgenic backgrounds. I: motor tests. Genes Brain Behav. 2004;3:206–15.

71. Nie T, Abel T Fear conditioning in inbred mouse strains: an analysis of the time course of memory. Behav Neurosci. 2001;115(4): 951–6.

72. Crawley JN, Belknap JK, Collins A, Crabbe JC, Frankel W, Henderson N, Hitzemann RJ, Maxson SC, Miner LL, Silva AJ, Wehner JM, Wynshaw-Boris A, Paylor R. Behavioral phenotypes of inbred mouse strains: implications and recommendations for molecular studies. Psychopharmacology (Berl). 1997 Jul;132(2):107–24.

73. Crawley JN, Paylor R. A proposed test battery and constellations of specific behavioral paradigms to investigate the behavioral phenotypes of transgenic and knockout mice. Horm Behav. 1997 Jun;31(3):197–211.

74. Maren S. Overtraining does not mitigate contextual fear conditioning deficits produced by neurotoxic lesions of the basolateral amygdala. J Neurosci. 1998 Apr 15;18(8):3088–97.

75. Bourtchouladze R, Abel T, Berman N, Gordon R, Lapidus K, Kandel ER. Different Training Procedures Recruit Either One or Two Critical Periods for Contextual Memory Consolidation,

Each of Which Requires Protein Synthesis and PKA Learn Mem. 1998;5:365–74.

76. Cordero MI, Sandi C. A role for brain glucocorticoid receptors in contextual fear conditioning: dependence upon training intensity. Brain Res.1998;786:11–7.

77. Pietersen CY, Bosker FJ, Postema F, den Boer JA. Fear conditioning and shock intensity: the choice between minimizing the stress induced and reducing the number of animals used. Lab Anim. 2006 Apr;40(2):180–5.

78. Quinn JJ, Wied HM, Ma QD, Tinsley MR, Fanselow MS. Dorsal hippocampus involvement in delay fear conditioning depends upon the strength of the tone-footshock association. Hippocampus. 2008;18(7):640–54.

79. Abrari K, Rashidy-Pour A, Semnanian S, Fathollahi Y. Administration of corticosterone after memory reactivation disrupts subsequent retrieval of a contextual conditioned fear memory: dependence upon training intensity. Neurobiol Learn Mem. 2008;89:178–84.

80. Wang SH, Teixeira CM, Wheeler AL, Frankland PW. The precision of remote context memories does not require the hippocampus. Nat Neurosci. 2009 Mar;12(3):253–5.

81. Scharf MT, Woo NH, Lattal KM, Young JZ, Nguyen PV, Abel T. Protein synthesis is required for the enhancement of long-term potentiation and long-term memory by spaced training. J Neurophysiol. 2002;87(6):2770–7.

82. Baldi E, Bucherelli C. The inverted "u-shaped" dose-effect relationships in learning and memory: modulation of arousal and consolidation. Nonlinearity Biol Toxicol Med. 2005 Jan;3(1):9–21.

83. Anagnostaras SG, Wood SC, Shuman T, Cai DJ, Leduc AD, Zurn KR, Zurn JB, Sage JR, Herrera GM. Automated assessment of pavlovian conditioned freezing and shock reactivity in mice using the video freeze system. Front Behav Neurosci 2010 Sep 30;4. pii: 158.

84. Marchand AR, Luck D, DiScala G. Evaluation of an improved automated analysis of freezing behaviour in rats and its use in trace fear conditioning. J Neurosci Methods. 2003;126(2): 145–53.

85. Godsil BP, Quinn JJ, Fanselow MS. Body temperature as a conditional response measure for pavlovian fear conditioning. Learn Mem. 2000; 7(5):353–6.

86. Jones SV, Heldt SA, Davis M, Ressler KJ. Olfactory-mediated fear conditioning in mice: simultaneous measurements of fear-potentiated startle and freezing. Behav Neurosci. 2005 Feb;119(1):329–35.

Chapter 12

The "Cut-Off Behavioral Criteria" Method: Modeling Clinical Diagnostic Criteria in Animal Studies of PTSD

Hagit Cohen, Michael A. Matar, and Joseph Zohar

Abstract

Posttraumatic stress disorder (PTSD) is clinically defined by exposure to a significantly threatening and/or horrifying event and the presence of a certain number of symptoms from each of three symptom clusters at least one month after the event. The procedures involved in defining clinical diagnostic criteria for mental disorders are lengthy and quit stringent. The APA and WHO (responsible for the DSM and ICD, respectively) periodically review the diagnostic criteria, a process which engenders heated discussion in the literature, with the aim of refining and improving their clinical, epidemiological, and research validity.

Animal behavioral studies, however, have generally tended to overlook this aspect and have commonly regarded the entire group of animals subjected to certain study conditions as homogeneous. The method to be described below was developed in an attempt to model diagnostic criteria in terms of individual patterns of response using behavioral measures and determining cut-off scores to distinguish between extremes of response or nonresponse, leaving a sizeable proportion of subjects in a middle group, outside each set of cut-off criteria.

This chapter will discuss the concept of the model and its background, provide detailed protocols for each of its components, and present a selection of studies employing and examining the model, alongside the underlying translational rationale of each.

Key words: Posttraumatic stress disorder (PTSD), Mouse, Anxiety, Fear, Predator odor, Predator-scent stress paradigm, Stress, Elevated plus maze, Acoustic startle

1. Background and Clinical Overview

Posttraumatic stress disorder (PTSD) is a pathological response to exposure to a traumatic event, with behavioral, emotional, functional, and physiological components, according to the diagnostic criteria of the Diagnostic and Statistic Manual, or DSM (1), since the inclusion of the diagnosis in 1980. A diagnosis of PTSD is conditional on a number of required symptoms being present 1 month or more after exposure to a triggering event: (1) intrusive reexperiencing of the traumatic event in the form of nightmares

Todd D. Gould (ed.), *Mood and Anxiety Related Phenotypes in Mice: Characterization Using Behavioral Tests, Volume II*, Neuromethods, vol. 63, DOI 10.1007/978-1-61779-313-4_12, © Springer Science+Business Media, LLC 2011

and flashbacks, with an exaggerated response to trauma-related reminders/cues; (2) persistent avoidance of stimuli associated with the trauma and emotional numbing; and (3) persistent symptoms of exaggerated startle response, increased physiological arousal, and sustained preparedness for an instant alarm response (1).

In the first hours to days following the experience, the vast majority of individuals exposed to an extreme event will demonstrate, to a varying degree, symptoms such as intense fear, helplessness, or horror followed by anxiety, depression, agitation, shock, or dissociation, and may have trouble functioning in their usual manner for a while (2–4). Retrospective and prospective epidemiological studies indicate that most individuals affected by a potential traumatic experience will adapt within a period of 1–4 weeks following exposure (2, 5). Epidemiological data from a range of events (from natural disasters to 9/11) indicate that only a proportion of persons exposed will develop long-term psychopathology (2, 5). In the United States, studies report that the rate of lifetime exposure to at least one "serious" traumatic event (excluding grief and mourning) is quite high; a conservative estimate reported 61% among men and 51% among women (6). Other studies have found similar rates (6–9). The lifetime prevalence of PTSD in the general population reaches about 7% overall (10), suggesting that about 20–30% of individuals exposed to severe stressors will develop PTSD (6). This figure varies depending on the type of trauma studied, where male rape victims suffer very high rates and populations exposed to natural disasters significantly less (11). The discrepancy between the proportion of the general population exposed to potentially traumatic experience and those who eventually fulfill criteria for the disorder suggests qualitative differences in vulnerability and/or resilience.

PTSD has severe, often debilitating, effects on widespread areas of functioning, severely compromising quality of life, affecting work, the family and social life, but most of all self-esteem, coping, and one's sense of mastery. Moreover, PTSD is often comorbid with other disorders such as depressive and anxiety disorders, drug and alcohol abuse, various degrees of cognitive and memory impairments, and sexual dysfunction (1). The development of PTSD is often a gradual process and extends over time through a series of stages ranging from relatively contained distress to severe disability (12). As the disorder evolves over time, pathological changes and debilitating comorbidity may become fixed and irreversible. Unlike processes in which exposure to repeated stimuli induces a process of learning or conditioning, implying increased efficiency in processing of data to produce the required response, the psychopathology underlying PTSD produces a paradoxical vulnerability to negative sequelae upon subsequent stress exposure (12).

The sequelae of exposure to a traumatizing stressor are subject to extensive clinical study. Clinical research often gives rise to important questions or hypotheses as to the pathogenesis, clinical course, and outcomes of such events. Among the issues raised are those relating to factors that may confer risk or resilience for the development of more severe stress-induced clinical outcomes, such as PTSD. By their nature, clinical studies raise issues concerning premorbid factors largely by means of extrapolating retrospectively. Prospective studies are near-impossible to conceive and would most probably be prohibitively expensive to put into practice.

An animal model provides a good approximation of certain aspects of the complex clinical disorder, enabling the study of questions raised in clinical research in a prospective study design and under far more controllable conditions. In order to achieve a satisfactory degree of validity and reliability, animal models of complex and intricate psychiatric disorders must fulfill certain criteria. For example, the behavioral responses must be observable and measurable and must reliably reflect clinical symptomatology, and pharmacological agents that are known to affect symptoms in human subjects should correct, with equal efficacy, measurable parameters that model symptoms of the disorder.

Developing an animal model for PTSD is not a trivial issue. Diagnosis in human patients relies heavily on personal reports of thoughts, dreams, and images, which cannot be studied in rodents. Furthermore, several of the typical symptoms of PTSD may be unique to humans and thus not be found in rodents. For example, intrusive memories of the traumatic event, one of the core symptoms for PTSD in humans, cannot be translated in animal behavioral models. Likewise, an important factor of the trauma in humans is the perception of the life-threatening potential of the situation. It is not clear whether rodents can make this judgment or which stressors will be most effective for animals. In addition, there is as yet no clearly effective pharmacological treatment for PTSD. It is thus difficult to test a potential rodent model for its pharmacological predictability in relation to PTSD or other traumatic stress-related disorders. Nevertheless, using animals to study PTSD holds advantages for several reasons. First, unlike many other mental disorders, the diagnostic criteria for PTSD specify an etiological factor, which is an exposure to a life-threatening, traumatic event (13). In a model for PTSD, variables such as the quality and intensity of the stressor and the degree of exposure to it can be carefully controlled, and the behavioral and concomitant physiological responses to a (valid) threatening stimulus can be studied. Second, little is known about pretrauma etiological aspects of the disorder, since, naturally, the studies so far have focused on retrospective assessments of the patients after the onset of PTSD. An animal model enables a prospective follow-up design, in which the disorder is triggered at a specified time and in a uniform manner, in controllable and statistically sound population samples, and enables the assessment of behavioral and gross physiological parameters. Moreover, unlike studies in human subjects, animal model studies enable the assessment of concomitant biomolecular changes in dissected brain areas and experimentation with pharmacological agents with potential therapeutic effects.

General recommendations: All protocols using live animals must first be reviewed and approved by an Institutional Animal Care and Use Committee – and must follow officially approved procedures for the care and use of laboratory animals (14).

Colony maintenance and handling: Handling and housing of the animals prior to all tests are crucial, but of particular importance when testing anxiety-like behaviors. It is critical only that the specific test conditions (and not other extraneous factors) determine the anxiety generated (14). Upon their arrival, new animals undergo acclimatization to the colony for about 7 days. All animals are handled daily for at least another 7 consecutive days prior to the experiment. Handling consists of picking the animals up with a gloved hand around the shoulder for 5 min/day.

The method had been applied mainly to rats based on the behavioral model for PTSD below, although mice have been involved in a number of studies, to be detailed below. In general, the mice have tended to respond in a manner almost identical to rat subjects. The descriptions below may thus be taken to hold true for mice as well, overall.

2. The Predator-Scent Stress Paradigm

Stress paradigms in animal studies aim to model criterion A of the DSM diagnostic criteria (1). They thus use extremely stressful experiences aimed at engendering a sense of threat and helplessness in the animal. Some of these have focused more on the intensity of the experience, whereas others have combined intensity with an attempt to design an ethologically valid experience, one which an animal might encounter in its natural environment.

The standard stressor in the following studies consists of exposure of rodents to the scent of the urine of their prime predator – the cat. Blanchard et al. (15–19), Adamec et al. (20–25), and others (14, 26, 27) have established the validity of this paradigm, in which adult rodents are inescapably exposed to urine-soiled substrate (cat litter) for 5–10 min in a closed environment, where both "fight" and "flight" options are ineffective. Predator stress has ecological validity in that it mimics brief intense threatening experiences inducing the expected range of behavioral and physiological responses. The potency of predator stimuli is comparable to that of

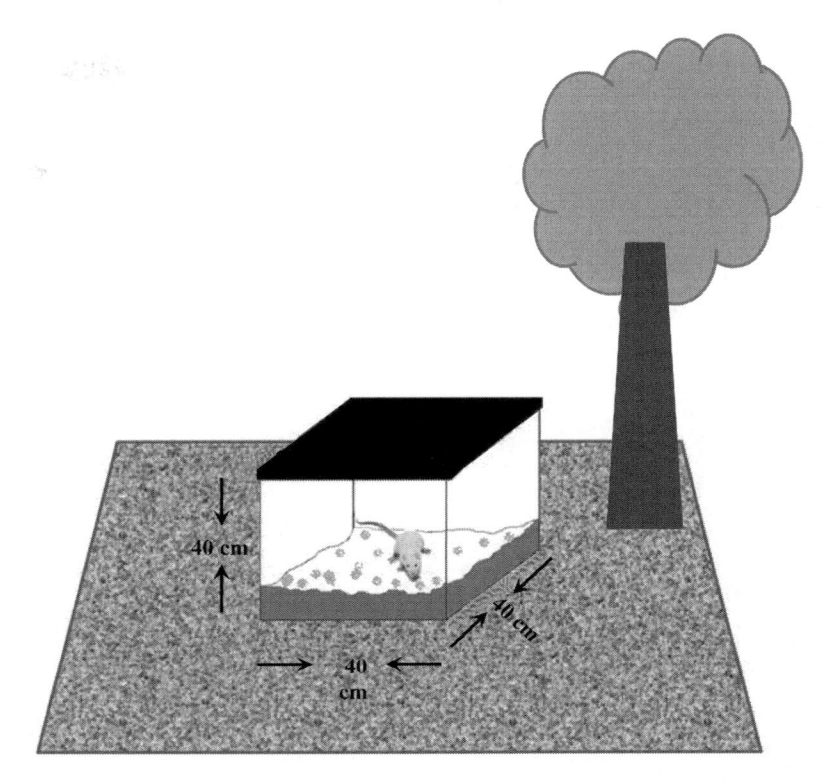

Fig. 1. *Predator-scent stress (PSS) test apparatus*: The apparatus consists of a 40 × 40 × 40-cm chamber of transparent Plexiglas with transparent roof. The floor is covered with a 5–6-cm layer of well-soiled cat litter (in use by the cat for 2 days, sifted for stools). The apparatus placed on a yard paving stone.

a variety of paradigms in which the threat is more tangible and immediate, such as paradigms based on inescapable pain or electric shock, swimming and near-drowning, a small raised platform, and even direct proximity to a kitten or a cat (separated by a mesh divide or a solid divide with an opening large enough for the rodent to slip through).

Predator-scent stress exposure: Stress consists of placing the test animals on well-soiled cat litter (in use by the cat for 2 days, sifted for stools) in a plastic cage (40×40×40 cm) (inescapable exposure) placed on a yard paving stone, for 10 min in a closed environment (Fig. 1). The soiled litter is changed between animals. Control animals are exposed to fresh, unused litter for the same amount of time and under the same conditions.

Critical factors: It is important that the soiled litter not be allowed into the animal-house or the area devoted to behavioral experiments at any time.

The predator stress paradigm has proven to be effective in inducing the expected range of behavioral and physiological responses (13, 21–23). These include freezing, avoidance, increased secretion of stress hormones, and changes in transmission from the hippocampus via the ventral angular bundle to the basolateral amygdala, and from the central amygdala to the lateral column of the periaqueductal gray (14, 15, 21, 22, 26, 28–39). These pathways are of interest because neuroplastic changes within them are associated with aversive learning.

3. Behavioral Assessments

A variety of mazes and open environments have been employed to assess changes in exploratory behavior resulting from stress exposure. These test environments assess behaviors whose disruption indicates anxiety-like fearful behaviors and behaviors reflecting avoidance. Various learning and memory tasks are employed in which both exploration and learned task performance can be assessed. Some studies have investigated social behavior in home cages and in challenge situations. The startle response, which characterizes many PTSD patients, has been employed as one of the more definitively measurable parameters for the hypervigilant/hyperalert component of the behavioral responses (14). In the studies presented below, exploratory behavior on the elevated plus maze (EPM) serves as the main platform for the assessment of overall behavior and the acoustic startle response (ASR) paradigm provides a precise quantification of hyperalertness, in terms of magnitude of response and habituation to the stimulus.

The elevated plus maze: The EPM is a well-characterized behavioral paradigm to investigate anxiety in rodents. The test is based upon the conflict between an innate aversion to exposed spaces and a tendency to explore new environments (14, 40). The EPM was designed to provide measures of anxiety that are relatively uncontaminated by changes in overall motor activity and has been extensively validated (40). The following sectors describe the maze for rats and mice.

Apparatus: The EPM consists of two open arms (sized for rats: 50×10 cm; for mice: 30×6 cm) and two enclosed arms (for rats: $50 \times 10 \times 40$ cm; for mice: $30 \times 6 \times 30$ cm), with an open roof, arranged such that the two open arms are opposite to each other. The maze is elevated from the floor (for rats: to a height of 50 cm and for mice: 30 cm) (Fig. 2).

Experimental procedures: Animals are placed on the central platform facing an open arm and allowed to explore the maze for 5 min. Arm entry is defined as entering an arm with all four paws. In a typical EPM set-up, a video camera is positioned directly above the maze. The videotape and the recorded images are scored by a

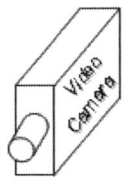

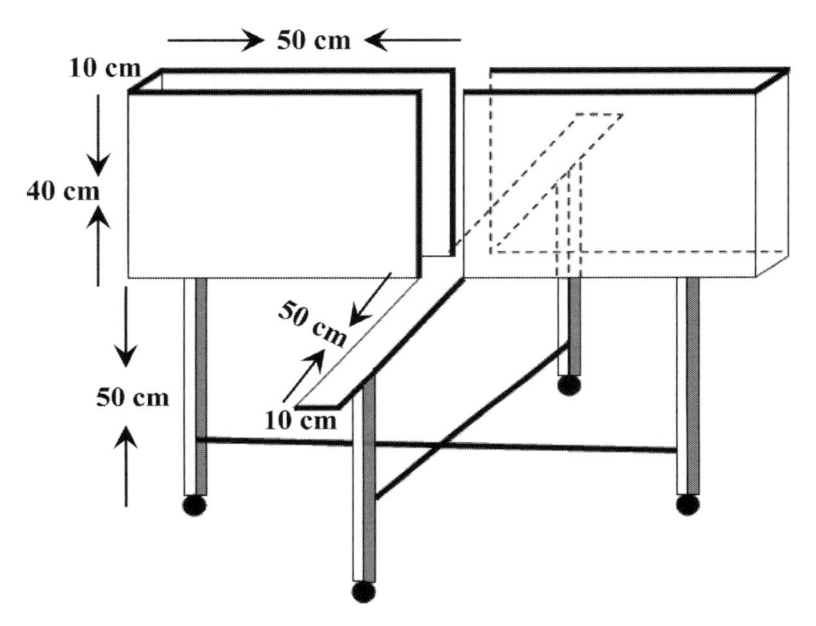

Fig. 2. *The elevated plus maze apparatus*: The elevated plus maze (sized for rats). The maze made of black Plexiglas, consists of four arms in form of a maze: two open arms (for rats: 50×10 cm; for mice: 30×6 cm) and two arms of the same size that enclosed by walls (for rats: 40 cm high; for mice: 30 cm). The open arms are opposite each other and converge into a central platform (for rats: 10×10 cm; for mice: 6×6 cm). A video camera mounted above the maze is used to observe the animal's behavior and record the trails for automatic computer analysis, for later additional scoring or both.

trained observer (in another room) unaware of the treatment conditions.

Critical factors: (1) Surfaces must be cleaned with a 5% ethanol/water solution and dried thoroughly between test sessions. (2) Should an animals fall off the maze, its data are excluded.

In our animal model, the EPM behaviors assessed are: time spent (duration) in open and closed arms and on the central platform; number of open and closed arm entries; and total exploration (entries into all arms). Total exploration is calculated as the number of entries into any arm of the maze in order to distinguish between impaired exploratory behavior, exploration limited to closed arms (avoidance), and free exploration. Moreover, we also calculated an "anxiety index" which integrated all the EPM measurements as follows:

Anxiety index =

$$1 - \left[\frac{\left(\dfrac{\text{time spent in the open arms}}{\text{total time on the maze}} \right) + \left(\dfrac{\text{number of entries to the open arms}}{\text{total activity on the maze}} \right)}{2} \right]$$

Anxiety index values range from 0 to 1 where an increase in the index expresses increased anxiety-like behavior. The anxiety index brings together the data for each of the individual parameters of exploratory behavior in the EPM and the accepted ratios used to date into a unified parameter, which reflects not only the absolute measures, but also indicates a (n) (overall) tendency.

The acoustic startle response: The adaptive response to a sudden and unexpected stimulus is to be startled by it, thus focusing attention and preparing the organism for a response, should it be required. Repeated stimuli ensuring in no perceivable danger ought to result in habituation. PTSD patients tend to display poor habituation and the phenomenon has been included in diagnostic criteria (1). Behaviorally, in rodent the startle response consists of rapid contraction of head, neck, trunk, and legs muscles in addition to the arrest of ongoing activity (41, 42). One of the most widely used stimuli is a series of repeated intense auditory signals eliciting an acoustic startle response (ASR) (65).

Apparatus: The acoustic startle apparatus is made of a Plexiglas cylinder (sized for rats: $180 \times 85 \times 90$ mm; for mice: $100 \times 60 \times 70$ mm) mounted on a Plexiglas platform and enclosed in a ventilated sound-attenuated cubicle equipped with high-frequency loudspeakers. A stabilimeter measures the whole-body flinch elicited by acoustic stimuli. Movements within the cylinder are detected and transduced by a piezoelectric accelerometer attached to the platform, digitized, and stored by the operating computer. The startle box is placed on a vibration isolation platform to minimize environmental vibration (Fig. 3). The system is computer-based and all stimulus parameters are specified by completing entries in software available from a single screen.

Experimental procedures: Animals are placed in the small chamber within the acoustic startle box. The door is closed and animals are allowed to acclimate. The startle response will depend on both subjective (individual) factors and on objective factors. The researcher must therefore determine a priori the parameters of the stimulus (duration, intensity, and especially the stimulus rise time) and take into account extraneous disruptive factors (background noise levels and intertrail intervals) (43). Startle magnitudes differ significantly among strain of laboratories rodents.

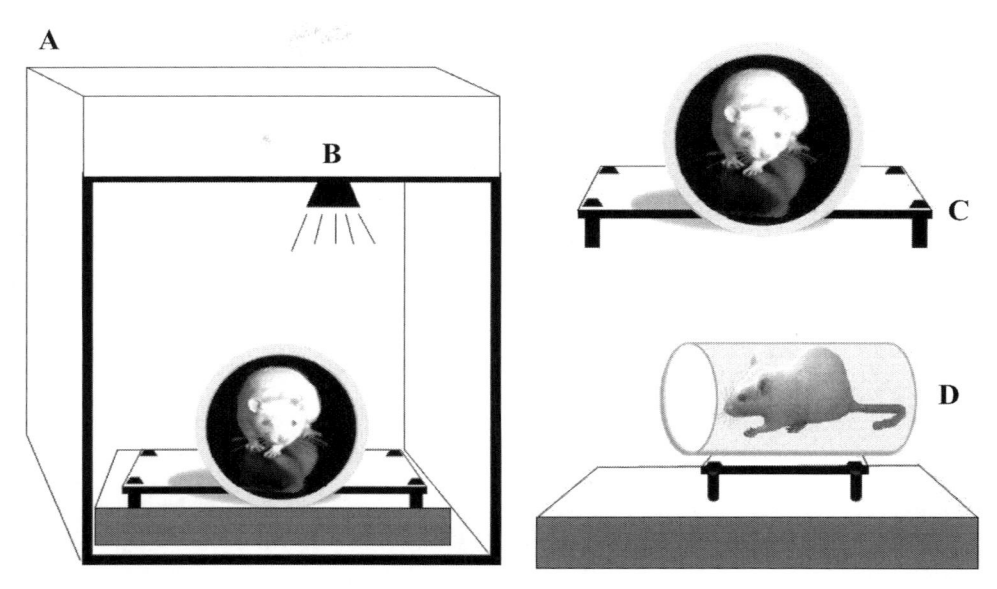

Fig. 3. *The acoustic startle response apparatus*: (**a**) The cabinet contains a complete sound generation system for white noise production, separately adjustable background noise levels, and accessory connections for optional stimuli. The audio source module is used to provide acoustic stimuli. For tone stimuli (**b**), amplitude, frequency, and duration are computer-controlled. Amplitude and duration of noise bursts are computer-controlled. Specialized amplifier circuitry, contained in the cabinet, permits the use of a dynamic standardization system that emulates actual startle response movements. The dynamic response sensor design ignores static animal weights enabling the full range of the transducer capacity to be available for response recording. (**d**) Animal enclosures are designed to locate the subject without using restraint so the animals do not suffer from restraint stress and confound the results of the startle testing.

Critical factors: (1) It is important to recalibrate the startle before each trail. (2) Animals must be of approximately equal weight and age. (3) Habituate individuals by placing them in test chamber for 30 min on each of 3–5 consecutive days. At the end of each habituation period, return animals to the colony. (4) The apparatus is cleaned with a 5% ethanol/water solution and dried thoroughly between test sessions.

In our animal model each test, session starts with a 5-min acclimatization period to background white noise of 68 dB, followed by 30 acoustic trial stimuli in six blocks (a 40-ms, 110-dB pulse of broadband noise, at intervals averaging 15 s (ranging from 12 to 30 s) between stimuli). The ASR assessment consists of: mean startle amplitude (averaged over all trials) and percent of startle habituation to repeated presentation of the acoustic pulse. Percent habituation – the percent change between the response to the first block of sound stimuli and the last – is calculated as follows:

Percent habituation =

$$\frac{100 \times \left[(\text{average startle amplitude in Block 1}) - (\text{average startle amplitude in Block 6})\right]}{(\text{average startle amplitude in Block 1})}$$

4. Timing of Behavioral Assessments

As to the timing of behavioral assessments, a large number of studies performed in a range of research centers indicate quite clearly that behavioral changes that are observed in rodents at Day 7 after stress exposure are unlikely to change significantly over the next 30 days (44). The average life expectancy for the domestic rat is between 2.5 and 3 years. Hence, behavioral patterns observed at Day 7 can reliably be taken to represent PTSD-like responses (i.e., "translating" a week for a rodent to a month for a human).

5. Individual Differences in Response to an Exposure to a Traumatic Experience

Researchers who work with animals have long been aware that individual study subjects tend to display a varying range of responses to stimuli, certainly where stress paradigms are concerned. This heterogeneity in responses was accepted for many years and regarded as unavoidable. Since humans clearly do not respond homogeneously to potentially traumatic experiences, any heterogeneity in animal responses might be regarded as confirming the validity of animal studies, rather than as a problem. It stands to reason that a model of diagnostic criteria for psychiatric disorders could be applied to animal responses to augment the validity of study data, as long as the criteria for classification are clearly defined, reliably reproducible, and yield results that conform to findings in human subjects.

6. Classification according to Cut-Off Behavioral Criteria

Data from a large series of studies had previously shown that 7 days after a single 10-min predator-scent exposure, the overall exposed population displayed significantly decreased time spent in the open arms and increased time in the closed arms of the EPM (which is translated to "avoidant" and "anxiety-like behavior"), higher mean startle responses, and lower startle habituation as compared to control animals (Fig. 4). It is important to note that the rodents' behavior was not uniformly disturbed, but rather demonstrated a broad range of variation in response severity (Fig. 5). Since the distribution of the individual variations in behavioral response was clearly nonnormal, tending towards a discontinuous, it was justifiable to classify the animals according to extent of behavioral change (Fig. 6).

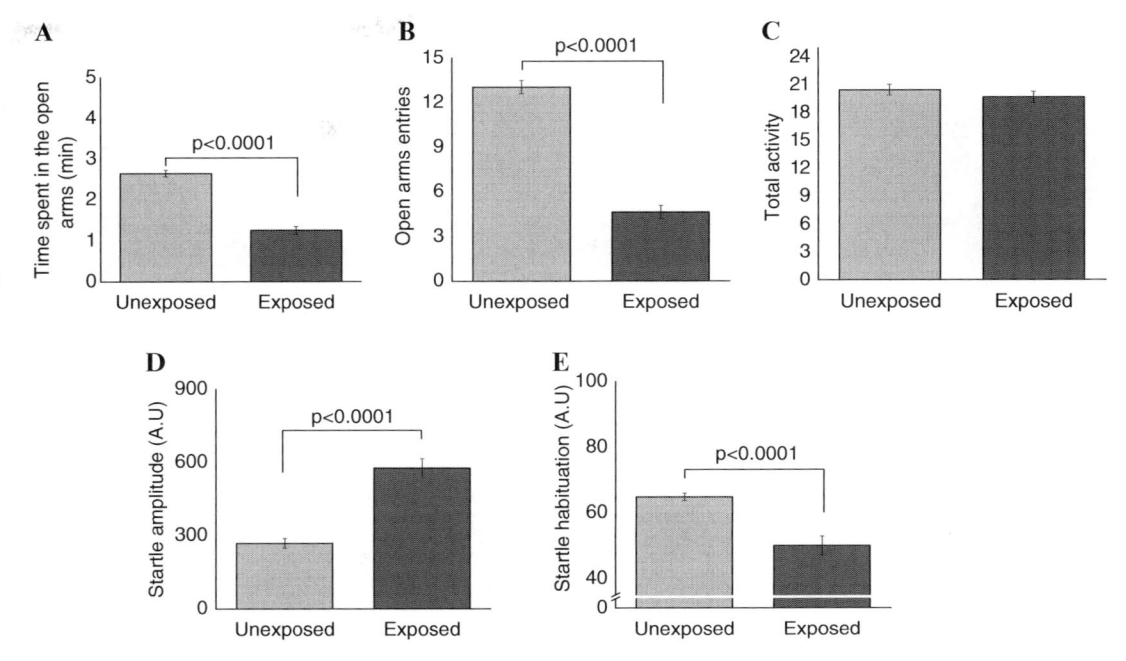

Fig. 4. The effect of the predator-scent stress (PSS) paradigm on overall anxiety-like behavior and acoustic startle response: A single ten-minute exposure to PSS significantly increased anxiety-like behavior/avoidance of open spaces as compared to unexposed controls. Time spent in the open arms (**a**) and entries into open arms (**b**) were significantly decreased after a single exposure to the stressor, as compared to control conditions ($F(1,298) = 126.4$, $p < 0.0001$ and $F(1,298) = 68.25$, $p < 0.0001$, respectively). There were no differences in total exploration (**c**) of the maze between groups. This result suggests overall anxiety and avoidance of exploration in the open arms, as opposed to an impairment of loco-motion/exploration. PSS exposure significantly increased the mean startle amplitude (**d**) and caused a significant deficit in the startle habituation of ASR (**e**) in exposed rats as compared to controls ($F(1,298) = 51.04$, $p < 0.0001$ and $F(1,298) = 45.25$, $p < 0.0001$, respectively). All data represent group mean ± SEM.

The behavioral measures for each of these groups on the EPM and ASR tests were employed to define the basic cut-off behavioral criteria (CBCs). The classification of individuals according to the degree to which their individual behavior is affected by a stressor is based on the premise that, in the natural environment, such extremely compromised behavior in response to the priming trigger may compromise behaviors essential for survival and is thus inadequate and maladaptive, representing a pathological degree of response. Since clinical diagnostic criteria require a sufficient number of symptoms from three symptom clusters in order to achieve satisfactory diagnostic specificity, the CBC response classification process requires that a given animal fulfill all criteria on both tests, performed in series. The standard algorithm for the CBC classification model also requires that prior to classification, a significant overall effect be demonstrated (Fig. 7).

The CBCs enable us to clearly classify a given animal as displaying extreme behavioral response (EBR) or minimal behavioral response (MBR) (i.e., extreme responses on both EPM and ASR

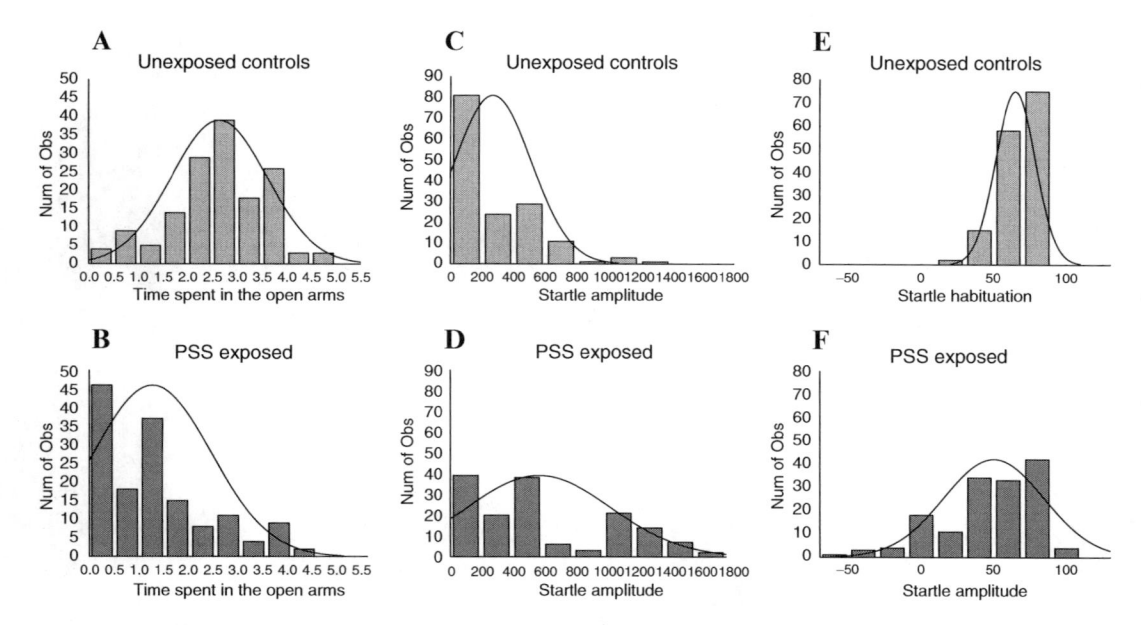

Fig. 5. *The distributions of the exposed groups and the control group*: The distributions of the exposed groups and the control group (time spent in the closed arms, startle response and startle habituation) were compared by a Kolmogorov–Smirnov statistic. The results showed that there are differences in the general shapes of the distributions curve in the two populations ($p < 0.001$). The distribution curve for exposed animals has a distinct shift (in the parameter of time spent in the open arms: shift to the left; startle response: a discontinuous nature, tending towards a bimodal distribution; startle habituation: shift to the left).

tests lead to classification as EBR, whereas minimal responses are defined as MBR) – both of which have been validated in a large series of studies. The remaining animals display clearly disrupted behavior patterns compared to controls, but the extent of the disruption does not cross the threshold for EBR. These are labeled partial behavioral responders (PBR) and have as yet not been further subclassified (45).

The pooled behavioral data for entire predator-scent stress (PSS)-exposed populations were reexamined according to the CBCs, revealing that the overall prevalence rate for EBR animals was approximately 21.3%, as compared to 1.3% in unexposed control populations. The prevalence of MBR animals in the PSS-exposed groups was 6.7% as compared to 20.0% in the control groups (Fig. 8).

The implication of this initial finding was that all prior study analyses must have included a significant proportion of animals whose behavior had not been affected by the stressor (MBR) and many animals whose response was of uncertain significance (PBR), alongside those whose response was unequivocally one of severely disrupted behavioral patterns (EBR). Hence, the method offered a feasible means for classifying animal response patterns to trauma, thereby increasing the conceptual accuracy of the data.

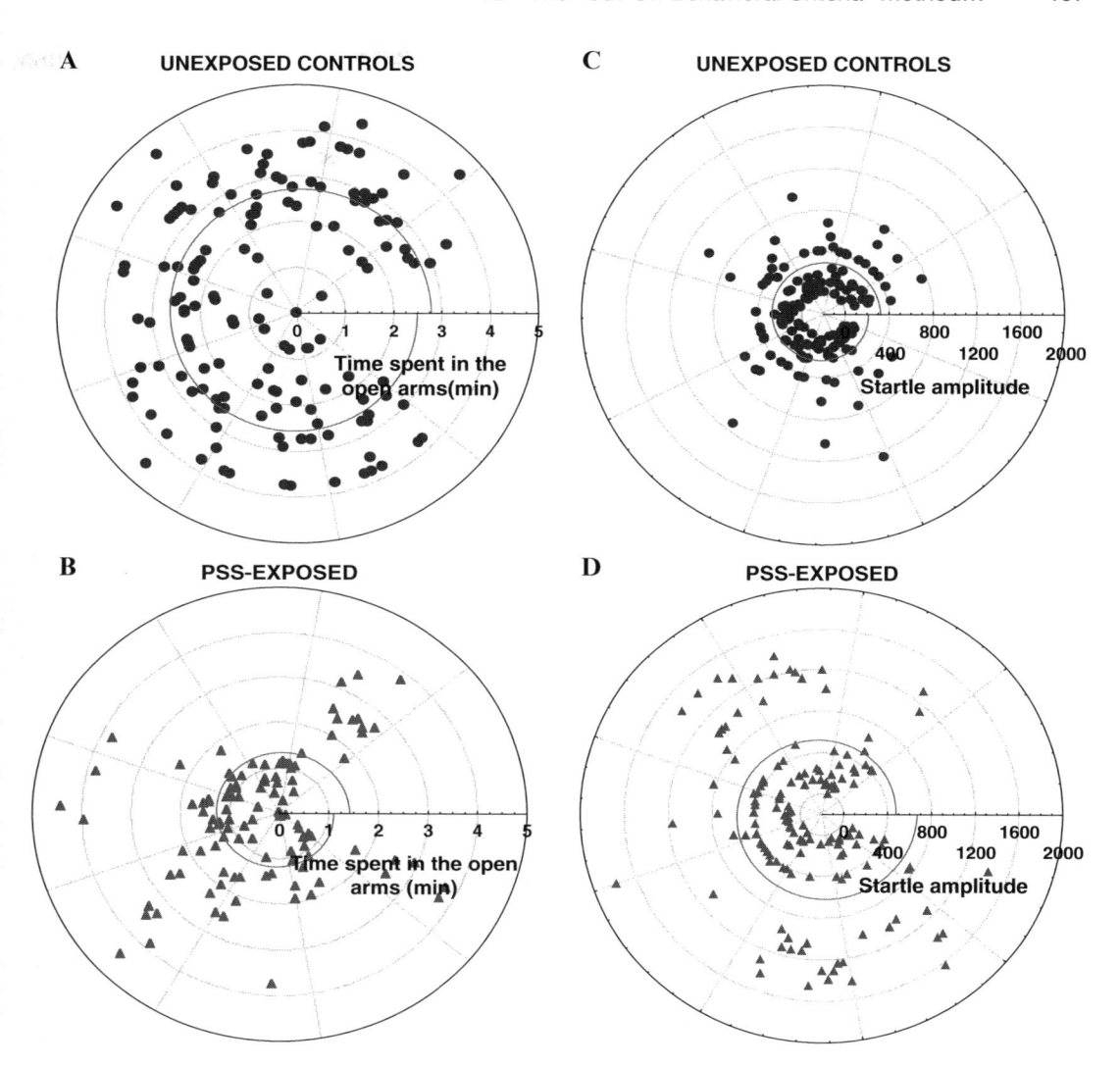

Fig. 6. *The effect of the predator-scent stress (PSS) paradigm on anxiety-like behavior and acoustic startle response*: The figures represent time spent in the open arms 6 (**a**, **b**) of the elevated plus maze and in the mean acoustic startle amplitude 6 (**c**, **d**) in rats exposed to a cat-scent, as compared to controls. Exposed rats spent less time in the open arms (*center*) and exhibited higher mean startle amplitude (circumference) and as compared to controls. Overall, a wide distribution in results was observed within the exposed rats with a broad range of variation in behavioral response. We thus hypothesized that the group is not homogeneous, and we may be dealing with several subgroups in this population. Based on the results of this phase of the study, the animals were subdivided into groups reflecting magnitude of response according to the CBCs, focusing selectively on EBR, PBR, and MBR.

It is of interest to note that the proportion of the entire exposed population fulfilling criteria for extreme responses (EBR) was compatible with epidemiological data for PTSD among trauma-exposed human populations (6), which report that between 15 and 35%

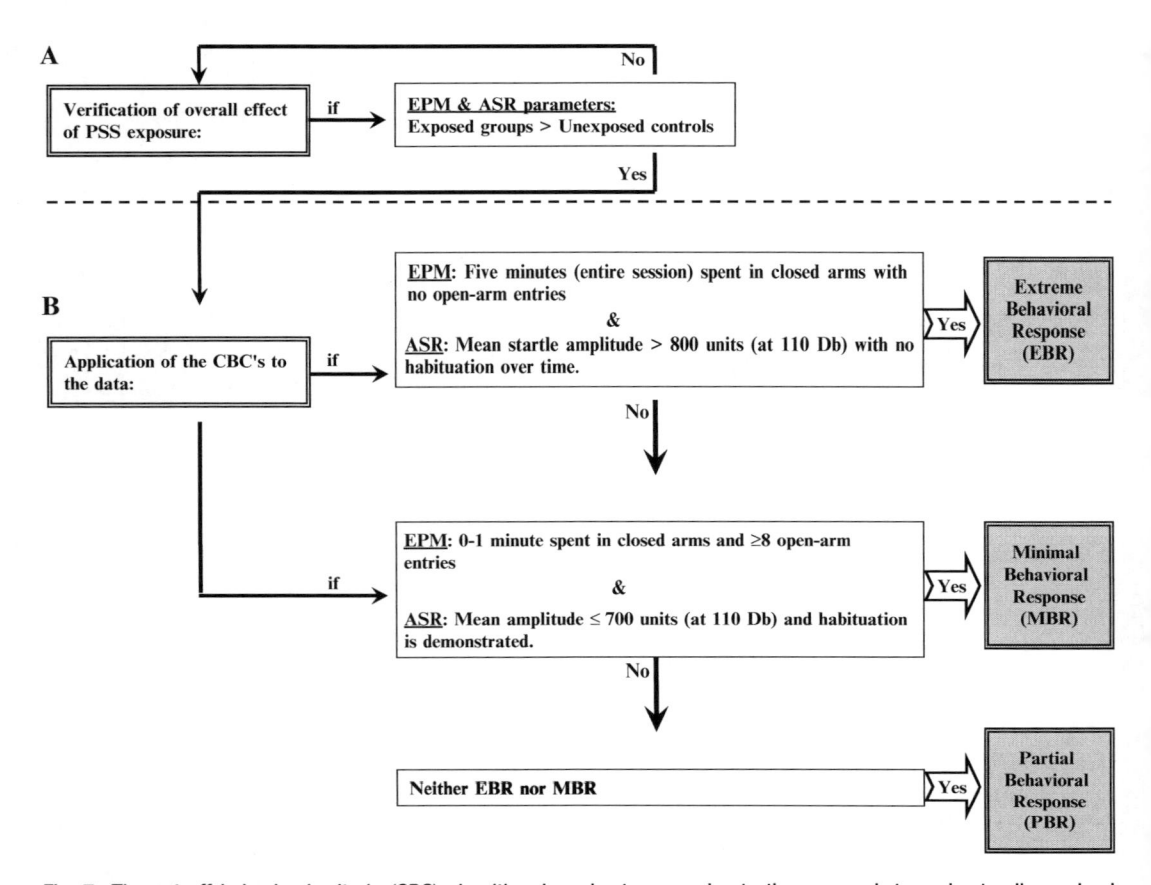

Fig. 7. *The cut-off behavioral criteria (CBC) algorithm:* In order to approximate the approach to understanding animal behavioral models more closely to contemporary clinical conceptions of PTSD, we use an approach that enables the classification of study animals into groups according to degree of response to the stressor, i.e., the degree to which individual behavior is altered or disrupted. In order to achieve this, behavioral criteria were defined and then complemented by the definition of cut-off criteria reflecting severity of response; this parallels inclusion and exclusion criteria applied in clinical research. The procedure requires the following steps: (**a**) Verification of global effect: The data must demonstrate that the stressor had a significant effect on the overall behavior of exposed vs. unexposed populations at the time of assessment. *Step A* is intended to test our zero-hypothesis, i.e., that exposure would have an overall effect on the rodents as a group compared to controls and yet that there would be individual differences in behavioral effects. This heterogeneity would form the basis for the definition of cut-off criteria for two groups of animals. We hypothesized that it would be possible to identify one group of rats, which demonstrated significant behavioral change and other with almost none. These three groups would then provide the data by which behavioral criteria could be defined for the second step. (**b**) Application of the CBCs to the data: In order to maximize the resolution and minimize false positives, extreme responses to both EPM and ASR paradigms, performed in sequence, were required for "inclusion" into the EBR group, whereas a negligible degree of response to both was required for inclusion in the MBR group. In order to focus on an elucidation of the criteria for a specific behavioral response, we looked into variation within the groups. Based on the severity of the PTSD-like behaviors, we divided the rats (within each group) into two dichotomous subgroups: extreme behavioral responders (EBR), minimal behavioral responders (MBR), and partial behavioral responders (PBR).

fulfill criteria for PTSD and that approximately 20–30% display partial or subsymptomatic clinical pictures (6, 8). This compatibility further supports the concept of criterion-based classification in terms of face validity.

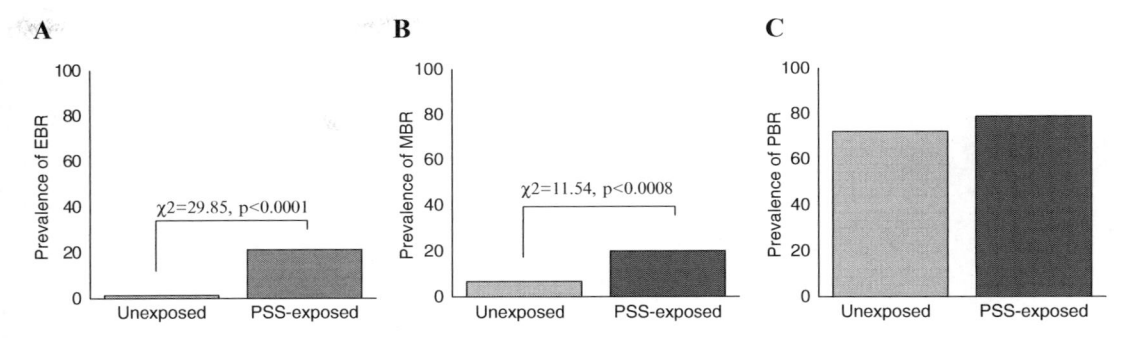

Fig. 8. *Relative prevalence rates according to CBCs*: Reanalysis of data applying cut-off behavioral criteria: (**a**) Prevalence of extreme behavioral response (EBR) rats. (**b**) Prevalence of minimal behavioral response (MBR) rats. (**c**) Prevalence of partial behavioral response (PBR) rats. There were significant differences in the prevalence rates of individuals displaying EBR among groups (Pearson $\chi^2 = 296.11$, df $= 3$, $p < 0.0001$). PSS exposure increased the prevalence of PTSD-like behavioral responses (EBR) ($x^2 = 29.85$, $p < 0.0001$) and concomitantly reduced prevalence rates of minimal behavioral response (MBR) ($x^2 = 11.54$, $p < 0.0008$), relative to unexposed controls.

7. Exposure to Trauma Cue

A disproportionate psycho-physiological response to trauma cues is integral to the clinical definition of PTSD. In order to model this component of PTSD, we required a stimulus which was not intrinsically threatening, yet acted as a clear-cut reminder of the traumatic stressor and a novel behavioral measure, in order to preclude biases (i.e., not the EPM or ASR).

7.1. Experimental Procedures

The trauma cue consists of placing the animals on fresh, unused cat litter for 10 min (i.e., identical to sham-exposure) (Fig. 9). Since the clinical symptom stands out because it occurs long after the actual event, the test must be timed accordingly. It is preferable to allow extended periods of time, certainly no less than 8 days. Many of our studies span a 30-day period and the trauma cue exposure is thus performed on day 31.

The behavioral outcome measure which we found reliable and valid is freezing behavior of the rodent placed on the litter. The effect of exposure to the trauma cue was recorded using an overhead video camera and scored for immobility (freezing) using the recorder images. Freezing behavior was defined as the absence of all movements except for those related to respiration (46–50). Total cumulative freezing time (total seconds spent freezing during each assessment period) was measured and calculated as a percentage of total time.

Freezing behavior indicates a sense of immediate threat and intense fear, and the fact that this behavior was engendered by a neutral reminder 8 days (or more) after stress exposure implies that a memory-related process and contextual association of the stimulus must have occurred.

Fig. 9. *Exposure to trauma cue*: The apparatus consists of a 40 × 40 × 40-cm chamber of transparent Plexiglas. The floor is covered with a 5–6-cm layer of unused cat litter. The apparatus placed on a yard paving stone.

7.2. Translational Validity of the Model

Behavioral response patterns vs. time: Time is an integral factor in traumatic stress-induced disorders. The prevalence rates of EBR were assessed among PSS-exposed animals on Days 1, 3, 5, 7, 30, and 90 after exposure. Initially (Day 1), almost all animals displayed extreme disruptions of behavior (EBR=90%). The proportion of EBR animals dropped rapidly over Days 1 and 3 and between Days 3 and 5 to about 21% at Day 7. This proportion remained stable till Day 30, dropping to about 17% by Day 90. The resulting time curve of EBR prevalence rates parallels the rates of stress-related symptoms in humans, culminating in acute and chronic traumatic stress disorders (Fig. 10) (44).

Physiological correlates: Physiological data were correlated with behavioral classification in a series of studies. These included the hypothalamus-pituitary-adrenal (HPA)-axis (circulating corticosterone, dehydroepiandrosterone, and its sulphate derivative dehydroepiandrosterone-sulphate levels), autonomic nervous system (heart rate and heart rate variability) (51, 52), and immune system (53). Although the gross population data had shown that the parameters in each study displayed significant responses to the stressor, CBC classification revealed that animals whose behavior conformed to EBR criteria were characterized by significantly more disturbances on all measures, whereas MBR animals displayed almost none.

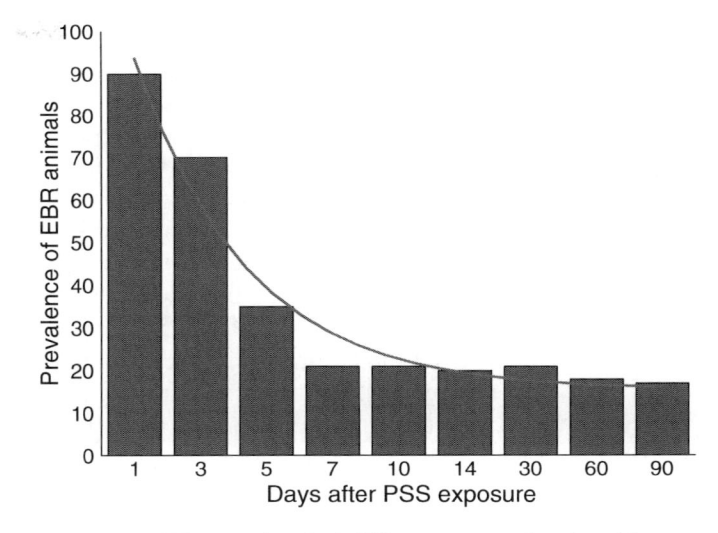

Fig. 10. Prevalence of EBR rats after single PSS exposure as a function of time.

Molecular neurobiological correlates. Selected brain areas, especially hippocampal substructures and frontal cortex, of animals classified according to the CBC procedure have been studied in correlation to both their behavioral and physiological response patterns. For example, the development of an EBR has been shown to be associated with a distinct pattern of long-term and persistent downregulation of brain-derived neurotrophic factor (BDNF) (mRNA and protein levels) and synaptophysin, and an upregulation of glucocorticoid receptor (GR) protein levels and tyrosine kinase receptor mRNA in the CA1 subregion of the hippocampus, compared to PBR and MBR animals and to unexposed controls (34, 54). The persistently higher levels of glucocorticoids are associated with attenuation of BDNF, synaptophysin, and expression of zif/268 and immediate early genes such as activity-regulated cytoskeletal-associated protein (Arc) in the EBR individuals; this suggests that glucocorticoids reflect or mediate the characteristic changes in neural plasticity and synaptic functioning underlying chronic, stress-induced behavioral disruption (35). Taken together, decreased hippocampal expression of these genes may have physiological consequences, for example, inducing damage to hippocampal neurons.

Strain/genetic studies. The CBC classification model was applied to genetically manipulated rodent strains in order to examine two aspects of PTSD. One study assessed the HPA-axis response in rat strains inbred to have either deficient or excessive HPA-axis responsiveness, compared to outbred rats. The other examined the heritability of vulnerability vs. resilience factors using inbred (near-isogenic) mouse strains exposed to PSS and classified according to the CBC method.

HPA-axis response in Lewis and Fischer rats. PTSD has been associated with disordered levels of circulating cortisol, an integral component of the stress response (increased levels according to some studies, and decreased in others) (55–61). The study examined whether low basal cortisol levels represent a consequence of traumatic exposure (i.e., possible neurotoxic effects of trauma) or a predisposing trait for pathological stress reactions in populations of inbred Lewis and Fischer rats compared to outbred Sprague–Dawley rats (controls). Lewis rats exhibit a reduced synthesis and secretion of corticotropin-releasing hormone, leading to reduced plasma adrenocorticotropic hormone and reduced corticosterone release from the adrenal cortex, whereas Fischer rats possess a hyperresponsive HPA-axis. After PSS exposure, prevalence rates of EBR individuals were significantly higher in Lewis (50%) than in Fischer rats (10%), or controls (25%) (62). However, exogenous administration of cortisol to Lewis rats before applying the stressor decreased the prevalence of EBR significantly (to 8%). These results suggest that a blunted HPA-axis response to stress may play a role in the susceptibility to experimentally induced PTSD-like behavioral changes, especially as these effects were reversed by preexposure administration of corticosterone (62).

Stress-induced behavioral responses in inbred mouse strains. Twin and family studies of PTSD patients raise questions as to a possible genetic predisposition to PTSD, although the relative contributions of genotype and environment to endophenotypic expression are unclear.

Six inbred strains of mice frequently employed in transgenic research were assessed at baseline and 7 days after PSS exposure (52). Inbred strains are expected to demonstrate ~97.5% homozygosity of loci, as the result of at least 20 generations of sibling matings. The results, however, revealed an unexpectedly high degree of within-strain individual heterogeneity at baseline and a high degree of heterogeneity in response to stress. This within-strain phenotypic heterogeneity might imply that environmental factors play a significant role in characterizing individual responses, in spite of the significant strain-related, i.e., genetic, underpinnings. This study suggests that heritable factors may be involved only in part of the endophenotypes associated with the PTSD-like behavioral phenotype and may be influenced indirectly by interactions with environmental variables (63).

8. A Final Comment

If the initial gross data show little or no effect for the PSS exposure, i.e., little or no difference between exposed and unexposed population:

1. Check whether the litter is scented or contains deodorizing granules to a degree which doesn't enable the urine scent to dominate. If yes, change litter and try again.

2. Try changing cat – there appear to be differences between individual cats of both genders.

3. Add some fur (from the cloth) to the litter. Fur can sometimes be helpful. Usually, there are sufficient levels of fur in the soiled litter.

9. Summary

The fact that it is possible to distinguish between affected and unaffected individuals reaffirms the fact that animals display a range of responses to stimuli, as do the humans they are intended to model. The inclusion of all exposed animals in data analysis overlooks the individual variability in their behavioral response to the stressor, and represents a source for potential bias. Drawing a distinction between animals whom we choose to regard as affected to a sufficient degree by a stimulus in order to be included in a study from those whom we do not consider to have been sufficiently affected may be seen as being conceptually similar to the application of inclusion and exclusion criteria when diagnosing patients or deciding whether to include them in a clinical study. This approach exposes patterns of responses which are significantly different from those seen when it is not applied, and which resemble patterns seen in human studies regarding the more severe type of long-term sequelae of exposure to stressors, including PTSD. Thus, just as one would define the study population for a clinical PTSD study as "meeting DSM IV criteria," it seems sensible to consider redefining the study population in animal models of stress and PTSD for those animals who were exposed to the paradigm and developed certain anxious/fearful behavioral changes.

The above-mentioned approach to the analysis of the data is conceptually equivalent to DSM criteria for PTSD and the criteria proposed by Yehuda and Antelman (64).

1. Regarding the conceptual validity of the model itself, "predator exposure trauma" is a potentially life-threatening situation (Criteria A) and it may represent a more "natural" setting than other types of stressors, such as electrical tail shocks and restraint on a teleological level (20), which might probably be related to extreme conditions like torture. The demonstrated reduction in the time spent by rats in the open arms of the plus maze may ever parallel the avoidance behavior seen in human PTSD (Criteria C). Since the total number of entries into any arms of the plus maze is the same, except that the MA rats keep to the closed arms only, the result is consistent with anxiety

and not with nonspecific impairment of locomotion. DSM-IV defines this symptom as a persistent avoidance of reminders of the trauma and numbing of responsiveness. Since the exposure took place in an open space, it is in keeping with the described definition.

2. The assessment of rats after 7 days has previously been established by Adamec et al. (20) and is taken to refer to PTSD and not to acute stress reaction. Our results (unpublished data) have since shown repeatedly that, as of day 7, the prevalence rate remains constant at about 25% until day 30. The DSM-IV defines PTSD as lasting more than a month after the exposure. So the persistence of anxiety symptoms on day 7 after the exposure fulfills the time element of PTSD (20).

3. One practical implication of the application of a method to distinguish between maladapted responders to well-adapted nonresponders to a stressogenic/traumatic paradigm is that trials using this method may consider including an interim step for the determination of the study's CBCs, between the stress/trauma paradigm and the measure of the selected parameters.

4. As the construct and face validity of study have been tuned, we hypothesize that the potential effect of this animal model on predictive validity of treatment paradigms would be better tuned as well. It is conceivable that the shift in emphasis onto the study of affected populations in animal models might improve the study of treatment modalities in this all too common incapacitating disorder.

10. Future Directions

Since the authors include both clinicians and scientists, we are acutely aware that what we have described is no more than a reliable compromise. We feel that the importance of the CBC model lies in the concept that animal subjects display a range of responses and that this is to be respected and taken into account. The stress paradigm and behavioral measures cannot possibly be the only ones applicable, nor are they necessarily the best.

The assumptions we have made regarding the behavioral response patterns are open to discussion. For instance, is the minimal response (MBR) indeed necessarily adaptive, or could the "unaffected" animals be displaying a behavior worthy of study, such as dissociation? What can we learn from the partial responders, the most common type of response, and how could one go about studying them?

11. Conclusions

The animal model presented here, which is a combination of exposure to a predator and a focus on setting apart the affected based on behavioral cut-off criteria, has demonstrated high face validity, construct validity, and predictive validity. The cumulative results of our studies indicate that the contribution of animal models can be further enhanced by classifying individual animal study subjects according to their response patterns. This approach enables researchers to test interventions that might be impossible or difficult to do in a clinical setting without any proper preclinical basis. The animal model also enables the researcher to go one step further and correlate specific anatomic biomolecular and physiological parameters with the degree and pattern of individual behavioral response.

Acknowledgments

We are grateful for funding from the National Institute for Psychobiology in Israel, funded by Charles E. Smith Family, the Israel Academy of Science and Humanities grant (416/09), and the Ministry of Health (3-0000-6086) grant to H.C.

References

1. Association AP. Diagnostic and Statistical Manual of Mental Disorders. 4th ed, text rev. ed. Washington, DC,: American Psychiatric Association; 2000.

2. Bryant RA, Creamer M, O'Donnell M, Silove D, McFarlane AC. A study of the protective function of acute morphine administration on subsequent posttraumatic stress disorder. Biol Psychiatry 2009;65(5):438–40.

3. Davidson JR. Pharmacologic treatment of acute and chronic stress following trauma: 2006. J Clin Psychiatry 2006;67 Suppl 2:34–9.

4. Yehuda R, McFarlane AC, Shalev AY. Predicting the development of posttraumatic stress disorder from the acute response to a traumatic event. Biol Psychiatry 1998;44(12):1305–13.

5. Foa EB, Stein DJ, McFarlane AC. Symptomatology and psychopathology of mental health problems after disaster. J Clin Psychiatry 2006;67(Suppl 2):15–25.

6. Young E, Breslau N. Cortisol and catecholamines in posttraumatic stress disorder: an epidemiologic community study. Arch Gen Psychiatry 2004;61:394–401.

7. Helzer J, Robins L, McEvoy L. Post-traumatic stress disorder in the general population. Findings of the epidemiologic catchment area survey. N Engl J Med 1987;317:1630–4.

8. Resnick HS, Yehuda R, Pitman RK, Foy DW. Effect of previous trauma on acute plasma cortisol level following rape. Am J Psychiatry 1995;152(11):1675–7.

9. Shore J, Vollmer W, Tatum E. Community patterns of posttraumatic stress disorders. J Nerv Ment Dis 1989;177:681–5.

10. Fairbank JA, Schlenger WE, Saigh PA, Davidson JRT. An Epidemiologic Profile of Post-Traumatic Stress Disorder: Prevalence, Comorbidity, and Risk Factors. In: M.J F, D.S C, A.Y D, eds. Neurobiological and clinical Consequences of Stress: From Normal Adaptation to PTSD. Philadelphia: Lippincott-Raven; 1995:415–27.

11. Peri T, Ben-Shakhar G, Orr SP, Shalev AY. Psychophysiologic assessment of aversive conditioning in posttraumatic stress disorder. Biological Psychiatry 2000;47(6):512–9.

12. Solomon Z, Shklar R, Mikulincer M. Frontline treatment of combat stress reaction: a 20-year

longitudinal evaluation study. Am J Psychiatry 2005;162:2309–14.

13. Nutt D, Davidson J. Post-Traumatic Stress Disorder Diagnosis, Management and Treatment. London: Taylor & Francis; 2000.

14. File SE, Zangrossi Jr H, Sanders FL, Mabbutt PS, . Dissociation between behavioral and corticosterone responses on repeated exposures to cat odor. Physiol Behav 1993;54:1109–11.

15. Blanchard DC, Griebel G, Blanchard RJ. Conditioning and residual emotionality effects of predator stimuli: some reflections on stress and emotion. Prog Neuropsychopharmacol Biol Psychiatry 2003;27(8):1177–85.

16. Blanchard RJ, Blanchard DC, Rodgers J, Weiss SM. The characterization and modelling of antipredator defensive behavior. Neurosci Biobehav Rev 1990;14(4):463–72.

17. Blanchard RJ, Griebel G, Henrie JA, Blanchard DC. Differentiation of anxiolytic and panicolytic drugs by effects on rat and mouse defense test batteries. Neurosci Biobehav Rev 1997; 21(6):783–9.

18. Blanchard RJ, Nikulina JN, Sakai RR, McKittrick C, McEwen B, Blanchard DC. Behavioral and endocrine change following chronic predatory stress. Physiol Behav 1998;63(4):561–9.

19. Blanchard RJ, Yang M, Li CI, Gervacio A, Blanchard DC. Cue and context conditioning of defensive behaviors to cat odor stimuli. Neurosci Biobehav Rev 2001;25(7–8):587–95.

20. Adamec R. Transmitter systems involved in neural plasticity underlying increased anxiety and defense--implications for understanding anxiety following traumatic stress. Neurosci Biobehav Rev 1997;21(6):755–65.

21. Adamec R, Head D, Blundell J, Burton P, Berton O. Lasting anxiogenic effects of feline predator stress in mice: sex differences in vulnerability to stress and predicting severity of anxiogenic response from the stress experience. Physiol Behav 2006;88(1–2):12–29. Epub 2006 Apr 19.

22. Adamec R, Muir C, Grimes M, Pearcey K. Involvement of noradrenergic and corticoid receptors in the consolidation of the lasting anxiogenic effects of predator stress. Behav Brain Res 2007;179(2):192–207. Epub 2007 Feb 6.

23. Adamec R, Strasser K, Blundell J, Burton P, McKay DW. Protein synthesis and the mechanisms of lasting change in anxiety induced by severe stress. Behav Brain Res 2006;167(2): 270–86.

24. Adamec RE, Blundell J, Burton P. Relationship of the predatory attack experience to neural plasticity, pCREB expression and neuroendocrine response. Neurosci Biobehav Rev 2006;30(3): 356–75. Epub 2005 Aug 22.

25. Adamec RE, Shallow T. Lasting effects on rodent anxiety of a single exposure to a cat. Physiol Behav 1993;54(1):101–9.

26. Diamond DM, Campbell AM, Park CR, Woodson JC, Conrad CD, Bachstetter AD, Mervis RF. Influence of predator stress on the consolidation versus retrieval of long-term spatial memory and hippocampal spinogenesis. Hippocampus 2006;16(7):571–6.

27. Griebel G, Blanchard DC, Jung A, Lee JC, Masuda CK, Blanchard RJ. Further evidence that the mouse defense test battery is useful for screening anxiolytic and panicolytic drugs: effects of acute and chronic treatment with alprazolam. Neuropharmacology 1995;34(12):1625–33.

28. Apfelbach R, Blanchard CD, Blanchard RJ, Hayes RA, McGregor IS. The effects of predator odors in mammalian prey species: a review of field and laboratory studies. Neurosci Biobehav Rev 2005;29(8):1123–44. Epub 2005 Aug 8.

29. Blundell J, Adamec R, Burton P. Role of NMDA receptors in the syndrome of behavioral changes produced by predator stress. Physiol Behav 2005;86(1–2):233–43.

30. Cohen H, Benjamin J, Kaplan Z, Kotler M. Administration of high-dose ketoconazole, an inhibitor of steroid synthesis, prevents posttraumatic anxiety in an animal model. Eur Neuropsychopharmacol 2000;10:429–35.

31. Cohen H, Friedberg S, Michael M, Kotler M, Zeev K. Interaction of CCK-4 induced anxiety and post-cat exposure anxiety in rats. Depress Anxiety 1996;4(3):144–5.

32. Lom B, Cohen-Cory S. Brain-derived neurotrophic factor differentially regulates retinal ganglion cell dendritic and axonal arborization in vivo. J Neurosci 1999;19(22):9928–38.

33. Endres T, Apfelbach R, Fendt M. Behavioral changes induced in rats by exposure to trimethylthiazoline, a component of fox odor. Behav Neurosci 2005;119(4):1004–10.

34. Kozlovsky N, Matar MA, Kaplan Z, Kotler M, Zohar J, Cohen H. Long-term down-regulation of BDNF mRNA in rat hippocampal CA1 subregion correlates with PTSD-like behavioural stress response. Int J Neuropsychopharmacol 2007:1–18.

35. Kozlovsky N, Matar MA, Kaplan Z, Kotler M, Zohar J, Cohen H. The immediate early gene Arc is associated with behavioral resilience to stress exposure in an animal model of posttraumatic stress disorder. Eur Neuropsychopharmacol 2007;18:107–16.

36. Mazor A, Matar M, Kozlovsky N, Zohar J, Kaplan Z, Cohen H. Gender-related qualitative

differences in baseline and post stress anxiety responses are not reflected in the incidence of criterion-based PTSD-like behavior patterns. The World Journal of Biological Psychiatry 2009;10:856–69.

37. Roseboom PH, Nanda SA, Bakshi VP, Trentani A, Newman SM, Kalin NH. Predator threat induces behavioral inhibition, pituitary-adrenal activation and changes in amygdala CRF-binding protein gene expression. Psychoneuroendocrinology 2007;32(1):44–55. Epub 2006 Nov 20.

38. Sullivan M, Gratton A. Relationships between stress-induced increases in medial prefrontal cortical dopamine and plasma corticosterone levels in rats: role of cerebral laterality. Neuroscience 1998;83:81–91.

39. Takahashi LK, Nakashima BR, Hong H, Watanabe K. The smell of danger: a behavioral and neural analysis of predator odor-induced fear. Neurosci Biobehav Rev 2005;29(8):1157–67. Epub 2005 Aug 10.

40. Pellow S, Chopin P, File SE, Briley M. Validation of open:closed arm entries in an elevated plus-maze as a measure of anxiety in the rat. J Neurosci Methods 1985;14(3):149–67.

41. B'aszczyk JW. Startle response to short acoustic stimuli in rats. Acta Neurobiol Exp 2003;63:25–30.

42. Graham FK. Distinguishing among orienting, defense, and startle reflexes. In: H.D. Kimmel EHvOaJFO, ed. The orienting reflex in humans. Hillsdale, New Jersey: Lawrence Erlbaum Associates Publishers; 1979:137–67.

43. Blaszczyk JW, Tajchert K. Effect of acoustic stimulus characteristics on the startle response in hooded rats. Acta Neurobiol Exp (Wars) 1997;57(4):315–21.

44. Cohen H, Zohar J, Matar MA, Zeev K, Loewenthal U, Richter-Levin G. Setting apart the affected: the use of behavioral criteria in animal models of post traumatic stress disorder. Neuropsychopharmacology 2004;29(11):1962–70.

45. Cohen H, Zohar J. Animal models of post traumatic stress disorder: The use of cut off behavioral criteria. The Annals New-York Academy of Sciences 2004;1032:167–78.

46. Cohen H, Kaplan Z, Matar M, Loewenthal U, Kozlovsky N, Zohar J. Anisomycin, a Protein Synthesis Inhibitor, Disrupts Traumatic Memory Consolidation and Attenuates Post Traumatic Stress Response in Rats. Biological Psychiatry 2006;60 (7):767–76.

47. Cohen H, Kozlovsky N, Matar MA, Kaplan Z, Zohar J. Mapping the brain pathways of traumatic memory: Inactivation of protein kinase M zeta in different brain regions disrupts traumatic memory processes and attenuates traumatic stress responses in rats. Eur Neuropsychopharmacol;20(4):253–71.

48. Cohen H, Matar M, Buskila D, Kaplan Z, Zohar J. Early post-stressor intervention with high dose corticosterone attenuates post traumatic stress response in an animal model of PTSD. Biol Psychiatry 2008;64:708–17.

49. Cohen H, Matar M, Richter-Levin G, Zohar J. The contribution of an animal model towards uncovering biological risk factors for PTSD. The Annals New-York Academy of Sciences 2006;1071:335–50.

50. Kozlovsky N, Matar MA, Kaplan Z, Zohar J, Cohen H. The role of the galaninergic system in modulating stress-related responses in an animal model of posttraumatic stress disorder. Biol Psychiatry 2009;65(5):383–91.

51. Cohen H, Joseph Z, Matar M. The relevance of differential response to trauma in an animal model of post-traumatic stress disorder. Biol Psychiatry 2003;53(6):463–73.

52. Cohen H, Kaplan Z, Matar MA, Loewenthal U, Zohar J, Richter-Levin G. Long-lasting behavioral effects of juvenile trauma in an animal model of PTSD associated with a failure of the autonomic nervous system to recover. Eur Neuropsychopharmacol 2007;17(6–7):464–677.

53. Cohen H, Ziv Y, Cardon M, Kaplan Z, Matar MA, Gidron Y, Schwartz M, Kipnis J. Maladaptation to mental stress mitigated by the adaptive immune system via depletion of naturally occurring regulatory CD4+CD25+ cells. J Neurobiol 2006;66(6):552–63.

54. Kozlovsky N, Matar MA, Kaplan Z, Zohar J, Cohen H. A distinct pattern of intracellular glucocorticoid-related responses is associated with extreme behavioral response to stress in an animal model of post-traumatic stress disorder. European Neuropsychopharmacology 2009;19(11):759–71.

55. Bremner JD, Licinio J, Darnell A, Krystal JH, Owens MJ, Southwick SM, Nemeroff CB, Charney DS. Elevated CSF corticotropin-releasing factor concentrations in posttraumatic stress disorder. Am J Psychiatry 1997;154(5):624–9.

56. Delahanty DL, Raimonde AJ, Spoonster E. Initial posttraumatic urinary cortisol levels predict subsequent PTSD symptoms in motor vehicle accident victims. Biol Psychiatry 2000;48(9):940–7.

57. Mason JW, Giller EL, Kosten TR, Ostroff RB, Podd L. Urinary free-cortisol levels in posttraumatic stress disorder patients. J Nerv Ment Dis 1986;174(3):145–9.

58. Pitman RK, Orr SP. Twenty-four hour urinary cortisol and catecholamine excretion in combat-related posttraumatic stress disorder. Biol Psychiatry 1990;27(2):245–7.

59. Rasmusson AM, Lipschitz DS, Wang S, Hu S, Vojvoda D, Bremner JD, Southwick SM, Charney DS. Increased pituitary and adrenal reactivity in premenopausal women with post-traumatic stress disorder. Biol Psychiatry 2001;50(12):965–77.

60. Yehuda R. Neuroendocrine aspects of PTSD. Handb Exp Pharmacol 2005(169): 371–403.

61. Yehuda R, Morris A, Labinsky E, Zemelman S, Schmeidler J. Ten-year follow-up study of cortisol levels in aging holocaust survivors with and without PTSD. J Trauma Stress 2007; 20(5):757–61.

62. Cohen H, Zohar J, Gidron Y, Matar MA, Belkind D, Loewenthal U, Kozlovsky N, Kaplan Z. Blunted HPA axis response to stress influences susceptibility to posttraumatic stress response in rats. Biol Psychiatry 2006;59(12): 1208–18.

63. Caspi A, Moffitt T. Gene-environment interactions in psychiatry: joining forces with neuroscience. Nat Rev Neurosci 2006;7:583–90.

64. Yehuda R, Antelman S. Criteria for rationally evaluating animal models of posttraumatic stress disorder. Biol Psychiatry 1993;33:479–86.

65. Ekman P, Friesen WV, Simons RC. Is the startle reaction an emotion? 1985;49(5):1416–1426.

Chapter 13

Measuring Variations in Maternal Behavior: Relevance for Studies of Mood and Anxiety

Becca Franks, James P. Curley, and Frances A. Champagne

Abstract

The assessment of variations in maternal behavior in laboratory rodents is challenging yet may provide an essential tool for understanding the mechanisms linking early life experiences to individual differences in stress responsivity and behavioral indices of depression and anxiety. In this chapter, the methodology for characterizing the quality and quantity of mother–pup interactions in mice as well as the strategy for analyzing this observational data is described in detail. Successful use of this approach is dependent on careful consideration of the wide variety of environmental variables that may influence maternal behaviors, such as pup licking/grooming, which have been demonstrated to be associated with a wide range of behavioral phenotypes in offspring. Maternal behavior can moderate the effect of genetic and neurobiological manipulations that are being developed in mice to study the etiology of psychopathology. The protocol described in this chapter can be applied to these studies to examine the interplay between genes and the environment.

Key words: Maternal behavior, Licking/grooming, Home-cage, Time-sampling, Postnatal

1. Background and Historical Significance

Adverse early life experiences have been demonstrated to be a significant predictor of later life psychopathology in humans, including risk of depression and anxiety (1). This risk is particularly evident among individuals who have experienced reduced or disrupted mother–infant interactions, as in the case of childhood neglect or abuse (2). These epidemiological findings have motivated the development of laboratory models in rodents where manipulation of the degree of contact between the dam and pups is altered during the postnatal period and the behavioral and neurobiological outcomes in offspring are assessed (3). Two classic approaches to inducing postnatal changes in the experience of

Todd D. Gould (ed.), *Mood and Anxiety Related Phenotypes in Mice: Characterization Using Behavioral Tests, Volume II*, Neuromethods, vol. 63, DOI 10.1007/978-1-61779-313-4_13, © Springer Science+Business Media, LLC 2011

rodent pups are neonatal handling and maternal separation. These paradigms have been used extensively within the literature and lead to significant effects on brain development, stress physiology, and behavior in adulthood, though the exact nature of the effect of these manipulations is dependent on the methodological approach taken and the strain/species being studied (4, 5).

The handling paradigm involves brief daily removal of pups from the home-cage after birth and was developed as a form of early life stimulation (6). The duration of separation between mothers and pups in handling studies can vary from 3 to 20 min and the condition of the pups during the separation, particularly whether the pups are provided with an external source of heat (i.e., placed under a heat lamp), also varies from study to study. The period during which the handling occurs typically includes the week following parturition and may extend across the entire preweaning period. Effects of brief maternal separation on corticosterone response to stress in adult offspring can also be observed among pups that remain in the home-cage while the mother is briefly removed (7). Though much of the early work on neonatal handling was conducted in rats, the long-term consequences of handling on emotionality in mice are also evident (8, 9). For example, among C57Bl/6 (B6) mice, the experience of postnatal handling reduces the motivation to escape a novel environment and reduces the corticosterone response to stress (9). Handling can also attenuate increased stress responsivity associated with genetically induced upregulation of the stress response in transgenic mice and illustrates the potential of the handling paradigm in studies of gene–environment interactions (10). The effectiveness of handling to induce behavioral and neurobiological change is significantly related to the background strain, genotype, and sex of the mouse being used, and these factors are an important methodological consideration when implementing this paradigm (11).

In contrast to the effects of handling, more prolonged separations between mother and pups have been used to enhance emotionality and behavioral indices of depression and anxiety (4). The maternal separation paradigm typically involves removing pups from the home-cage for at least 1 h daily during the postnatal period. Similar to the handling manipulation, there are a wide range of variations in methodology used within the maternal separation paradigm, including the duration of separation (ranging from 1 to 24 h), the timing of the separation within the postnatal period (i.e., daily separations during the first week postnatal, throughout the preweaning period, or a single separation within the postnatal period), and the condition of mother and pups during separation (i.e., both placed in novel cages, pups kept warm vs. at ambient temperatures). Maternal separation in mice has been demonstrated to have acute effects on hypothalamic-pituitary-adrenal (HPA) function leading to increased stress-induced glucocorticoids and reduced hippocampal glucocorticoid receptor gene expression (12). In B6 maternally separated male mice, increased risk assessment behaviors (stretch-attend postures) during open-field testing increased immobility during forced swim, and memory deficits have been reported (13). However, as in the case of neonatal handling, variation in methodological parameters as well as the strain, genotype, and sex of the mice being assessed will influence the effectiveness of this early-life manipulation to induce long-term neurobehavioral outcomes (8, 11).

Neonatal handling and maternal separation are both considered early life stressors as they both lead to increases in plasma corticosterone levels in pups when they are separated from the dam (14). However, the reduced stress response that is observed in adults who experienced brief maternal separation in the form of neonatal handling has led to the hypothesis that the effect of handling may be due to separation-induced increases in the maternal behavior of dams toward the pups following the handling procedure (5, 15). In laboratory rodents, there are increased levels of maternal care, particularly licking/grooming (LG), observed in dams toward pups among handled litters (16). If mouse dams are treated with anxiolytic drugs during the separation period, there are no increases in maternal care directed toward pups following reunion and no long-term reductions in emotionality observed in the handled offspring (17). Though it may not be necessary to induce changes in maternal care in order to observe handling effects in offspring (18, 19), these studies suggest that variation in the quality and quantity of maternal care may be a critical aspect of the developmental experience that shapes the emotionality of offspring.

Natural variations in maternal care observed in laboratory environments among inbred and outbred rodents can be used as a predictor of offspring neurobiology and behavior (20, 21). Though this approach has been more thoroughly explored in rats, there are emerging data from studies in mice illustrating the persistent effects on emotionality of variations in maternal care (22–24). The maternal behavior of rodents varies across the postnatal period and consists of crouching over pups, nursing, licking pups, and nest-building (NB) in addition to time off the nest spent eating, drinking, and self-grooming (20, 21). There are considerable between-strain and within-strain differences in the frequency of the various aspects of maternal behavior. Studies of the long-term impact of maternal care indicate that the frequency of LG may be a critical feature of the mother–infant interaction that shapes emotionality and HPA response to stress as well as multiple dimensions of behavior and physiology (cognition, social behavior, and reproduction) correlated to changes in region-specific gene expression within the brain (25). These findings are consistent with the handling literature, which has identified increased LG in handled dams. The frequency of LG can be influenced by the environmental experiences of the dam prior to the postpartum period (26–30) and by the conditions of the postnatal rearing environment (23, 31) and thus exhibits a high degree of plasticity. The genotype of the dam can also influence the frequency of LG (32), which is an important consideration in studies where the phenotypic consequences of a transgenic manipulation are being assessed. In rodents, cross-fostering of offspring between strains that differ in the frequency of postnatal LG can lead to shifts in the development of offspring (33, 34), an effect that can also be observed in within-strain cross-fostering designs (20).

Assessment of variations in maternal behavior in mice is challenging but if successful can yield exciting new insights into the pathways linking early life experience to molecular, neurobiological, and behavioral outcomes relevant to studies on mood and anxiety. Traditionally, measures of maternal behavior in mice have focused on motivation to retrieve pups or the latency or degree of maternal aggression displayed toward an intruder (35–37). While informative regarding the mechanisms of maternal behavior (38), these approaches are likely not relevant for understanding the rearing experience of laboratory mice, since disruption to the nest or introduction of unfamiliar mice into the home-cage of a postparturient female is not typical in laboratory rearing environments. Thus, methodological approaches involving home-cage maternal observations and assessment of individual differences in maternal care toward offspring are essential in studies examining the direct impact of maternal care on offspring development as well as the moderating effect of these experiences on the influences of environment, genotype, and strain. In the subsequent sections, we provide a detailed description of this methodology, outline strategies for analyzing maternal data, and discuss tips for troubleshooting this protocol.

2. Equipment, Materials, and Setup

2.1. Animals

Both inbred and outbred strains of laboratory mice are suitable for measuring home-cage maternal behavior. However, as mentioned in the previous section, strains vary considerably in the frequency of mother–pup interactions, particularly in LG behavior. For example, Balb/c and 129Sv mice engage in significantly reduced levels of LG compared to B6 and Swiss. If the goal of the study is to assess offspring who have received low vs. high levels of LG in a within-strain model, then selecting a strain with a high average level of LG (i.e., B6) would be recommended. Alternatively, a cross-fostering design could be used between strains that exhibit low (i.e., Balb/c) vs. high (i.e., B6) levels of LG (33, 39). To generate litters for these studies, female and male mice can be successfully mated at 8 weeks of age with mating success declining after 32 weeks. Within a cohort of mating females, it is best to use animals that are similar in age.

2.2. Animal Facility

The maternal behavior and rearing experiences of laboratory animals can be easily disturbed by the routine traffic and mainte-nance activities that occur in all animal facilities. Ideally, these stud-ies should be conducted in smaller animal housing rooms separate from large vivariums. Though investigators often do not control the light cycle within animal housing rooms, studies should avoid transferring mice to and from rooms that differ in this cycle. For example, if offspring are going to be tested on measures that require reverse lighting (i.e., dark during the day, lights on at night), it would be preferable that they are reared under these same lighting conditions. There has been no systematic study of whether maternal data collected exclusively under light vs. dark conditions are better predictors of the overall pattern of maternal care, and both strategies have been successful approaches in studying mother–pup interactions in mice (22, 31, 34). If observations are being conducted during the dark phase of the cycle, lamps with red light bulbs can be arranged within the observation area to provide suitable illumination.

2.3. Housing and Husbandry

The observation of home-cage maternal behavior will require that mice are housed in Plexiglas cages that permit a clear view of the dam and litter. Though the shoe-box cages that are typically used for housing mice will be sufficient, larger cages permit the observa-tion of more dynamic interactions between the dam and pups. The bedding material is also an important consideration. Corncob bed-ding does not provide sufficient nesting material for lactating females and if using this type of bedding, nestlets will need to be provided. However, the nestlets allow mice to build very elaborate nests which will prevent reliable observations of mother–pup inter-actions. An alternative is to use pine shavings as the bedding mate-rial. Mice can build nests from this bedding yet still be viewed from the exterior of the cage, making nestlets unnecessary.

2.4. Mating

It is recommended that if mice are being used from a commercial breeder, a 2-week period of acclimatization to the animal facility should be implemented to promote reduced stress at the time of mating. Multiple females (2–3) can be housed together with one male to promote the breeding of multiple litters. If the intention of the study is to examine the effects of within-strain variations in maternal care, 30–40 litters will be necessary to generate sufficient numbers of "low" vs. "high" litters. Since the determination of the relative level of maternal care exhibited by a female will be dependent on comparison to a cohort mean and standard deviation (SD) (see Sect. 4.2), it will be ideal to breed the females as a cohort so that litters will be born at roughly the same time. Males can be removed after a 2-week period to avoid any contact with postnatal pups which will reduce the risk of infanticide and prevent the male from influencing the postnatal development of offspring.

2.5. Postpartum Monitoring and Husbandry

Mated females should be routinely monitored and singly housed a few days prior to parturition. Though it is often difficult to ascertain when a female will give birth (particularly if the litter is small), the weight gain of pregnant females will be evident 2 weeks after mating. Once singly housed, females should be monitored daily to establish the date of birth of the litter. The process of parturition may take over an hour to complete, particularly if the litter is large, and it is best to leave the female undisturbed until this process is completed. After birth, the dam will retrieve her pups to the nest and clean them to remove remnants of the placenta. Once this process is complete, pups can be weighed and counted and placed with the dam into a clean cage, mixing in some of the soiled bedding material from the old cage to provide olfactory continuity and reduce rehousing stress. During the postnatal observations of maternal care, cage cleaning is very disruptive. Cleaning the cage on the day of birth (day 0) will allow for undisturbed assessment of home-cage behavior from postnatal days 1–6 and cages should not be cleaned until after observations are completed on postnatal day 6. We recommend cleaning the cage every 7 days thereafter until weaning. Cleaned cages can then be positioned on the racks in the animal housing room to permit viewing of activity within the cage (see Fig. 1). Observations of maternal behavior can then commence the following day on postnatal day 1.

Fig. 1. Photo of home-cage observation set-up (*left*) with unique cage identifiers (*A.1, A.2, A.3*) and a close-up of a lactating female B6 mouse with a litter of pups (*right*).

3. Procedure

The acquisition of behavioral data from the home-cage can be achieved either through video recordings or by observers who document the behavior in the home-cage from within the animal housing room. The later approach will be preferable when there are numerous cages to be observed and when viewing the mother–pup interactions require dynamic changes in the position of the observer (as is often the case). In both cases, raters/observers will need to be trained to a high degree of inter-rater reliability. The description of maternal behaviors that can be observed is provided in Table 1 and includes nursing postures, LG pups, NB, self-grooming, eating, and drinking. These behaviors are not mutually exclusive and can occur in a variety of combinations. There may be a wide variety of nursing postures that can be observed, and a more detailed account of these postures has previously been described (22). However, for most studies, the behaviors outlined in Table 1 will capture the features of mother–pup interactions that are predictive of long-term outcomes in offspring. Each individual behavior should be assigned a unique alphanumeric identifier (e.g., nursing = N, nest-building = B) and raters/observers should be provided with

Table 1
Description of home-cage maternal behaviors in mice

Behavior	Description	Code
Nursing (crouch)	Dam is positioned over the pups to permit sucking or thermo-regulation with a low to moderate arch in her back	N
Arched nursing	Dam is positioned over the pups with a high arch in her back to permit sucking and pup movement	A
Passive nursing	Dam is lying on her side with her ventrum exposed to the sucking pups	P
Licking/grooming	Dam is licking pups (anogenital or body region)	G
Nest-building	Dam is picking up pieces of bedding and retrieving these to the nest or moving bedding in the nest with her snout	B
Self-grooming	Dam is licking herself (often occurs during bouts of pup licking)	S
Eating	Dam is eating	E
Drinking	Dam is drinking	D
Contact with pups	Dam is in contact with the pups but not in a posture that promotes sucking (i.e., sitting next to pups)	C
No contact with pups	Dam is off the nest and not in contact with any pups (and not engaging in any of the other behaviors noted above)	X

a detailed legend outlining these identifiers. The schedule of observations during the day relative to the light cycle should be consistent across days and for each litter. Furthermore, it is important to avoid conducting observations within an hour before or after the light–dark transition in room lighting. It is recommended that a minimum of 4 h of observations (e.g., 11 am, 12 pm, 3 pm, 4 pm) be conducted each day for each litter. Across the postpartum period, there is a significant decline in maternal behavior (particularly LG (20, 21)) and it is generally accepted that the critical window for many long-term developmental effects will require assessment of maternal care from the time of birth to at least 6 consecutive days.

3.1. Time-Sampling Observation of Maternal Behavior

In rodent studies examining the long-term consequences of maternal behavior for offspring development, a typical data collection technique involves time-sampling the home-cage behavior of the dam during each observation session (20, 23). At the start of the observation session, observers should acquaint themselves with the location of each cage to be observed. Data collection sheets should be organized such that there is a row for each "time of sampling" and a column for each litter to be observed (for an example observation sheet, see Fig. 2a). For data collection and analysis, it will be necessary for each cage to have a unique alphanumeric identifier code (see Fig. 1). A stopwatch will also be necessary to time the intervals between observations. Once the observer is ready to commence the data collection, the stopwatch is started and the observer notes the behavior of the dam in the first cage. For example, if the dam is crouched over the pups in a nursing posture and licking the pups, the observer can write "NG." If the dam is eating and not in contact with the pups, the observer can write "E" on the observation sheet. It is important to take a visual "snap-shot" of behavior (lasting 1–2 s) within the cage since the behavior is likely to change if the cage is observed continuously. Once the behavior in the first cage is noted, the observer should move to the next cage and repeat the rating process. Once all of the cages have been observed and the maternal behavior recorded, the observer returns to the first cage and repeats the entire process. A good time-sampling interval is 3 min. When using a 3-min interval, series of observations are conducted at 0, 3, 6 min and so on until the 1-h session is completed. This will result in 20 observations of each cage within the session. With a schedule of four sessions per day across 6 consecutive days, a resulting 480 observations (20 observations × 4 sessions × 6 days) per litter will be obtained.

3.2. Focal Observations of Maternal Behavior

In cases where there are fewer cages to be observed, maternal behavior can also be measured through continuous observation of a single cage. In this case, the observer also needs a stopwatch to note the time at which behaviors emerge or change and data collection sheets can be arranged with multiple rows in which the

A

DATE	10-Oct	
TIME	11AM	
OBSERVER	Frances	
Litter ID	A.1	A.2
0	X	N
3	X	N
6	X	N
9	X	NG
12	X	NG
15	E	NG
18	E	NS
21	E	NS
24	D	NG
27	D	N
30	X	C
33	X	X
36	X	X
39	CB	X
42	C	X
45	C	E
48	N	E
51	NG	E
54	N	B
57	A	B

B

DATE	10-Oct-10
TIME	11AM
OBSERVER	Frances
Litter ID	A.1
TIME	BEHAVIOR
0	off nest
0:56	contact with pups
1:48	pup licking starts
2:38	licking pups stops
2:59	nursing starts
8:09	nursing & nest-building
9:35	nest-building stops
10:20	arched nursing starts
12:34	contact (not nursing)
15:22	crouch nursing
20.41	pup licking starts
28.34	pup licking stops
29.12	self-grooming starts
36.44	self-grooming stops
45.34	off nest
52.11	eating
56.17	eating stops
57.56	nest-building starts
59.33	nest-building stops

Fig. 2. Example observation sheets and maternal behavior ratings. (**a**) Time-sampling observation data of two litters (*A.1* and *A.2*). (**b**) A focal observation of litter *A.1* indicating the start and stop times of maternal behavior.

time and behaviors can be noted (e.g., observation sheet, see Fig. 2b). At the start of the session, the observer will start the stopwatch and note the behaviors that are evident. When there is a transition to another behavior or a current behavior stops, this should also be recorded. This continues for 1 h until the session is completed. This data collection strategy is ideal for looking at the duration and number of bouts of maternal behavior (20) and provides a rich source of information on the temporal patterns of mother–pup interactions.

**3.3. Observing
Communal Maternal
Behavior**

Communal rearing of pups is an early life experience that can have a significant impact on behavioral and neurobiological phenotypes related to anxiety and depression (40). Observation of mother–pup interactions in a communal nest can be conducted using a time-sampling procedure as outlined for a single dam and litter. However, this will require individual marking of each dam in the nest. For albino mice, black bands marked on the tail (i.e., 1-band, 2-bands, 3-bands) can clearly distinguish each female (31). For B6 mice, ear punches or tags might be necessary to distinguish between females.

**3.4. Cross-Fostering,
Handling, and
Maternal Separation**

Experimental approaches to studying environmental and/or genetic influences on phenotype in rodents often utilize a manipulation of the early rearing experiences. In this context, cross-fostering can be used to alter rearing experiences, and can lead to varying developmental trajectories among offspring (33). However, though most laboratory mice will readily accept foster pups and provide maternal care toward these pups, the genotype or treatment history of the pups may influence the frequency of maternal behavior (34). To maintain significant strain differences in maternal behavior when cross-fostering is being used may require an increased sample size to account for these pup-induced effects. When cross-fostering is done on the day of birth, the behavioral sampling protocol that has been outlined can be used to verify the frequency of maternal care. If the fostering is being done at some point later in the postnatal period, the timing of the manipulation relative to the observation session should be consistent over days and litters and be noted as a potential influence on the observational data collected immediately after the fostering procedure. This strategy should also be used if daily handling or maternal separation is being conducted.

4. Data Analysis

The observation protocols outlined in the previous section result in numerous individual data points that can be used in multiple ways depending on the overall experimental design of the study. There are multiple statistical approaches to using this data and the choice of approach will depend on the independent or dependent variables associated with maternal behavior.

**4.1. Calculating
Average or Daily
Frequencies of
Maternal Behavior**

Time sampling observational data collected for each litter can be entered into a spreadsheet in which the litter ID, time, and date are noted. Frequencies of the occurrence of each combination of behaviors (e.g., nursing and licking (NG), nursing and eating (NE), nursing and NB) or of a single behavior of interest (e.g., licking – G) can then be calculated. The frequency of a behavior will be determined by calculating the number of times a behavior occurs as a proportion of the total number of observations conducted. For example, if pup licking (G) is observed to occur 48 times during the 480 observations of a dam, the average frequency of pup licking for that dam would be 10%. A similar approach can be taken if calculating average daily maternal behavior, with the numerator and denominator reflecting the frequency of behavior observed during a single day and the total number of observations conducted during this particular day.

4.2. Determining Low vs. High Maternal Behavior

When observational data is available on numerous dams (i.e., 30–40), a normal distribution of maternal behavior is likely to be evident. As such, females can be selected as engaging in low vs. high levels of a particular aspect of maternal behavior and this grouping may be a useful strategy in subsequent analysis of offspring characteristics. For example, Fig. 3a illustrates a distribution of pup LG behavior collected during postnatal days 1–6 in observations of B6 dams ($n = 40$). The values are the average frequency of observed LG during this 6-day period. Low LG dams are defined as engaging in this behavior at a frequency that is less that 1 SD below the cohort average whereas high LG dams engage in this behavior at a frequency that is greater than 1 SD above the mean. In the sample cohort illustrated in Fig. 3a, the average frequency of LG is 9.46% with a SD of 3.16%. Using this information, low (<6.30%; $n = 7$) and high (>12.62%; $n = 6$) licking grooming dams can be identified, with remaining females classified as "mid" ($n = 27$), as illustrated in Fig. 3b. Offspring born to these dams can undergo subsequent testing to determine the relationship between maternal care and behavioral phenotypes. For example, in the sample cohort described, one male from each litter was assessed in the open-field test at 60 days of age and the duration of time spent in the inner area of the apparatus was measured as an indication of anxiety-like behavior (41). In this sample, using the average LG frequency for each dam yields a significant correlation with time (seconds) spent in the inner area ($r = 0.75$, $p < 0.001$; illustrated in Fig. 3c). One-way ANOVA using the group classifications (low, mid, high) of LG behavior also indicates a significant effect ($F(2,39) = 16.25$, $p < 0.001$), and Tukey's post hoc analysis indicates a significant difference between all three groups ($p < 0.01$). Offspring behavioral data can be presented as a function of the group classification of maternal LG (Fig. 3d).

4.3. Calculating Maternal Behavior Bout Lengths

Focal observational data can be used to assess the frequency and duration of bouts of maternal behavior. This assessment can also be done with time-sampling protocols; however, if the behavior is very short in duration (i.e., less than the interval between observations) then the duration estimates will not be accurate. Assessment of bout duration requires determining the start and stop times of

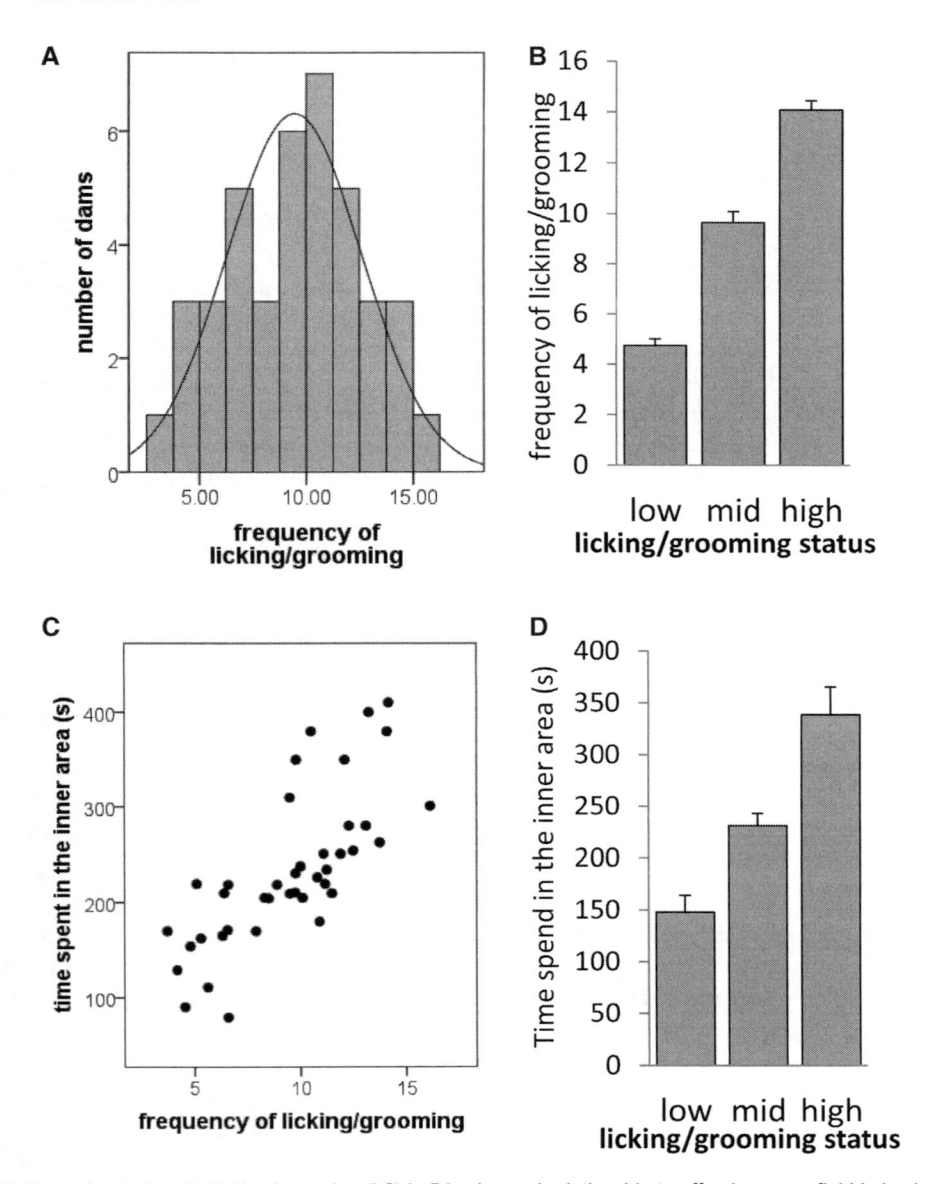

Fig. 3. Analysis of variations in licking/grooming (LG) in B6 mice and relationship to offspring open-field behavior. (a) In a cohort of 40 B6 females, average LG behavior observed from postnatal days 1–6 is normally distributed. (b) Based on the cohort mean, B6 females can be classified as low, mid, or high in LG behavior. (c) Average LG behavior can be used as a continuous variable to permit analysis of the linear correlation with offspring open-field performance (time (s) spent in the inner area of a 10-min open-field test). (d) Categorical maternal data (low, mid, high) can likewise be used to compare mean differences in open-field behavior.

the behavior (based on the noted times on the observation record – see Fig. 2b) and calculating the length of time of each start/stop occurrence. The average bout duration for a dam would be the sum of the individual bout durations divided by the total number of bouts. Mice strains differ significantly in the duration and frequency of bouts of nursing and LG, likely reflecting differences in activity levels between strains (21).

4.4. Using Multilevel Modeling in Studies of Maternal Behavior

To analyze change over time, to parse the relationship between two co-occurring variables, or to examine how early maternal care affects later litter neurobiology and behavior, a more sophisticated statistical methodology can be useful. For example, moment to moment, the pups' corticosterone may increase/decrease while interacting with dams, but this lower level effect can be distinct from the higher level effect, i.e., pups with high corticosterone in general tend to have mothers that provide more/less care. Traditional statistical methods would be ill equipped to differentiate between these two effects, but they are easily identifiable with multilevel modeling (42). Multilevel (hierarchical or random effects) modeling is now well established as a standard technique in the statistical toolkit and eminently suited to answer these questions (43, 44). By accounting for repeated observations within a single group (i.e., multiple observations of a single litter), multilevel modeling allows researchers to retain the bulk of their data in raw unaggregated form, thus gaining statistical power and permitting more refined experimental questions. To use this approach, data should be arranged in a spreadsheet so that all similar observations are stacked within a single column/one variable, while observation type (e.g., postnatal day, time) and litter ID are indicated with two other variables. The data are then analyzed as with GLM or regular regression but using a multilevel error structure. These techniques are easily applied in most current statistical packages (43, 45).

5. Experimental Variables and Troubleshooting

5.1. Low Levels of Maternal Behavior with Minimal Variation

Though this may be due to specific characteristics of the strain or genotype being used, a common cause of reduced maternal behavior in rodents are stressors within the physical or social environment. Fluctuations in temperature, humidity, and noise can disturb patterns of mother–pup interactions. Collaborating with animal care staff is essential to maintaining appropriate conditions for assessing home-cage behaviors and observations should be conducted when there are few if any ongoing activities in the housing room. Schedule observations during the day so as to avoid times when routine cage maintenance/cleaning are being conducted in the room.

5.2. Outliers in Maternal and Offspring Behavior

It is not uncommon for there to be dams or offspring that exhibit extremes of behavior. In some cases, there are predictors of when this will occur. Very small litter (i.e., fewer than four pups) or very large litter (i.e., greater than 14 pups) sizes can influence the pattern of maternal behavior, average pup weight, and affect the experience of individual pups within the litters. One approach to address this issue is to cull litters to a standard size (i.e., eight pups) and to not use very small litters. However, litter size and average pup weight can also be used as covariates in data analyses and may yield interesting interactions between life-history variables and outcome measures. In general, it is best to record as much information as possible on each litter, particularly any adverse experiences that may alter the data.

5.3. High Levels of Mortality

There are often periods of time during the course of the year where breeding success is very low and the mortality of litters is very high. Reproduction in rodents is highly dependent on olfactory cues and any change in the olfactory environment may lead to a reduction in the number of litters produced that survive to weaning age. If the source of the reduced reproductive success cannot be identified, it may be recommended to use a different mouse strain, as there are likely to be strain difference in resilience to these environmental fluctuations.

5.4. Avoiding Observer Effects

Variability in behavioral data can often be linked to the particular characteristics of individual experimenters/observers. In addition to appropriate training to obtain an inter-rater reliability greater than 90%, observers should be provided with detailed instructions regarding the importance of not disturbing the litters while conducting the observations. Cell phones, loud noises, lighted computer screens, and heavy perfumes/aftershaves are definitely to be avoided.

5.5. Making Fine-Tuned Behavioral Distinctions

Training observers is a difficult task. In addition to using video recordings to generate inter-rater reliability, accompanying observers during sessions and providing feedback will improve the quality of the maternal data. This will be particularly important for difficult behavioral categories, such as the distinction between contact and nursing or between self-grooming and pup-grooming.

6. Concluding Remarks

Across species, the quality of the early life environment can shape development leading to increased risk or resilience to later-life disorder. Measures of variation in maternal behavior have proven to be valuable tools in understanding the mechanisms of these developmental effects, and the successful use of mice as a model for these studies is dependent on the appropriate assessment of home-cage maternal behavior. The plasticity of maternal behavior in response to the environment contributes to the challenge of such an approach but with detailed observational data, a meaningful characterization of mother–pup interactions is possible.

This methodology can be applied to studies of induced genetic and neurobiological modification that are being developed in mice and may be a critical methodological approach for examining the interplay between genes and the environment.

References

1. Nemeroff, C.B., *Early-Life Adversity, CRF Dysregulation, and Vulnerability to Mood and Anxiety Disorders.* Psychopharmacol Bull, 2004. **38 Suppl 1**: p. 14–20.

2. Trickett, P. and C. McBride-Chang, *The developmental impact of different forms of child abuse and neglect.* Developmental Reviews, 1995. **15**: p. 11–37.

3. Sanchez, M.M., C.O. Ladd, and P.M. Plotsky, *Early adverse experience as a developmental risk factor for later psychopathology: evidence from rodent and primate models.* Dev Psychopathol, 2001. **13**(3): p. 419–49.

4. Pryce, C.R., et al., *Comparison of the effects of early handling and early deprivation on conditioned stimulus, context, and spatial learning and memory in adult rats.* Behav Neurosci, 2003. **117**(5): p. 883–93.

5. Pryce, C.R. and J. Feldon, *Long-term neurobehavioural impact of the postnatal environment in rats: manipulations, effects and mediating mechanisms.* Neurosci Biobehav Rev, 2003. **27**(1–2): p. 57–71.

6. Levine, S., et al., *Physiological and behavioral effects of infantile stimulation.* Physiol Behav, 1967. **2**(1): p. 55–59.

7. Thoman, E.B. and S. Levine, *Role of maternal disturbance and temperature change in early experience studies.* Physiol Behav, 1969. **4**: p. 143–145.

8. Anisman, H., et al., *Do early-life events permanently alter behavioral and hormonal responses to stressors?* Int J Dev Neurosci, 1998. **16**(3–4): p. 149–64.

9. Parfitt, D.B., et al., *Differential early rearing environments can accentuate or attenuate the responses to stress in male C57BL/6 mice.* Brain Res, 2004. **1016**(1): p. 111–8.

10. Zanettini, C., et al., *Postnatal handling reverses social anxiety in serotonin receptor 1A knockout mice.* Genes Brain Behav, 2010. **9**(1): p. 26–32.

11. Millstein, R.A. and A. Holmes, *Effects of repeated maternal separation on anxiety- and depression-related phenotypes in different mouse strains.* Neurosci Biobehav Rev, 2007. **31**(1): p. 3–17.

12. Schmidt, M.V., et al., *The postnatal development of the hypothalamic-pituitary-adrenal axis in the mouse.* Int J Dev Neurosci, 2003. **21**(3): p. 125–32.

13. Murgatroyd, C., et al., *Dynamic DNA methylation programs persistent adverse effects of early-life stress.* Nat Neurosci, 2009. **12**(12): p. 1559–66.

14. Levine, S., et al., *Time course of the effect of maternal deprivation on the hypothalamic-pituitary-adrenal axis in the infant rat.* Dev Psychobiol, 1991. **24**(8): p. 547–558.

15. Denenberg, V.H., *Commentary: is maternal stimulation the mediator of the handling effect in infancy?* Dev Psychobiol, 1999. **34**(1): p. 1–3.

16. Liu, D., et al., *Maternal care, hippocampal glucocorticoid receptors, and hypothalamic-pituitary-adrenal responses to stress.* Science, 1997. **277**(5332): p. 1659–62.

17. D'Amato, F.R., et al., *Long-term effects of postnatal manipulation on emotionality are prevented by maternal anxiolytic treatment in mice.* Dev Psychobiol, 1998. **32**(3): p. 225–34.

18. Macri, S., G.J. Mason, and H. Wurbel, *Dissociation in the effects of neonatal maternal separations on maternal care and the offspring's HPA and fear responses in rats.* Eur J Neurosci, 2004. **20**(4): p. 1017–24.

19. Tang, A.C., et al., *Programming social, cognitive, and neuroendocrine development by early exposure to novelty.* Proc Natl Acad Sci U S A, 2006. **103**(42): p. 15716–21.

20. Champagne, F.A., et al., *Variations in maternal care in the rat as a mediating influence for the effects of environment on development.* Physiol Behav, 2003. **79**(3): p. 359–71.

21. Champagne, F.A., et al., *Natural variations in postpartum maternal care in inbred and outbred mice.* Physiol Behav, 2007. **91**(2–3): p. 325–34.

22. Coutellier, L., et al., *Effects of foraging demand on maternal behaviour and adult offspring anxiety and stress response in C57BL/6 mice.* Behav Brain Res, 2009. **196**(2): p. 192–9.

23. Coutellier, L., et al., *Variations in the postnatal maternal environment in mice: effects on maternal behaviour and behavioural and endocrine responses in the adult offspring.* Physiol Behav, 2008. **93**(1–2): p. 395–407.

24. Pedersen, C.A. and M.L. Boccia, *Oxytocin and mothers' developmental effects on their daughters*, in *Neurobiology of the Parental Brain*, R. Bridges, Editor. 2008, Academic Press.

25. Meaney, M.J., *Maternal care, gene expression, and the transmission of individual differences in stress reactivity across generations*. Annu Rev Neurosci, 2001. **24**: p. 1161–92.

26. Boccia, M.L. and C.A. Pedersen, *Brief vs. long maternal separations in infancy: contrasting relationships with adult maternal behavior and lactation levels of aggression and anxiety*. Psychoneuroendocrinology, 2001. **26**(7): p. 657–72.

27. Champagne, F.A. and M.J. Meaney, *Stress during gestation alters postpartum maternal care and the development of the offspring in a rodent model*. Biol Psychiatry, 2006. **59**(12): p. 1227–35.

28. Champagne, F.A. and M.J. Meaney, *Transgenerational effects of social environment on variations in maternal care and behavioral response to novelty*. Behav Neurosci, 2007. **121**(6): p. 1353–63.

29. Gonzalez, A., et al., *Intergenerational effects of complete maternal deprivation and replacement stimulation on maternal behavior and emotionality in female rats*. Dev Psychobiol, 2001. **38**(1): p. 11–32.

30. Moore, C.L. and K.L. Power, *Prenatal stress affects mother-infant interaction in Norway rats*. Dev Psychobiol, 1986. **19**(3): p. 235–45.

31. Curley, J.P., et al., *Social enrichment during postnatal development induces transgenerational effects on emotional and reproductive behavior in mice*. Front Behav Neurosci, 2009. **3**: p. 25.

32. Champagne, F.A., et al., *Paternal influence on female behavior: the role of Peg3 in exploration, olfaction, and neuroendocrine regulation of maternal behavior of female mice*. Behav Neurosci, 2009. **123**(3): p. 469–80.

33. Priebe, K., et al., *Maternal influences on adult stress and anxiety-like behavior in C57BL/6J and BALB/cJ mice: a cross-fostering study*. Dev Psychobiol, 2005. **47**(4): p. 398–407.

34. Curley, J.P., et al., *Developmental shifts in the behavioral phenotypes of inbred mice: the role of postnatal and juvenile social experiences*. Behav Genet, 2010. **40**(2): p. 220–32.

35. Carlier, M., P. Roubertoux, and C. Cohen-Salmon, *Differences in patterns of pup care in Mus musculus domesticus l-Comparisons between eleven inbred strains*. Behav Neural Biol, 1982. **35**(2): p. 205–10.

36. Wainwright, P.E., *Maternal performance of inbred and hybrid laboratory mice (Mus musculus)*. J Comp Physiol Psychol, 1982. **94**: p. 694–707.

37. Gammie, S.C. and R.J. Nelson, *cFOS and pCREB activation and maternal aggression in mice*. Brain Res, 2001. **898**(2): p. 232–41.

38. Gammie, S.C., *Current models and future directions for understanding the neural circuitries of maternal behaviors in rodents*. Behav Cogn Neurosci Rev, 2005. **4**(2): p. 119–35.

39. Caldji, C., et al., *Maternal behavior regulates benzodiazepine/GABAA receptor subunit expression in brain regions associated with fear in BALB/c and C57BL/6 mice*. Neuropsychopharmacology, 2004. **29**(7): p. 1344–52.

40. Branchi, I., *The mouse communal nest: investigating the epigenetic influences of the early social environment on brain and behavior development*. Neurosci Biobehav Rev, 2009. **33**(4): p. 551–9.

41. Crawley, J.N., *Exploratory behavior models of anxiety in mice*. Neurosci Biobehav Rev, 1985. **9**(1): p. 37–44.

42. van de Pol, M.V. and J. Wright, *A simple method for distinguishing within- versus between-subject effects using mixed models*. Animal Behaviour, 2009. **77**(3): p. 753–758.

43. Gellman, A. and J. Hill, *Data Analysis Using Regression and Multilevel/Hierarchical Models*. 2006, Cambridge: Cambridge University Press.

44. Singer, J. and J. Willet, *Applied Longitudinal Data Analysis: Modeling Change and Event Occurrence*. 2003, New York, NY: Oxford University Press.

45. Rabe-Hesketh, S. and A. Skrondal, *Multilevel and Longitudinal Modeling Using Stata* 2nd ed. 2008, College Station, TX: Stata Press.

Chapter 14

Emotionality-Related Consequences of Early Weaning in Mice and Rats

Takefumi Kikusui

Abstract

Among all mammalian species, pups are the most highly dependent on their mothers not only for nutrition but also for physical interaction. Therefore, disruption of the mother–pup interaction changes the physiology and behavior of pups. We reviewed the experimental procedure of early weaning, which brings about changes in the behavior and neuronal systems of the offspring of rats and mice. Thus far, we have demonstrated that early weaning results in a persistent increase in anxiety-like and aggressive behavior in adult mice. Early weaned mice also showed higher hypothalamic-pituitary-adrenal activity in response to novelty stress. Neurochemically, early weaned male mice, but not female mice, showed precocious myelination in the amygdala, decreased brain-derived neurotrophic factor protein levels in the hippocampus and prefrontal cortex, and reduced bromodeoxyuridine immunoreactivity in the dentate gyrus. These results are similar to those of human psychological disorders, and suggest that deprivation of the mother–infant interaction during the late lactation period in mice and rats provides an interesting animal model to understand the neuromolecular basis of and psychopharmacological treatments for human psychological disorders related to anxiety and higher emotionality.

Key words: Early weaning, Anxiety, Stress response

1. Background and Historical Overview

The importance of mammalian mother–infant relationships in developing infants, including the roles of nutrition and physical contact, has been characterized from behavioral, endocrine, and neurochemical perspectives. This idea was first suggested in humans by Bowlby's attachment theory (1). Thereafter, many psychological research studies have reported that child abuse or childhood neglect is correlated with severe, deleterious, long-term effects on the child's cognitive, socioemotional, and behavioral development (2). The developmental effects of mother–infant bonding have also been demonstrated experimentally in nonhuman primates (3).

Todd D. Gould (ed.), *Mood and Anxiety Related Phenotypes in Mice: Characterization Using Behavioral Tests, Volume II*, Neuromethods, vol. 63, DOI 10.1007/978-1-61779-313-4_14, © Springer Science+Business Media, LLC 2011

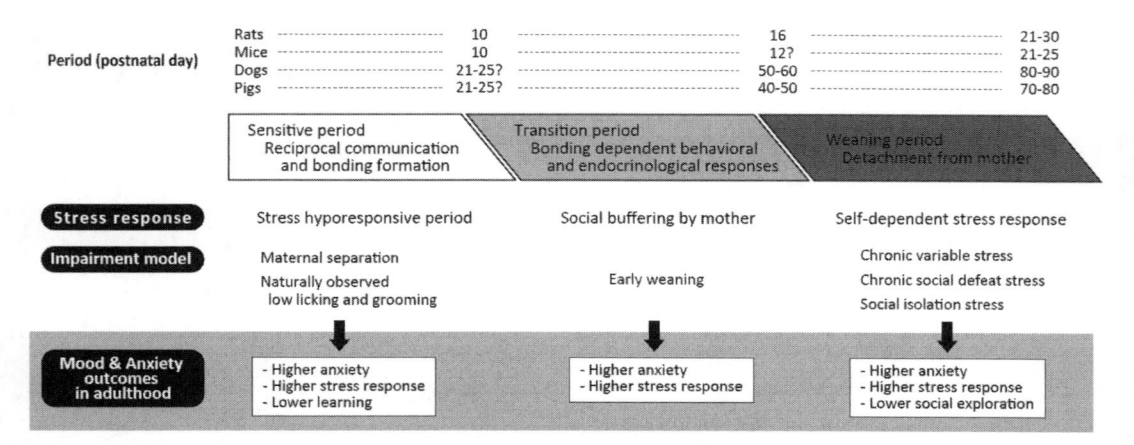

Fig. 1. Representative animal models for investigating the effects of social environments during the neonatal and juvenile periods on neurobehavioral development. During the very first neonatal period, that is, the first 2 weeks in rats and mice, maternal separation or naturally observed less maternal licking and grooming is associated with higher anxiety, higher stress response, and lower learning in adulthood. The second period is a transition period in which maternal factors can inhibit the neuroendocrine stress response. In this period, early weaning also leads to higher anxiety, as well as a higher stress response in the offspring. After weaning, chronic variable stress, social defeat stress, or social isolation stress also increases anxiety and stress response, while decreasing social exploration.

In these primate studies, maternally deprived infants showed antisocial behavior and aggression, and sometimes displayed higher emotional responses to other group members in their adulthood (4). Some of them also showed depression-like behavior (5), which is similar to the psychological symptoms of depressive, posttraumatic stress disorder (PTSD), and bipolar human patients. Based on these findings, neurobehavioral studies examining the impacts of neonatal mother–infant relationship have been widely focused on and well documented.

Maternal separation/natural variation of maternal licking and grooming model: There are a number of rodent models that show interesting behavioral and neurochemical outcomes induced by manipulations of social environments in the developmental period (Fig. 1). The first neonatal period, encompassing postnatal days 0–10 in mice and rats, in which the pups do not show a corticosterone response to novelty stress, is called the "stress hyporesponsive period" (6). In this period, the pups are highly dependent on the mother, and the mother shows intensive care to the pups. One of the well-documented models is maternal separation, in which neonatal pups are separated from the mother 1–3 h a day for 10–14 days. Complete maternal deprivation during the first 10 days of the neonatal period has been shown to have serious consequences for the pups later in life, including increased anxiety-like behavior (7) and enhanced neuroendocrine response to stressors (8). These behavioral and neuroendocrine changes persist throughout life, and associated epigenetic changes are also found in the central nervous system, indicating that maternal presence during the neonatal period has significant effects on the development of neural networks (9).

Another pioneer study was done by Michael Meaney's group at McGill University. They discovered that variation in maternal licking and grooming modulates the regulation of stress response through changing the epigenetic codes of glucocorticoid receptors in the hippocampus (10). When pups were reared by a highly licking and grooming mother, they developed hypomethylation of the regulatory region of the glucocorticoid receptor in the hippocampus, accompanied by a higher expression of the glucocorticoid receptor. In contrast, when pups were reared by a less licking and grooming mother, they showed both hypermethylation and lower expression of the glucocorticoid receptor (11, 12).

Early weaning model: In a natural setting, the domestic rat mother stays with her pups for more than 12 h a day and shows intense nursing behavior until weaning, which generally occurs between 25 and 33 days postpartum. Rat pups are able to eat, maintain body temperature, and evacuate after approximately 2 weeks of life (13), and weaning within this period has received scientific attention in terms of its neurobehavioral consequences. Similarly, in domestic mice (*Mus musculus*), a study based on seminatural observations described separation from the dam to have occurred when the offspring reached puberty, at the age of 4–5 weeks (14). After weaning, a key event marking the end of the early developmental stage, offspring must become nutritionally and behaviorally independent from their dam. Thus, it is highly likely that the weaning process influences the physiological and neurobehavioral development of rat pups. In support of this postulate, we and other researchers have identified behaviors that are influenced by early weaning (15, 16). For example, the precocious weaning of rats and mice enhances anxiety and aggression for a long period (17–19) and results in decreased play–fighting behaviors (20), suggesting that parent–pup interaction during the later lactation period is important for behavioral development (21). Moreover, early weaned male rats have higher autonomic responses to novel stressors (22) and early weaned mice exhibit an increased stress-endocrine response to mild stress (23). Therefore, early weaning may alter the development of the neural circuits for the stress response (Fig. 2).

In addition to behavioral changes, neurochemical development of the brain is also affected by early weaning. Early weaned male mice showed precocious myelination in the amygdala, decreased brain-derived neurotrophic factor (BDNF) protein levels in the hippocampus and prefrontal cortex, and reduced bromodeoxyuridine immunoreactivity in the dentate gyrus.

Neuro-behavioral outcomes induced by early-weaning	
Elevated plus-maze test (rats, mice)	Increased anxiety
Hole board test (rats)	Increased anxiety
Autonomic response to stress (rats)	Enhanced
Forced swim test (mice)	Increased immobility time
Maternal behavior (female mice)	Decreased
Sexual behavior (male mice)	Inhibited
Aggression after food deprivation (mice)	Enhanced
Aggression after social instigation (mice)	Decreased
Aggression toward cage-mates (mice)	Enhanced
Social play behavior (rats)	Decreased
Basal corticosterone (mice)	Increased
Corticosterone response to stress (mice)	Prolonged
Hippocampal glucocorticoid receptor (mice)	Decreased
Brain weight (mice)	Decreased
Myelin formation (mice)	Precocious
Hippocampal BDNF protein (mice)	Decreased
Hippocampal neurogenesis (mice)	Decreased

Fig. 2. Summary of the effects of early weaning. Early weaning causes higher anxiety (*top* four behavioral tests), decreased social behavior (*middle* six behavioral tests), enhanced stress responses (*middle* three parameters), and brain neurochemical changes (*bottom* four parameters).

In addition, it was revealed that early weaning decreases the mRNA expression level of the glucocorticoid and 5HT-1B receptors in the hippocampus (Fig. 2). Increased neurogenesis in the hippocampus in response to antidepressant treatments (24) or decreased neurogenesis due to irradiation of the hippocampus (25) is observed in parallel with behavioral changes, suggesting that neurogenesis in the hippocampus is related to anxiety/fear behavior. Regarding the effect of BDNF on behavior, mice genetically deficient in the BDNF–TrkB system failed to show antidepressant behavioral outcomes, suggesting that hippocampal BDNF modulates neurogenesis and behavior in these mice (26). In support of this hypothesis, BDNF itself causes antidepressant-like behavioral changes (27). Recently, Tsankova et al. (28) demonstrated that chronic social defeat decreases BDNF expression in the hippocampus through DNA modification, and can be recovered with antidepressant treatments. Taken together, these results demonstrate that not only the early neonatal period but also the late lactation period are important for the development of the nervous system, which is involved in emotional behavior and stress response.

2. Procedure

A significant number of rat and mouse studies regarding weaning have been demonstrated in our laboratory. In mice of the strains BALB/c and ICR, the day of birth was designated as postnatal day 0 (PD0) for each litter. On PD2, the litters were culled to 6–10 individuals, in a 1:1 male-to-female ratio, to standardize the litter size. Throughout the nursing period, the animals were not disturbed, except for a brief cleaning of the cage each week. All pups were kept with both parents until weaning. On PD14, half of the litter was separated from the dam and housed in same-sex groups of 2 or 3 mice in $12.5 \times 20 \times 11$ cm cages (early weaned group). The early weaned mice were fed ordinary adult pellets, as were the normally weaned mice after weaning. The other half of each litter was weaned on PD21, the natural weaning time, and housed in same-sex groups of 2 or 3 mice in $12.5 \times 20 \times 11$ cm cages (normally weaned group). The cages were cleaned once a week, on a day on which all of the litters were not necessarily at the same postnatal stage. Both male and female mice were used in the following experiments and were housed in groups of 2–3 per cage with animals of the same sex and same treatment until behavioral test and sampling. Two or three male and female mice from each litter were used for each group to minimize the interlitter differences, and each mouse was tested once in the elevated plus maze test and underwent blood sampling at either age.

We also tried to duplicate the experiment with the use of C57BL/6 mice, and discovered that the day of weaning needed to be slightly modified. In C57BL/6, the day of early weaning was PD16 and the day of normal weaning was PD25 because this mouse strain grows more slowly (29). Half of the C57BL/6 mice that were weaned on PD14 died.

In rats, the weaning procedure is slightly different from that in mice; the weaning time point was changed from PD21 (for mice) to PD30 (for rats), considering the difference in somatic growth rates between the two species, based on previous findings (30). For each litter, the day of birth was designated as PD0. On PD2, the litters were culled to eight, with half being male and the other half female, to standardize the litter size. Throughout the nursing period, care was taken not to disturb the animals, except for a brief cage cleaning each week. On PD16, half of each litter was separated from the dam and housed in same-sex groups of 2 or 3 mice in $41 \times 26 \times 18$ cm cages. The early weaned rats were fed ordinary adult pellets, as were the normally weaned mice after weaning. The other half of each litter was weaned on PD30, the natural time of weaning, and housed together in $41 \times 26 \times 18$ cm cages ($n = 2-3$ same-sex siblings per cage).

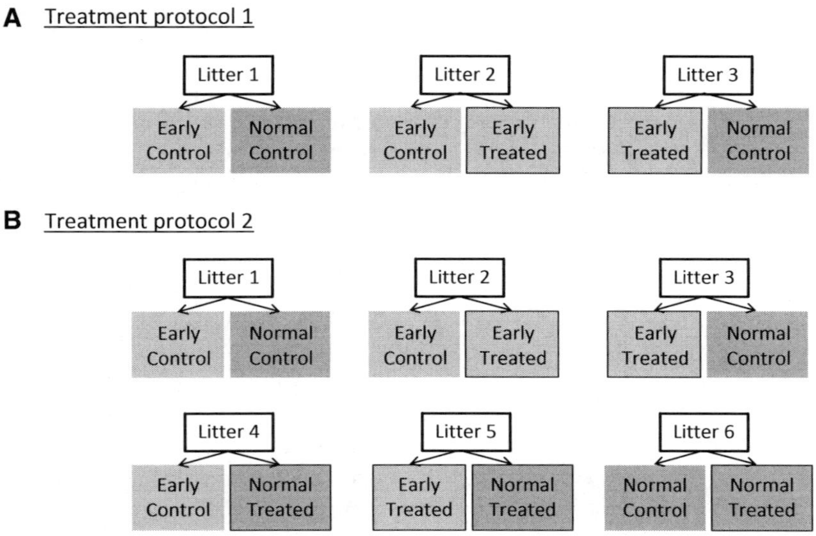

A Treatment protocol 1

B Treatment protocol 2

Fig. 3. Preparation of experimental groups. The least protocol is shown in (a), in which three groups are presented: early control, early treated, and normal control. Each litter should be assigned as 2, rather than 3 or 4, groups. The experimental groups are counterbalanced by preparing other litters. Another protocol is that six litters are prepared and each litter is assigned into two groups (b). In this case, a normally weaned group and a treatment group can also be obtained. These protocols can minimize the litter effects, and relatively stable data can be obtained.

Preparing experimental groups: In conducting experimental manipulations on early and normally weaned animals, such as drug treatments or environmental enrichments, the "litter effect" must be taken into consideration. Each litter should be assigned as 2, rather than 3 or 4, groups. Experimental groups are counterbalanced by preparing other litters. The least protocol is shown in Fig. 3a, in which three groups are presented: early control, early treated, and normal control. In this protocol, the treatment effects that can recover the behavioral or neurochemical changes induced by early weaning can at least be evaluated. The ideal result is that the early weaned but drug-treated animals show a difference from the early weaned control animals but are equivalent to the normal controls. Another protocol is that six litters are prepared and each litter is assigned into two groups (Fig. 3b). In this case, a normally weaned group and a treatment group can also be obtained. For example, if one wants to examine the interaction of two environments (weaning time and later environmental enrichments), a comparison of four groups – early weaned and standard environment, early weaned and enriched environment, normally weaned and standard environment, and normally weaned and enriched environment – can be performed. These protocols can minimize the litter effects, and relatively stable data can be obtained.

3. Data Analysis and Anticipated Results

Usually, the main purpose of this kind of study is to compare early and normally weaned animals. In this case, the data comparison is simple; the obtained parameters are compared by a multivariate analysis of covariance (MANCOVA), depending on the parameters measured in the experiments. For example, if the elevated plus maze test is conducted using early and normally weaned animals, the behavioral parameters, such as open arm entry, open arm duration, head dipping, stretched-out behavior, defecation, and distance moved, should be analyzed using a multivariate analysis of variance (MANOVA), because these parameters are related to each other and MANOVA has the advantage of controlling Type I errors and providing a multivariate analysis of effects by taking into account the correlations between dependent variables.

Another issue to consider is sex difference, which, in response to early weaning stress, is also dependent on the strain of the animals. The early weaning of male, but not female, Wistar rats was associated with increased anxiety, as assessed by the elevated plus maze, and exacerbated autonomic responses, as assessed by the novelty test (22, 31). In addition, early weaned ICR male, but not female, mice showed increased basal levels of circulating corticosterone and decreased levels of hippocampal glucocorticoid receptor mRNA compared to standard-weaned mice (23). Early weaned ICR male, but not female, mice also showed precocious myelination (32). However, the response to early weaning was observed to be different between strains: both male and female BALB/c mice showed increased anxiety (18); however, in the case of ICR mice, only the males displayed increased anxiety (32). We observed other sex-dependent effects of early weaning. Early weaned C57BL/6 males, but not females, showed a reduction of pup-killing behavior toward unfamiliar pups as compared to the normally weaned ones (Fig. 4). Moreover, early weaned C57BL/6 females, but not males, showed an increase of immobility time, as well as a decrease of struggling time, in the forced swim test (Fig. 5). Therefore, experimenters need to take sex difference into consideration, based on the strain they use and the outcomes they measure.

4. Troubleshooting

Mixed housing: One of the most frequent mistakes in experiments is the mixed housing of mice treated with other manipulations. If early and normally weaned mice are housed together after weaning (at 3 weeks of age), the difference in behavioral phenotypes

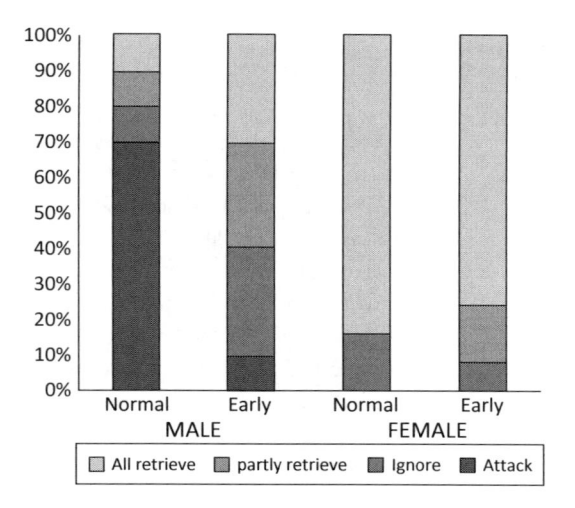

Fig. 4. Maternal retrieving behavior. C56BL6 male and female mice were subjected to early weaning (weaned on PD16) or normal weaning (PD25). Upon reaching adulthood, the mice were housed singly in a home cage for 24 h, and then three alien pups were introduced into the cage at its corners. A 30-min observation was performed with video recording, and the behaviors of the mice were separated into four categories: all retrieved in 30 min, partly retrieved, ignored, and attacking the pups.

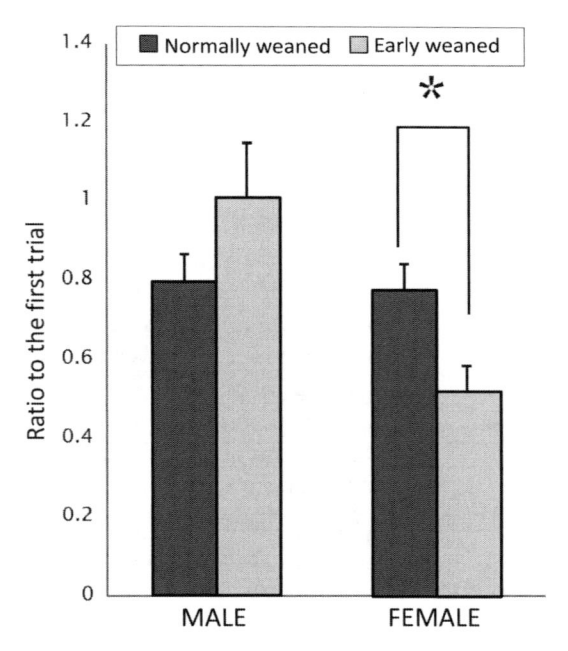

Fig. 5. Struggling ratio in the forced swim test. C56BL6 male and female mice were subjected to early weaning (weaned on PD16) or normal weaning (PD25). Upon reaching adulthood, the subject mice underwent the forced swim test, 5 min a day for 3 consecutive days. The immobilization time and struggling were scored, and the ratio of the third trial to the first trial was calculated. In the early weaned females, but not the males, the struggling ratio decreased as compared to that of the normally weaned ones.

disappears. The reason for this may be that mixed housing after weaning was found to lead to alterations in adult anxiety behavior (33). Therefore, the same group of animals should always be separated from other groups of animals.

Bedding materials and cage cleaning: Maternal behavior itself is very sensitive to the housing environment, lighting, temperature, and bedding materials. The preferred lighting is around 50–70 lux above the cage top, and the preferred temperature is 22–24°C. If the room temperature is higher, mother mice frequently step away from the nest, which will change the mother–pup interactions. The nest materials are also important. The parental behaviors of the male C57BL/6 strain were influenced by the type of cage bedding; the mice showed increased parental behavior when the bedding was purified paper chips (Alpha-Dri; Shepherd Specialty Papers, Watertown, TN) compared to autoclaved wood chips (Beta Chip; NEPCO, Warrensburg, NY; unpublished observation) (34). Thus far, we have used Corn Cob bedding (Shepherd's Cob: SHEPHERD) with Alpha-Dri nest materials. Too frequent cage cleaning must also be avoided, as it will lead to a decrease in maternal behavior and increase in stress response of the pups, especially in the first 2 weeks of the neonatal period. In this period, half of the bedding should be changed once a week, and it is preferable to allow the nesting area to remain as it is.

References

1. Bowlby J (1969) *Attachment and loss. Vol. 1, Attachment*. Hogarth, London.

2. Hildyard KL & Wolfe DA (2002) Child neglect: Developmental issues and outcomes. Child Abuse Negl 26: 679–695.

3. Rilling JK, Winslow JT, O'Brien D, Gutman DA, Hoffman JM & Kilts CD (2001) Neural correlates of maternal separation in rhesus monkeys. Biol Psychiatry 49: 146–157.

4. Suomi SJ (1997) Early determinants of behaviour: Evidence from primate studies. Br Med Bull 53: 170–184.

5. Suomi S, Delizio R & Harlow H (1976) Social rehabilitation of separation-induced depressive disorders in monkeys. Am J Psychiatry 133: 1279–1285.

6. Sapolsky RM & Meaney MJ (1986) Maturation of the adrenocortical stress response: Neuroendocrine control mechanisms and the stress hyporesponsive period. Brain Res 396: 64–76.

7. Lehmann J, Pryce CR, Bettschen D & Feldon J (1999) The maternal separation paradigm and adult emotionality and cognition in male and female wistar rats. Pharmacology, Biochemistry & Behavior. 64: 705–715.

8. Ogawa T, Mikuni M, Kuroda Y, Muneoka K, Mori KJ & Takahashi K (1994) Periodic maternal deprivation alters stress response in adult offspring: Potentiates the negative feedback regulation of restraint stress-induced adrenocortical response and reduces the frequencies of open field-induced behaviors. Pharmacol Biochem Behav 49: 961–97.

9. Holmes A, le Guisquet AM, Vogel E, Millstein RA, Leman S & Belzung C (2005) Early life genetic, epigenetic and environmental factors shaping emotionality in rodents. Neurosci Biobehav Rev 29: 1335–1346.

10. Liu D, Diorio J, Tannenbaum B, Caldji C, Francis D, Freedman A, Sharma S, Pearson D, Plotsky PM & Meaney MJ (1997) Maternal care, hippocampal glucocorticoid receptors, and hypothalamic-pituitary-adrenal responses to stress. Science 277: 1659–1662.

11. Weaver IC, La Plante P, Weaver S, Parent A, Sharma S, Diorio J, Chapman KE, Seckl JR, Szyf M & Meaney MJ (2001) Early environmental regulation of hippocampal glucocorticoid receptor gene expression: Characterization of intracellular mediators and potential genomic target sites. Mol Cell Endocrinol 185: 205–18.

12. Weaver ICG, Szyf M & Meaney MJ (2002) From maternal care to gene expression: DNA methylation and the maternal programming of stress responses. Endocr Res 28: 699–699.

13. Cramer CP, Thiels E & Alberts JR (1990) Weaning in rats: I. maternal behavior. Dev Psychobiol 23: 479–493.

14. Crowcroft P (1966) *Mice all over,* (G.T. Foulis & Co. Ltd, London).

15. Janus K (1987) Effects of early separation of young rats from the mother on their open-field behavior. Physiol Behav 40: 711–715.

16. Terranova ML & Laviola G (2001) Delta-opioid modulation of social interactions in juvenile mice weaned at different ages. Physiol Behav 73: 393–400.

17. Nakamura K, Kikusui T, Takeuchi Y & Mori Y (2003) The influence of early weaning on aggressive behavior in mice. Journal of Veterinary Medical Science 65: 1347–1349.

18. Kikusui T, Takeuchi Y & Mori Y (2004) Early weaning induces anxiety and aggression in mice. Physiol Behav 81: 37–42.

19. Kanari K, Kikusui T, Takeuchi Y & Mori Y (2005) Multidimensional structure of anxiety-related behavior in early-weaned rats. Behav Brain Res 156: 45–52.

20. Shimozuru M, Kodama Y, Iwasa T, Kikusui T, Takeuchi Y & Mori Y (2007) Early weaning decreases play-fighting behavior during the postweaning developmental period of wistar rats. Dev Psychobiol 49: 343–350.

21. Kikusui T & Mori Y (2009) Behavioural and neurochemical consequences of early weaning in rodents. J Neuroendocrinol 21: 427–431.

22. Ito A, Kikusui T, Takeuchi Y & Mori Y (2006) Effects of early weaning on anxiety and auto-nomic responses to stress in rats. Behav Brain Res 171: 87–93.

23. Kikusui T, Nakamura K, Kakuma Y & Yuji M (2006) Early weaning augments neuroendo-crine stress responses in mice. Behav Brain Res 175: 96–103.

24. Malberg JE, Eisch AJ, Nestler EJ & Duman RS (2000) Chronic antidepressant treatment increases neurogenesis in adult rat hippocam-pus. Journal of Neuroscience 20: 9104.

25. Santarelli L, Saxe M, Gross C, Surget A, Battaglia F, Dulawa S, Weisstaub N, Lee J, Duman R, Arancio O, Belzung C & Hen R (2003) Requirement of hippocampal neuro-genesis for the behavioral effects of antidepres-sants. Science 301: 805–809.

26. Saarelainen T, Hendolin P, Lucas G, Koponen E, Sairanen M, MacDonald E, Agerman K, Haapasalo A, Nawa H, Aloyz R, Ernfors P & Castren E (2003) Activation of the TrkB neu-rotrophin receptor is induced by antidepressant drugs and is required for antidepressant-induced behavioral effects. J Neurosci 23: 349–357.

27. Shirayama Y, Chen ACH, Nakagawa S, Russell DS & Duman RS (2002) Brain-derived neurotrophic factor produces antidepressant effects in behavioral models of depression. Journal of Neuroscience 22: 3251.

28. Tsankova NM, Berton O, Renthal W, Kumar A, Neve RL & Nestler EJ (2006) Sustained hip-pocampal chromatin regulation in a mouse model of depression and antidepressant action. Nat Neurosci 9: 519–525.

29. Lambert R (2007) Breeding strategies for maintaining colonies of laboratory mice. A Jackson Laboratory Resource Manual. http://jaxmice.jax.org/manual/breeding_strategies_manual.pdf).

30. Rosenblatt JS (1965) *The basis of synchrony in the behavioral interaction between the mother and her offspring in the laboratory rat,* ed Foss BM in Determinants of Infant Behavior, Vol. 3, (Methuen and Co Ltd, London).

31. Kodama Y, Kikusui T, Takeuchi Y & Mori Y (2008) Effects of early weaning on anxiety and prefrontal cortical and hippocampal myelina-tion in male and female wistar rats. Dev Psychobiol 50: 332–342.

32. Kikusui T, Kiyokawa Y & Mori Y (2007) Deprivation of mother-pup interaction by early weaning alters myelin formation in male, but not female, ICR mice. Brain Res 1133: 115–122.

33. Curley J, Rock V, Moynihan A, Bateson P, Keverne E & Champagne F (2010) Developmental shifts in the behavioral phenotypes of inbred mice: The role of postnatal and juvenile social experiences. Behav Genet 40: 220–232.

34. Kuroda KO, Meaney MJ, Uetani N, Fortin Y, Ponton A & Kato T (2007) ERK-FosB signal-ing in dorsal MPOA neurons plays a major role in the initiation of parental behavior in mice. Mol Cell Neurosci 36: 121–131.

Chapter 15

Separation-Induced Depression in the Mouse

Richard E. Brown, Alison L. Martin, and Rhian K. Gunn

Abstract

This chapter outlines the experimental methods for inducing Depression-like behavior in mice by separating group-housed mice to individual housing. This is a model of loneliness in mice. Loneliness-induced depression is measured using the forced swim test (FST) and tail suspension test (TST). The first part of the chapter describes the methods for inducing separation-induced depression. It summarizes the experimental protocols including behavioral testing of the animals. The second part describes some of the issues to be concerned within this paradigm. An up-to-date collection of protocols and procedures for troubleshooting are provided.

Key words: Depression, Social stress, Mouse model, Anxiety, FST, TST, Behavior, Social isolation

1. Background and Historical Significance

Loneliness is the sadness of being alone or the feeling of dejection arising from the lack of social interaction. In humans, loneliness can lead to depression (1, 2), high blood pressure, poor sleep patterns (3, 4), and obesity (5, 6). Loneliness can also facilitate the onset of life-threatening diseases including cancer and Alzheimer's disease (7, 8). Loneliness has been thought of as a stressor, leading to elevated cortisol levels (7), and as a "social pain response," leading to the activation of pain-related areas in the brain (9–11). This chapter discusses an animal model for the study of loneliness-induced depression that we have developed (12).

There is a strong demand for the development of new pharmacological treatments for depression, but the etiology and physiological mechanisms underlying depression are largely unknown (13). In order to develop new and more effective treatments for depression, an animal model needs to be identified that addresses

Todd D. Gould (ed.), *Mood and Anxiety Related Phenotypes in Mice: Characterization Using Behavioral Tests, Volume II*, Neuromethods, vol. 63, DOI 10.1007/978-1-61779-313-4_15, © Springer Science+Business Media, LLC 2011

the symptom profile, the etiology, and the neurochemical processes underlying depression (14, 15). Depression involves a heterogeneous set of symptoms, including low mood and anhedonia, and often symptoms of anxiety may be present (14, 16). Although women are almost twice as likely to become depressed and exhibit more severe symptoms of depression than men (17–19), very little research on depression is done on women (20–22).

In an animal model, the procedure for inducing depression must be independent of the procedure for measuring depression, and the behavioral tests for measuring depression must be able to dissociate depression from other psychological states such as anxiety, fear, and sensory-motor dysfunction (23, 24). The two most reliable tests for measuring Depression-like behavior in rodents are the forced swim test (FST) (25, 26) and the tail suspension test (TST) (27).

There are numerous rodent models of depression, but most focus on one or a few symptoms of depression and not the overall disorder. In addition, the majority of these models fail to investigate female animals, even though women are more likely to develop depression than men (14, 19, 28). A number of procedures have been advocated for inducing depression in animal models. These involve genetic modification, surgical intervention, environmental stressors, and social stressors (23). Procedures such as exposure to chronic restraint stress (29), learned helplessness (30), and corticosterone (CORT) challenge (31), focus on face and predictive validity, limiting their effectiveness for understanding depression and developing new antidepressants (17, 32–34).

One of the most reliable methods for inducing depression in animal models is the chronic mild stress procedure, which focuses on the etiology of depression, subjecting mice to mild stressors over a prolonged period of time (35). These stressors can be physical, such as damp bedding, tilting the cage, and food deprivation, or social, such as home-cage intruders and social defeat in aggressive encounters (36, 37). Because depression in humans is believed to be elicited by social stress rather than physical stress (38), there has been a shift toward the social stressor for inducing depression (39, 40).

One procedure that seems particularly amenable to the study of depression in females is the social separation model. Isolation rearing has a number of neural, endocrine, and behavioral effects in mice and has been used to model a number of neuropsychiatric disorders (41, 42). Long-term social isolation also causes stress and anxiety-like behavior in a modified open field test in female CD-1 mice (17). Antidepressants, such as fluoxetine, reduce depressive-like behavior in mice (43–47), including those who have been socially isolated (42). The neuro-behavioral effects of social isolation, however, may depend on the age at which isolation occurs and the duration of the separation. Long-term isolation starting at puberty (21 days of age) and continuing until adulthood (70 days of age) may have very different neural and behavioral consequences than short-term separation/isolation that begins in adulthood.

Isolation rearing prevents animals from forming social relationships and developing social interactions whereas short-term separation in adulthood removes an animal from a social group to which it has become accustomed. Short-term social separation in adults may provide a model of depression. As loneliness is a significant precursor of depression in humans (48, 49), social separation may be an etiological factor for depression in rodents. Social separation produces Depression-like behaviors, such as decreased sucrose intake (anhedonia) and increased plasma CORT in response to a stressor in rodents (14, 50). Short-term isolation or separation during adulthood appears to produce depressive-like and some anxiety-like symptoms in female rodents as we and others have shown (12, 17, 50).

2. Equipment, Materials, Setup

2.1. Animals

In our experiments, we used female C56BL/6J mice from the Jackson Laboratory (Bar Harbour, ME, USA), but this protocol can be used with any outbred or inbred mouse strain as well as mutant, knockout, and transgenic mice.

2.2. Housing Environment

All mice were housed in clear, plastic cages ($18.8 \times 28.0 \times 12.5$ cm) with stainless steel tops. Wood shavings were used for bedding and all cages included a PVC tube (4.0 cm inside diameter and 7.5 cm long) for shelter. Mice were given ad libitum access to Purina rodent chow (#5001) and tap water. The colony room was maintained at a constant temperature (22 ± 2°C) and was on a 12:12 reversed light:dark cycle with the lights off at 9:30 am.

3. Procedure

The procedure for separation-induced depressive behavior is detailed below. This procedure can readily be adapted to investigate the interactions between environment, mouse strain, gene manipulations, and/or chronic drug treatment on Depression-like behavior and depression-like changes in hormones and neural function as measured by histological and biochemical changes.

3.1. Grouped, Isolated, and Separated Mice

Upon arrival in our laboratory at 8 weeks of age, mice were housed in same-sex groups of four or placed in individual housing (isolated). Socially housed mice were separated into individual housing after

8 weeks of social housing. All mice were tail marked for individual recognition. If ear punching is used for individual recognition, it should occur at the same time for all mice and at least 1 week before separation. Restraint and ear punching caused elevated heart rate and body temperature for at least 1 h following the procedure, as well as increased motor activity, but these effects were not present after 6 h (51).

3.2. Duration of Social Housing Before Separation

We housed 8-week-old mice in social groups for 15 weeks before separating them into individual housing. These mice were housed in social groups by the breeder (Jackson Laboratories, Bar Harbor, ME) from weaning to 8 weeks of age, when we purchased them. Likewise, group-housed mice were housed in groups for 8 weeks (8–16 weeks of age) or 16 weeks (8–24 weeks of age) before testing.

3.3. Duration of Isolation or Separation Before Testing

Mice were housed in isolation for 8 weeks (8–16 weeks of age) or 16 weeks (8–24 weeks of age) before testing. Mice housed in groups from 8 to 23 weeks of age were separated for 1 week (23–24 weeks of age) before testing.

4. Dependent Measures

As depression is a multi-faceted disease, a paradigm that aims to model it effectively should address multiple behavioral measures. We used both the FST and the TST as measures of Depression-like behavior, as they have been shown to have different sensitivities to antidepressant treatments and it is thought that they measure changes in different neurotransmitter systems (52). We also measured weight gain, anxiety (in the light/dark box), and CORT levels.

4.1. Body Weight

Because depression can lead to overweight and obesity in humans (5, 6), all mice were weighed each morning during the study using an Ohaus Scout II digital scale.

4.2. Forced Swim Test

The FST was performed using the method of Porsolt et al. (25). A 2 L glass beaker was filled to 13 cm with tap water (1,600 mL) and allowed to sit overnight to achieve room temperature ($25 \pm 2°C$) (Fig. 1a). The water was deep enough so that mice were not able to touch the bottom of the beaker with their tails. Each mouse was tested in one 6-min trial in which the number of bouts of immobility and the duration of immobility in seconds was scored live using Hindsight (MS-DOS Version 1.5) computer program and video recorded by a camcorder. The duration of immobility was then converted to a percentage of trial time. Immobility was defined as the cessation of all movements except

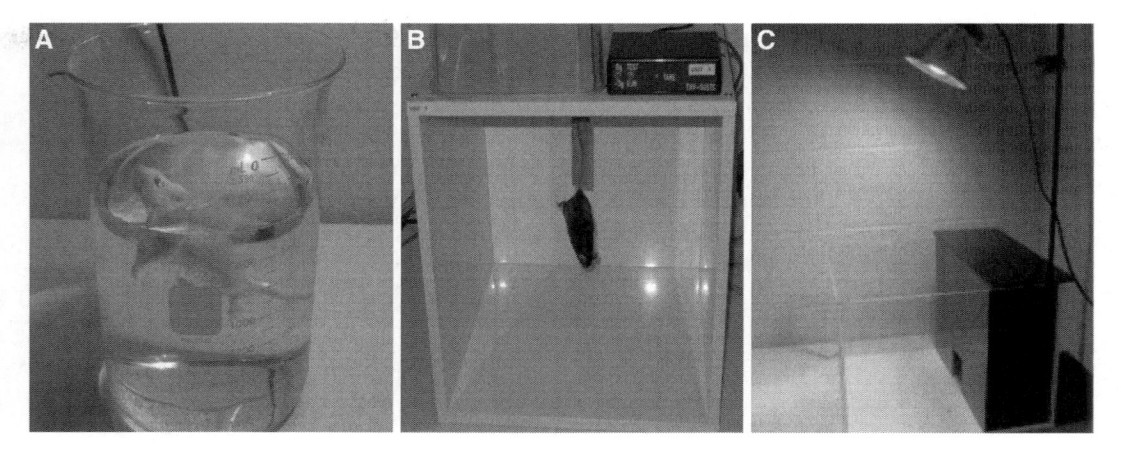

Fig. 1. Apparatus used for: (**a**) the forced swim test; (**b**) tail suspension test; (**c**) light/dark box.

those necessary to stay afloat, such as paddling lightly with one foot (23, 53, 54). The apparatus and procedure are described in detail by Hascoët and Bourin (26).

4.3. Tail Suspension Test

In the TST, the mouse was attached by the tail, using Scotch Brand Super Tape, to an aluminum strip ($11.5 \times 2.2 \times 0.15$ cm) suspended vertically from a Med Associates test apparatus (St. Albans, VT, USA) (Fig. 1b). This apparatus consisted of a box ($32 \times 33 \times 33$ cm) open on one side to allow the observer to view the subjects and for videotaping. Each mouse was tested in one 6-min trial in which the number of bouts of immobility (defined as the cessation of all movements) and the duration of immobility in seconds was scored live using Hindsight (MS-DOS Version 1.5) and recorded by a camcorder. The duration of immobility was then converted to a percentage of trial time. The apparatus and procedure are described in detail by O'Leary and Cryan (27).

4.4. Light/Dark Box

Anxiety-like behavior was measured in a light/dark box ($45 \times 27 \times 27$ cm), consisting of a light compartment (27×27 cm) and a smaller dark compartment (18×27 cm) connected by a door (7.5×7.5 cm) in the center of the wall separating the two compartments (Fig. 1c). The floor, walls, and lid of the light compartment were made of clear Plexiglas and walls, floor, and lid of the dark compartment were made of black Plexiglas. A 60-W bulb located 40 cm above the center of the light compartment provided bright illumination (640 lux, Sekonic Flashmate).

Using the test procedure of Costall et al. (55) for anxiety sensitivity, mice were placed in the center of the light compartment facing the door and were allowed to explore the box for 5 min. Each mouse's behavior was scored by an experimenter using Hindsight (MS-DOS Version 1.5) and video-recorded. After the

5-min test, the mouse was returned to its home cage and the apparatus was thoroughly cleaned with Sparkleen detergent (Fisher Brand) as a soiled LD box removes the neophobia associated with the box (56).

Four behaviors were scored for each mouse: Time (sec) in the light compartment, time spent in the dark compartment, the number of transitions between the compartments, and the number of head pokes (also called "head outs") from the dark into the light compartment. The percentage of time spent in the light compartment was analyzed. Transitions were defined as the number of times the mouse passes into the opposite compartment with all four paws, and head pokes were defined as when only the head and forefeet of a mouse in the dark compartment pass into the light compartment and then retreat back into the dark compartment. The apparatus and procedure are explained in detail by Hascoët and Bourin (57).

4.5. Corticosterone Levels

Baseline CORT can be measured using trunk blood at the end of study. Mice were euthanized using carbon dioxide and then decapitated and trunk blood was collected. Euthanasia and blood collection were done at a consistent time in the light/dark cycle because of the circadian rhythm of CORT secretion. The procedure should be done in a room away from the testing room, and mice should be closely monitored during the procedure to avoid unnecessary pain or stress. The order in which blood was taken was randomized for group and treatment to minimize differences due to the cyclic release of CORT. Following the procedure outlined by Mashoodh et al. (58), all blood samples were collected into heparinized tubes (Becton Dickinson, NJ, USA) and centrifuged at $6,000 \times g$ at 4°C. Resulting plasma was retained and stored at −80°C until assay. Levels of CORT were determined using a Correlate-EIA CORT immunosorbent assay kit (Assay Designs, MI, USA). Samples were diluted 1:40 prior to assay.

4.6. Antidepressant Drug Administration

There is the option of chronically administering antidepressant drugs such as fluoxetine during this protocol (26). Half of the isolated mice and half of the group-housed mice were given 10 mg/kg/day of fluoxetine in their drinking water (59) and half were given plain water. The mice were separated at 8 weeks of age and drug treatment began at 13 weeks of age, 5 weeks after separation, and was maintained until the end of behavioral testing. Treated water was kept in opaque bottles (wrapped in duct tape) to protect from light degradation. This method of dosing eliminates the stress that would be induced by daily intraperitoneal injections and is more valid with respect to human dosing.

The average daily water intake for each mouse was recorded for 1 week prior to drug administration and was monitored throughout the duration of drug treatment. For group-housed mice, it was assumed that each mouse consumed an amount consistent with its body weight (59). The weight of the mice was monitored and correlated with water consumption and drug dosage. The effects of a number of antidepressant drugs on Depression-like behavior in the FST are summarized by Hascoët and Bourin (26).

5. Data Analysis

For comparing the behavior of group-housed, isolated, and separated mice, one-way ANOVAs with post-hoc (Tukey HSD) tests and student t-tests were used. For comparing group housing and isolation in conjunction with chronic drug administration, two-way ANOVAs and post-hoc student t-tests were used.

6. Anticipated Results

We theorized that if social separation induced a depression-like state in mice, that this would be detected by changes in weight, by increased immobility in the FST and TST, and by changes in corticosteroid levels. If anxiety increased, this would be detected by changes in behavior in the light/dark box. If the separated mice were depressed, we proposed that treatment with an antidepressant would reverse the Depression-like behavior.

6.1. Body Weight

The body weight of isolated mice generally increased in comparison with the group-housed mice over the duration of separation. Increases in body weight in socially isolated mice became significant 4–6 weeks after separation (Fig. 2) (12).

6.2. Forced Swim Test

Socially isolated mice showed an increase in percent time spent immobile compared to group-housed mice and the percentage of time spent immobile increased with increased duration of isolation, as mice separated for 1 week were immobile longer than group-housed mice but not as long as mice housed in isolation for 16 weeks (Fig. 3a). The number of bouts of immobility in the FST did not show an increase with length of separation (Fig. 3b).

6.3. Tail Suspension Test

Separated mice had significantly more bouts of immobility in the TST than group-housed mice. Separated mice also spent more time immobile than group-housed mice but this was not significant (Fig. 4).

6.4. Light/Dark Box

Neither isolated nor separated mice spent less time in the light compartment than group-housed mice, but isolated mice made fewer transitions between the light and dark compartments than grouped mice (Fig. 5). This suggests that the isolated mice showed a slight increase in anxiety.

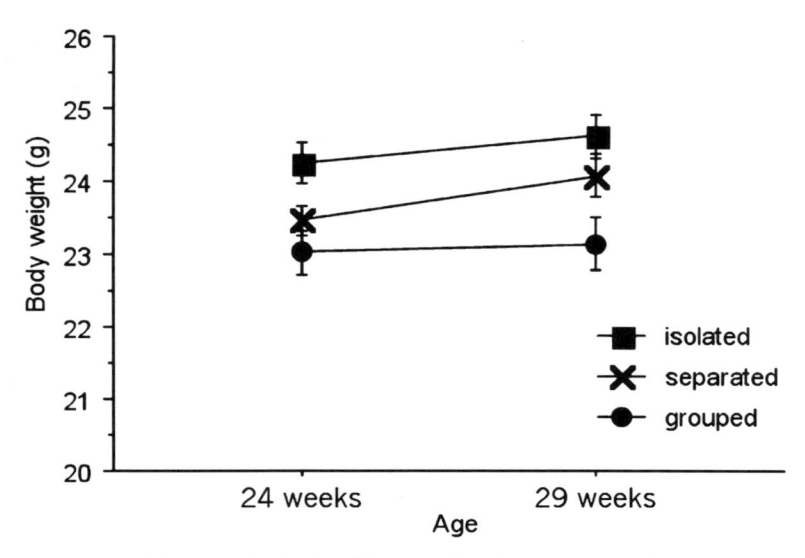

Fig. 2. Mean (±s.e.m.) body weight. At the beginning of behavioral testing (24 weeks of age), isolated mice were heavier than grouped and separated mice ($F(2,33) = 5.51$, $p < 0.01$). Five weeks later, isolated mice weighed significantly more than grouped mice ($F(2,33) = 5.69$, $p < 0.01$) and separated mice showed an increase in weight compared to grouped mice, but this did not reach statistical significance. Reprinted from Martin and Brown (12), with permission from Elsevier.

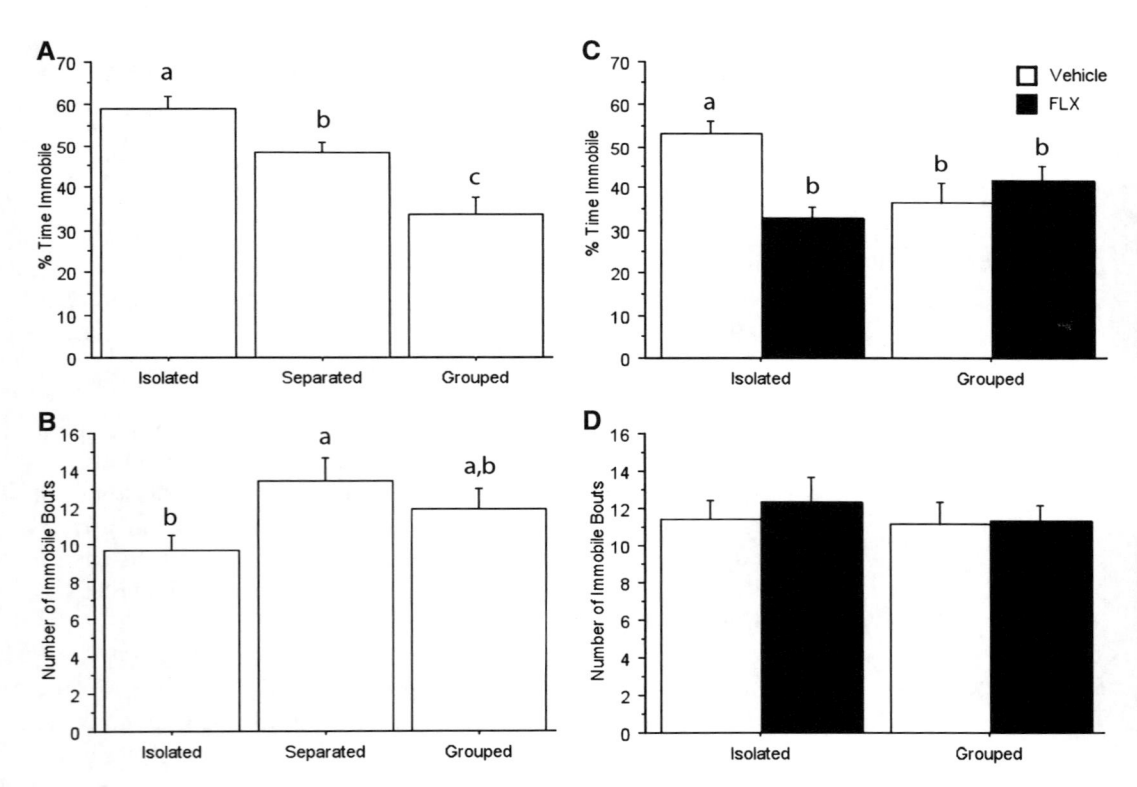

Fig. 3. Forced swim test. (a) Isolated and separated mice spent more time immobile (mean ± s.e.m.) than group-housed mice ($p < 0.00001$). (b) Isolated mice had significantly fewer bouts of immobility (mean ± s.e.m.) than separated mice ($p < 0.05$), but there was no difference between isolated and group-housed mice. (c) The effects of isolation were reversed with fluoxetine (FLX) ($p < 0.01$), as mice receiving drug were immobile significantly less than those receiving vehicle. (d) Neither fluoxetine nor housing condition affected the number of bouts of immobility. Groups with different letters are significantly different ($p < 0.05$). Reprinted from Martin and Brown (12), with permission from Elsevier.

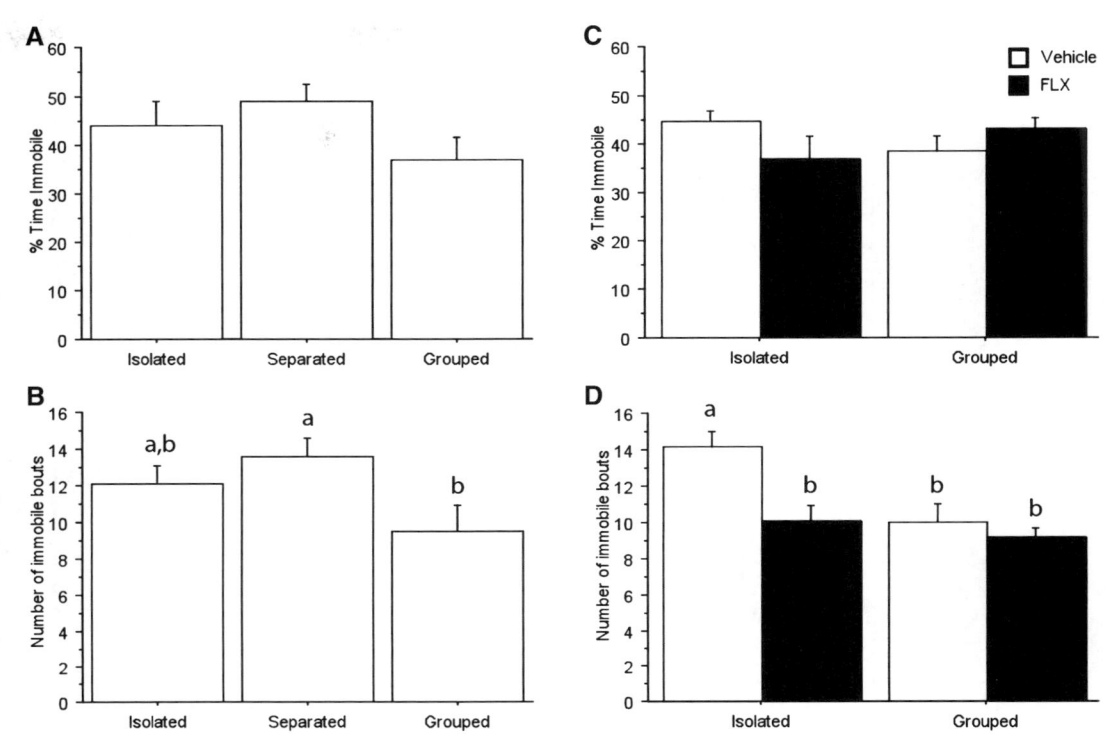

Fig. 4. Tail suspension test. (**a**) There was no effect of housing on mean (±s.e.m.) percent time spent immobile. (**b**) Separated mice had significantly more bouts of immobility (mean±s.e.m.) than group-housed mice ($p<0.05$). (**c**) There was no effect of housing or fluoxetine treatment (FLX) on mean (±s.e.m.) percent time spent immobile, however (**d**) fluoxetine treatment significantly reduced the mean (±s.e.m.) number of bouts of immobility in individually housed mice compared to control treatment ($p<0.01$). Groups with different letters are significantly different ($p<0.05$). Reprinted from Martin and Brown (12), with permission from Elsevier.

6.5. Corticosterone Levels

Although an increase in CORT was expected in individually housed mice, we found that group-housed mice had significantly higher baseline plasma CORT levels than separated mice (12). Similar results were reported by Beitia et al. (60), who found that socially stressed male OF1 mice had lower baseline CORT levels than controls, and Konkle et al. (61), who observed that individually housed Long Evans and Sprague-Dawley female rats had a significantly lower level of baseline CORT than group-housed females. It has been suggested that there is a gradual decrease in the basal adrenal activity in response to continued exposure to stressors (60). The relationship between separation and corticosteroid levels bears further investigation.

6.6. Effects of Antidepressant Drugs

In our study, chronic oral fluoxetine treatment (10 mg/kg/day) did not affect weight gain. In the FST, there was a significant fluoxetine by housing interaction as separated mice receiving vehicle were immobile significantly longer than group-housed mice and separated mice receiving fluoxetine (Fig. 3c). There was no effect

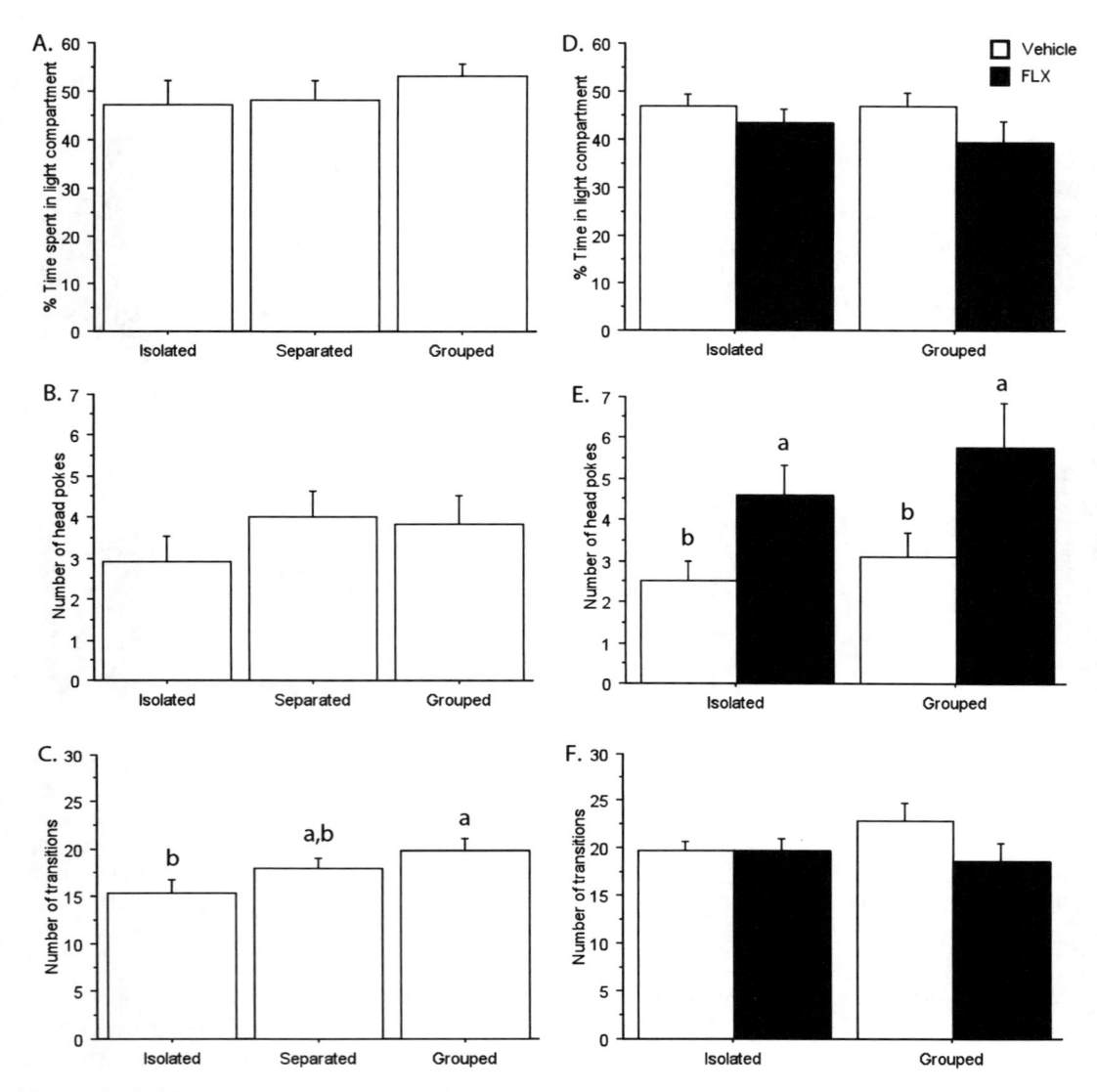

Fig. 5. Light/dark box. (**a**) There was no effect of housing on mean (±s.e.m.) percent time spent in the light compartment or (**b**) mean (±s.e.m.) number of head pokes from the dark compartment into the light compartment. (**c**) Isolated mice showed fewer transitions (mean ± s.e.m.) than group-housed mice ($p < 0.05$). (**d**) There was no effect of housing or fluoxetine treatment on mean (±s.e.m.) time spent in the light compartment. (**e**) Fluoxetine increased the mean (±s.e.m.) number of head pokes for both individually housed and group-housed mice ($p < 0.01$). (**f**) There was no effect of housing or drug on the mean (±s.e.m.) number of transitions between compartments. Groups with different letters are significantly different ($p < 0.05$). Reprinted from Martin and Brown (12), with permission from Elsevier.

of fluoxetine on the number of bouts of immobility (Fig. 3d). In the TST, fluoxetine reduced the percent time immobile in separated mice more than group-housed mice, but these results were not statistically significant (Fig. 4c). Fluoxetine did, however, significantly reduce the number of bouts of immobility in separated mice in the TST (Fig. 4d). In the L/D box, fluoxetine increased the frequency of head pokes by both separated and group-housed mice (Fig. 5e), but there was no effect on percent time spent in the light zone or transitions between the light and dark zones (Fig. 5d, f) (12). More work is thus required to investigate the effects of antidepressants on Depression-like behavior following social separation.

7. Experimental Variables and Troubleshooting: What Can Go Wrong?

7.1. Social/Isolation Housing Parameters

7.1.1. Duration of Group Housing Before Separation

We found that mice housed together for 8 weeks and for 16 weeks both showed increased Depression-like behavior when separated, but it is not known whether the duration of group-housing before separation has a linear effect on Depression-like behavior. Male mice show an increase in active social behavior with age, but the age of group formation may strongly affect social stability (62, 63). As long as the mice are late adolescents (8 weeks of age) or adults (10 weeks or older) when moved to individual housing, the same trend in depressive behavior has been observed. No parametric studies have been done on the effects of the duration of social housing before separation.

7.1.2. Duration of Separation Before Testing

There is a relation between the duration of separation and the severity of symptoms as measured by the FST but this appears to reach a ceiling, as there was no difference between 8 and 16 weeks of separation. No parametric studies have been done on the duration of separation before testing. As depression is a precursor to a number of other neuropsychiatric symptoms, there may be a linear relation between duration of social separation and the severity of symptoms (64). There may also be long-term behavioral effects that occur when mice separated in early adulthood reach old age (65).

7.1.3. Age at Separation

Isolation rearing prevents animals from developing social interactions whereas short-term separation in adulthood removes an animal from a social group, which may explain why isolation rearing beginning at puberty (21 days) and continuing until adulthood (70 days) has different neural and behavioral consequences than separation/isolation that begins in adulthood. Short-term separation during adulthood appears to produce depressive-like and some anxiety-like symptoms in female rodents as we and others have shown (12, 17, 50). On the other hand, isolation rearing from weaning results in a wide range of behavioral abnormalities in social behavior, communication, cognitive, and emotional behavior (41).

7.2. Sex Differences

Females and males have different behavioral reactions to social isolation. Abramov et al. (66) observed that 3 weeks of social isolation increased anxiety in female 129sv/C57BL6 mice while males showed an increase in aggression and exploration. Karolewicz and Paul (67), however, found that individual housing of adult male NIH Swiss mice for 8 weeks reduced levels of immobility in the FST and TST. Palanza (17) found that individual housing for 1 week reduced exploratory behavior and increased anxiety in female CD-1 mice, but reduced anxiety and increased exploratory behavior in males. At this time, the specific sex differences in our paradigm have not been tested.

7.3. Estrous Cycle

Although the cyclic fluctuations of gonadal hormones over the estrous cycle can affect behavior in female mice, these effects are small, especially in C57BL/6J females, and differences between groups can be detected irrespective of estrous cycle variation (68). Testing can be randomized over the estrous cycle and data pooled to control for the effects of estrous cycle on behavior, but the effects of gonadal and other hormones on separation-induced Depression-like behavior have yet to be studied.

7.4. Tests That Require Separation

Some behavioral tests require short-term separation of mice that are normally group housed. The effects of separation for a few days have not yet been investigated, but we did see behavioral changes with as little as 1 week of separation. The long-term effects of brief social separation on Depression-like behavior are unknown.

7.5. Drug Treatments

A wide variety of antidepressant (26) and anti-anxiety drugs (57) could be examined for their effects on the behavior of separated mice in the TST, FST, light/dark box, or elevated plus-maze. The effects of type of drug, route of administration, and dose on separation-induced depression have yet to be investigated.

7.6. Sensitivity of Dependent Measures

There are many behavioral tests for anxiety- and Depression-like behavior in mice, and these may have subtle differences in ethological validity and in their underlying neurochemical basis. For example, the FST and TST have different sensitivities to antidepressant treatments, and it is thought that these two tests measure changes in different neurotransmitter systems (53). The FST is primarily associated with the dopaminergic and serotonergic systems while the TST is associated with the adrenergic system (53, 69). Thus, these two measures might be used to dissociate the neurochemical basis of Depression-like behavior.

7.7. The Comorbidity of Depression and Anxiety

When we designed our studies of separation-induced depression, we thought that we could dissociate measures of depression from those of anxiety, but in human patients, especially women, there is an 80% comorbidity between anxiety and depression (18, 70). The reduced number of transitions in the L/D box may be related to immobility in the FST and TST (Fig. 5). Bourin and Hascoët (56) concluded that an increase in transitions in the L/D box without an increase in spontaneous locomotion could reflect increased anxiety. But it is hard to make conclusions about the effect of isolation on anxiety, as we have so far only tested C57BL/6J mice, and this strain is known have low baseline anxiety levels (71). The L/D box is thought to best measure depressive anxiety caused by uncontrollable stress, such as exposure to a novel and/or aversive environment (56). The L/D box is as sensitive as the elevated plus maze and the open field in detecting strain differences in anxiety (72). In a free-exploration paradigm, Palanza (17) found that compared to group-housed females, individually housed female CD-1 mice spent less time exploring the open field than group-housed females, showed less rearing and decreased activity, all indices of high anxiety. The increase in both anxiety and depressive behavior in socially isolated mice lends validity to this model of major depressive disorder as human females are often affected by comorbid depression and anxiety (18, 70).

8. Concluding Remarks

The separation-induced depression model allows for the comprehensive study of novel treatments of depression at behavioral, neurochemical, and genetic levels. The added benefit of this model is that it is simple and does not require any special resources or skills, nor does the administration of the antidepressant produce a stress artifact in the behavior or physiology of the animals (59). This makes the social separation model of depression very accessible to research labs and will hopefully allow for ease of investigation into the interaction of depression and anxiety, as well as studies with other mouse models and genetic phenotypes.

Acknowledgments

This research was funded by NSERC of Canada.

References

1. Cacioppo JT, Hawkley LC, Thirsted RA (2010) Perceived social isolation makes me sad: 5-Year cross-lagged analyses of loneliness and depressive symptomatology in the Chicago Health, Aging, and Social Relations study. Psychol Aging 25:453–463

2. Cacioppo JT, Hughes ME, Waite LJ et al (2006) Loneliness as a specific risk factor for depressive symptoms: Cross-sectional and longitudinal analyses. Psychol Aging 21: 140–151

3. Hawkley LC, Burleson MH, Berntson GG, Cacioppo JT (2003) Loneliness in everyday life: Cardiovascular activity, psychosocial context, and health behaviors. J Pers Soc Psychol 85:105–120

4. Cacioppo JT, Hawkley LC, Crawford LE, et al (2002) Loneliness and health: Potential mechanisms. Psychosom Med 64:407–417

5. Luppino FS, de Wit LM, Bouvy PF, et al (2010) Overweight, obesity and depression. A systematic review and meta-analysis of longitudinal studies. Arch Gen Psychiatry 67:220–229

6. Stunkard AJ, Faith MS, Allison KC (2003) Depression and obesity. Biol Psychiatry 54:330–337

7. Hawkley LC, Cacioppo JT (2003) Loneliness and pathways to disease. Brain Behav Immun 17:S98–S105

8. Wilson RS, Krueger KR, Arnold SE, et al (2007) Loneliness and risk of Alzheimer's disease. Arch Gen Psychiatry 64:234–240

9. Eisenberger NI, Lieberman MD, Williams, KD (2003) Does rejection hurt? An fMRI study of social exclusion. Science 302:290–292

10. Eisenberger NI, Taylor SE, Gable SL et al (2007) Neural pathways link social support to attenuated neuroendocrine stress responses. NeuroImage 35:1601–1612

11. Panskepp J (2003) Feeling the pain of social loss. Science 302:237–239

12. Martin AL, Brown RE (2010) The lonely mouse: Verification of a separation-induced model of depression in female mice. Behav Brain Res 207:196–207

13. Belmaker RH, Agam G (2008) Major depressive disorder. N Engl J Med 358:55–68

14. Anisman H, Matheson K (2005) Stress, depression, and anhedonia: caveats concerning animal models. Neurosci Biobehav Rev 29:525–546

15. Markou A, Chiamulera C, Geyer MA, et al (2009) Removing obstacles in neuroscience drug discovery: The future path for animal models. Neuropsychopharmacology 34:74–89

16. Nestler EJ, Barrot M, DiLeone RJ, et al (2002) Neurobiology of depression. Neuron 34:13–25

17. Palanza P (2001) Animal models of anxiety and depression: how are females different? Neurosci Biobehav Rev 25:219–233

18. Young EA, Abelson JL, Cameron OG (2004) Effect of comorbid anxiety disorders on the Hypothalamic-Pituitary-Adrenal axis to a social stressor in major depression. Biol Psychiatry 56:113–120

19. Patten SB, Wang JL, Williams JVA, et al (2006) Descriptive epidemiology of major depression in Canada. Can J Psychiatry 51:84–90

20. Accortt EE, Freeman MP, Allen JJ (2008) Women and major depressive disorder: clinical perspectives on causal pathways. J Womens Health (Larchmt) 17:1583–90

21. Check Hayden E (2010) Sex bias blights drug studies. Nature 464:332–333

22. Beery AK, Zucker I (2011) Sex bias in neuroscience and biomedical research. Neurosci Biobehav Rev 35:565–572

23. Fuchs E, Flügge G (2006) Experimental animal models for the simulation of depression and anxiety. Dialogues Clin Neurosci 8:323–33

24. Brown RE, Wong AA (2007) The influence of visual ability on learning and memory performance in 13 strains of mice. Learning Mem 14:134–144

25. Porsolt RD, Pichon LE, Jalfre M (1977) Depression: a new animal model sensitive to antidepressant treatments. Nature 266:730–732

26. Hascoët M, Bourin M (2009) The forced swimming test in mice: a suitable model to study antidepressants. In: Gould TD (ed) Mood and Anxiety Related Phenotypes in Mice, Neuromethods. Humana Press, New York, p. 85–118

27. O'Leary OF, Cryan JF (2009). The tail-suspension test: a model for characterizing antidepressant activity in mice. In: Gould TD (ed) Mood and Anxiety Related Phenotypes in Mice, Neuromethods. Humana Press, New York, p. 119–137

28. Leuner B, Mendolia-Loffredo S, Shors TJ (2004) Males and females respond differently to controllability and antidepressant treatment. Biol Psychiatry 56:964–970

29. Strekalova T, Steinbusch H (2009) Factors of reproducibility of anhedonia induction in a chronic stress depression model in mice. In: Gould TD (ed) Mood and Anxiety Related Phenotypes in Mice, Neuromethods. Humana Press, New York, p. 153–176

30. Anisman H, Merali M (2009) Learned helplessness in mice. In: Gould TD (ed) Mood and Anxiety Related Phenotypes in Mice, Neuromethods. Humana Press, New York, p. 177–196

31. Johnson SA, Fournier NM, Kalynchuk LE (2006) Effect of different doses of corticosterone on depression-like behavior and HPA axis responses to a novel stressor. Behav Brain Res 168:280–288

32. Cabib S, Puglisi-Allegra S (1996) Stress, depression and the mesolimbic dopamine system. Psychopharmacology 128:331–342

33. Müller MB, Keck ME (2001) Genetically engineered mice for studies of stress-related clinical conditions. J Psychiatr Res 36:53–76

34. Vollmayr B, Henn FA (2003) Stress models of depression. Clin Neurosci Res 3: 245–251

35. Willner P (2005) Chronic mild stress (CMS) revisited: Consistency and behavioural-neurobiological concordance in the effects of CMS. Neuropsychobiology 52:90–110

36. Dalla C, Antoniou K, Kokras N, et al (2008) Sex differences in the effects of two stress paradigms on dopaminergic neurotransmission. Physiol Behav 93:595–605

37. Mineur YS, Belzung C, Crusio WE (2006) Effects of unpredictable chronic mild stress on anxiety and depression-like behavior in mice. Behav Brain Res 175:43–50

38. Blazer DG 2nd, Hybels CF (2005) Origins of depression in later life. Psychol Med 35:1241–52

39. Bartolomucci A, Fuchs E, Koolhaas JM, Ohl F (2009) Acute and chronic defeat: stress protocols and behavioural testing. In: Gould TD (ed) Mood and Anxiety Related Phenotypes in Mice, Neuromethods. Humana Press, New York, p. 261–275

40. Malatynska E, Pinhasov A, Knapp RJ (2009) Reduction of submissive behavior model for antidepressant drug testing in mice. In: Gould TD (ed) Mood and Anxiety Related Phenotypes in Mice, Neuromethods. Humana Press, New York, p. 277–296

41. Brain P (1975) What does individual housing mean to a mouse? Life Sci 16:187–200

42. Koike H, Ibi D, Mizoguchi H, et al (2009) Behavioral abnormality and pharmacologic response in social isolation-reared mice. Behav Brain Res 202:114–21

43. Cervo L, Canetta A, Calcagno E, et al (2005) Genotype-dependent activity of tryptophan hydroxylase-2 determines the response to citalopram in a mouse model of depression. Neurobiol Dis 23:8165–8172

44. Elizalde N, Gil-Bea FJ, Ramirez MJ, et al (2008) Long-lasting behavioral effects and recognition memory deficit induced by chronic mild stress in mice: effect of antidepressant treatment. Psychopharmacology 199:1–14

45. Norcross M, Poonam M, Enoch AJ, et al (2008) Effects of adolescent fluoxetine treatment on fear-, anxiety- or stress-related behaviours in C57BL/6J or BALB/cJ mice. Psychopharmacology 200:413–424

46. Ripoll N, David DJP, Dailly E, et al (2003) Antidepressant-like effects in various mice strains in the tail suspension test. Behav Brain Res 143:193–200

47. Stone EA, Lin Y, Quartermain D (2008) Evaluation of the repeated open-space swim model of depression in the mouse. Pharmacol Biochem Behav 91:190–5

48. Bekhet AK, Zauszniewski JA, Nakhla WE (2008) Loneliness: a concept analysis. Nursing Forum 43:207–13

49. Heinrich LM, Gullone E (2006) The clinical significance of loneliness: a literature review. Clin Psychol Rev 26:695–718

50. Grippo AJ, Cushing BS, Carter S (2007) Depression-like behavior and stressor-induced neuroendocrine activation in female prairie voles exposed to chronic social isolation. Psychoneuroendocrinology 32:966–980

51. Cinelli P, Rettich A, Seifert B, et al (2007) Comparative analysis and physiological impact of different tissue biopsy methodologies used for the genotyping of laboratory mice. Lab Anim 41:174–184

52. Cryan JF, Valentino RJ, Lucki I (2005) Assessing substrates underlying the behavioral effects of antidepressants using the modified rat forced swimming test. Neurosci Biobehav Rev 29:547–569

53. Renard CE, Dailly E, David DJP, et al (2003) Monoamine metabolism changes following the mouse forced swimming test but not the tail suspension test. Fundam Clin Pharmacol 17:449–455

54. Roy M, David NK, Danao JV, et al (2005) Genetic inactivation of melanin-concentrating hormone receptor subtype 1 (MCHR1) in mice exerts anxiolytic-like behavioral effects. Neuropsychopharmacology 31:112–120

55. Costall B, Jones BJ, Kelly ME, et al (1989) Exploration of mice in a black and white test box: Validation as a model of anxiety. Pharmacol Biochem Behav 32:777–785

56. Bourin M, Hascoët M (2003) The mouse light/dark box test. Eur J Pharmacol 463:55–65

57. Hascoët M, Bourin M (2009) The mouse light-dark box test. In: Gould TD (ed) Mood and Anxiety Related Phenotypes in Mice, Neuromethods. Humana Press, New York, p. 197–223

58. Mashoodh R, Wright LD, Hébert K, Perrot-Sinal TS (2008) Investigation of sex differences in behavioural, endocrine, and neural measures following repeated psychological stressor exposure. Behav Brain Res 188: 368–379

59. Holick KA, Lee DC, Hen R, Dulawa SC (2008) Behavioral effects of chronic fluoxetine in BALB/cJ mice do not require adult hippocampal neurogenesis or the serotonin 1A receptor. Neuropsyhopharmacology 33:406–417

60. Beitia G, Garmendia L, Azpiroz A, et al (2005) Time-dependent behavioral, neurochemical, and immune consequences of repeated experiences of social defeat stress in male mice and the ameliorative effects of fluoxetine. Brain Behav Immun 19:530–539

61. Konkle ATM, Baker SL, Kentner AC, et al (2003) Evaluation of the effects of chronic mild stressors on hedonic and physiological responses: Sex and strain compared. Brain Res 992:227–238

62. Van Loo PLP, Van de Weerd HA, Van Zutphen LFM, Baumans V (2004) Preference for social contact versus environmental enrichment in male laboratory mice. Lab Anim 38:178–188

63. Olsson IAS, Westlund K (2007) More than numbers matter: The effect of social factors on

behaviour and welfare of laboratory rodents and non-human primates. Appl Anim Behav Sci 103:229–254

64. Brenes JC, Fornaguera J (2009) The effect of chronic fluoxetine on social isolation-induced changes on sucrose consumption, immobility behavior, and on serotonin and dopamine function in hippocampus and ventral striatum. Behav Brain Res 198:199–205

65. Kercmar J, Büdefeld, Grgurevic N, et al (2011) Adolescent social isolation changes social recognition in adult mice. Behav Brain Res 216:647–651

66. Abramov U, Raud S, Kõks J et al (2004) Targeted mutation of CCK_2 receptor gene antagonises behavioural changes induced by social isolation in female, but not in male mice. Behav Brain Res 155:1–11

67. Karolewicz B, Paul IA (2001) Group housing of mice increases immobility and antidepressant sensitivity in the forced swim test and tail suspension test. Eur J Pharmacol 415:197–201

68. Meziane H, Ouagazzal A-M, Aubert L, et al (2007) Estrous cycle effects on behavior of C57BL/6J and BALB/cByJ female mice: implications for phenotyping strategies. Genes Brain Behav 6:192–200

69. Cryan JF, Page ME, Lucki I (2005) Differential behavioral effects of the antidepressants reboxetine, fluoxetine, and moclobemide in a modified forced swim test following chronic treatment. Psychopharmacology 182: 335–344

70. Gorwood P (2004) Generalized anxiety disorder and major depressive disorder comorbidity: an example of genetic pleiotropy? Eur Psychiatry 19:27–33

71. Rogers RJ, Boullier E, Chatzimichalaki P, et al (2002) Contrasting phenotypes of C57BL/6JO1aHsd, 129S2/SvHsd and 129/SvEv mice in two exploration-based tests of anxiety-related behaviour. Physiol Behav 77:301–310

72. Milner LC, Crabbe JC (2007) Three murine anxiety models: results from multiple inbred strain comparisons. Genes Brain Behav 7:496–505

Chapter 16

Induction of Persistent Depressive-Like Behavior by Corticosterone

Shannon L. Gourley and Jane R. Taylor

Abstract

Multiple biological processes are implicated in the neurobiology of depression based primarily on the characterization of antidepressant efficacy in naïve rodents rather than on models that recapitulate the protracted feelings of anhedonia and helplessness that typify depression. In order to address this issue, the authors developed a protocol utilizing chronic oral exposure to the stress-associated adrenal hormone, corticosterone (CORT), in mice to induce anhedonic- and other depressive-like behaviors that are persistent for a significant duration of the animals' lifespan, yet reversible by chronic antidepressant treatment. As we will discuss in this chapter, prior chronic CORT exposure has multiple behavioral consequences relevant to stress-related mood disorders despite normalization of blood serum CORT levels after weaning. An additional example (for which data are also provided) is persistently disrupted locomotor activity, suggestive of neurovegetative malaise and early waking, both characteristics of depression in humans. In sum, prior chronic CORT exposure provides an alternative method to chronic mild stress models of depression that is easily replicable and persistent, thereby modeling the chronic depressive-like state in humans.

Key words: Corticosterone, Depression, Anhedonia, Antidepressant, Stress, Malaise

1. Background and Historical Significance

Depression is a debilitating and chronic illness characterized by feelings of helplessness, anhedonia, and loss of motivation to perform even everyday tasks. Although the relationship between stress and Major Depressive Disorder is incompletely understood, stress-induced cortisol release from the adrenal glands and subsequent activation of glucocorticoid receptors in brain likely plays a crucial role, as stressful life events are potent factors that trigger, induce, or exacerbate major depressive episodes (1, 2). While no stress-induced depression-like behavioral phenotype in rodents can fully

Todd D. Gould (ed.), *Mood and Anxiety Related Phenotypes in Mice: Characterization Using Behavioral Tests, Volume II*, Neuromethods, vol. 63, DOI 10.1007/978-1-61779-313-4_16, © Springer Science+Business Media, LLC 2011

recapitulate the human condition, a model that produces a *persistent, varied* behavioral sequela that is sensitive to chronic antidepressant treatment would be a major advance for understanding the neurobiology of depression (3, 4). Standard models of depression that rely on environmental stress manipulations, such as learned helplessness, chronic mild stress (CMS), and repeated social defeat, are hampered by protocol variability and reported difficulties in replication [(5–8); but see (4)], highlighting the need for a reliable, easily replicable depression model.

In this chapter, we will describe a depression model based on chronic oral exposure to the stress hormone corticosterone (CORT) in hemisuccinate form in rodents. Dissolution of CORT hemisuccinate in the drinking water has been used in the past to administer CORT to nursing pups via the lactating dam (9, 10), and to evaluate the reinforcing properties of the steroid (11) and its contribution to vulnerability to drugs of abuse (12). The CORT hemisuccinate exposure protocol described here is a chronic exposure method optimized for use in modeling the persistent depressive-like state in rodents, allowing for multiple behavioral tests in the same animals using an etiologically relevant model of depression that is easily replicable between and within laboratories. It also does not require CORT to be on-board during behavioral testing, which could confound responding in certain testing situations (e.g., anxiety-like responses to novelty soon after animals consumed CORT were reported in (13)).

The use of CORT hemisuccinate for oral administration provides an alternative to dissolving CORT in small amounts of EtOH (e.g., (14, 15)). While either of these methods allows for the identification of how CORT contributes to stress-related changes in neuronal morphology and behavior (12, 14, 15), as well as putative molecular and physiological correlates of depressive-like behavior (16–21), CORT dissolution in EtOH obviously requires treating control animals with EtOH, potentially confounding certain experiments, particularly those involving chronic exposure. The model as described here also has the added advantage that it carries face, predictive, and construct validity: For example, in line with CMS models and several cases of human depression, circulating CORT levels do not differ in prior CORT-exposed and control animals after exposure, suggesting stress-related insult persists long after HPA axis challenge (18). Prior CORT exposure induces anhedonic-like insensitivity to sucrose in both rats and mice, modeling a major symptom of depression, and hedonic sensitivity to sucrose can be restored by chronic, but not subchronic, antidepressant administration (18–20, 22). Moreover, locomotor activity during the dark cycle is disrupted for several weeks after CORT exposure, suggestive of disturbances in circadian cycling that include early waking, another common symptom of depression (Fig. 2).

Because our aims have largely been to evaluate the long-term consequences of chronic CORT exposure, the behavioral data presented here were collected from mice up to 5 weeks after the exposure period ended, although it is likely that some behavioral effects emerge *during* the exposure period; this topic is discussed in greater detail in Sect. 4. Optimized dosing protocols are provided, as are blood serum CORT levels during and after exposure in adult male C57bl/6 mice, and typical anhedonic-like insensitivity to sucrose after CORT is illustrated. Reversal by chronic, but not subchronic, antidepressant treatment is also shown, and protocols for use in replicating these experiments are provided in Sect. 6.

2. Equipment, Materials, and Setup

2.1. Animals

Our studies have historically used group-housed male C57bl/6J mice at least 8 weeks of age (e.g., Charles River Laboratories, Kingston, NY), although other mouse (and rat) strains would be expected to show similar sensitivities. CORT hemisuccinate is dissolved in regular tap water in a concentration of 25 or 100 µg/mL for mice and administered chronically with a weaning period described below (Sect. 3) at the end of the exposure period. Animals in our studies were maintained on a 12-h light cycle (0700 on) and experimentally naïve. Note that all protocols using live animals must first be reviewed and approved by an Institutional Animal Care and Use Committee (IACUC) or conform to governmental regulations regarding the care and use of laboratory animals.

2.2. Test Environment

Mice are administered CORT in the drinking water for a 20-day period as described below in Sect. 3. Water containing CORT replaces animals' regular drinking water, is available in the home cage for 24 h/day, and provides the only source of hydration. Otherwise, standard laboratory housing, light cycle regulation, and humidity control are maintained.

2.3. Setup

The following materials are necessary:

Corticosterone hemisuccinate (4-pregnen-11β 21-DIOL-3 20-DIONE 21-hemisuccinate) (Steraloids, Inc., Newport, RI)

10 N NaOH and HCl for pHing solutions (e.g., Sigma)

pH meter

Disposable Pasteur pipettes (for pHing solutions, e.g., Fisher Scientific)

Stir plate, magnetic stir bar

Standard laboratory rodent water bottles

Metric balance

1 or 2-L clean glass bottles with lids for solution storage

Tap water

3. Procedure

3.1. CORT Hemisuccinate Dissolution and Administration

1. Weigh out the appropriate amount of CORT hemisuccinate, depending on the concentration and desired volume, correcting for the added molecular weight of the salt by multiplying the desired concentration by a factor of 1.289. In other words, for a desired concentration of 25 μg/mL, add 32.22 mg CORT hemisuccinate to 1 L of tap water. For 100 μg/mL, dissolve 128.9 mg CORT hemisuccinate into 1 L of tap water, etc. The molecular weight of CORT is 346.46; the molecular weight of CORT hemisuccinate is 446.53. Concentrations in this chapter are expressed as free-base values.

2. Add CORT to 1 or 2 L of tap water in a clean glass storage container. Increase the pH to ~12 by adding 10 N NaOH using a sterile disposable Pasteur pipette. Stir at room temperature or at 4°C until dissolved (3–7 h). Stirring at 4°C is preferable because it will slow CORT decay, which is discussed in further detail in Sect. 5.1. If the experimenters choose not to adjust the pH, CORT takes significantly longer to dissolve (~30 h). In either case, no precipitate should be identifiable when CORT is dissolved.

3. Following dissolution, neutralize the pH to 7.0–7.4 with 10 N HCl. Add HCl slowly, as the pH will rise and fall dramatically because the solution is not buffered.

4. Once the CORT solution is prepared, present animals with the CORT *in place of* normal drinking water for 14 days so CORT provides the only source of hydration throughout the exposure period. To calculate the daily dose of CORT achieved, weigh bottles and mice daily and divide consumption values by the total body weight of the animals in the cage, then adjust for the CORT concentration in the drinking water. In the authors' experience, adult male C57bl/6 mice typically consume 4–6 mL/day. Over the course of the initial "full-dose" 2-week period, the average CORT doses achieved by mice consuming a 25 μg/mL concentration typically range from 5.7 to 6.9 mg/kg/day (see also Fig. 1a); these values are comparable to adult rats consuming 50 μg/mL, as rats consume less fluid per gram of body weight than mice.

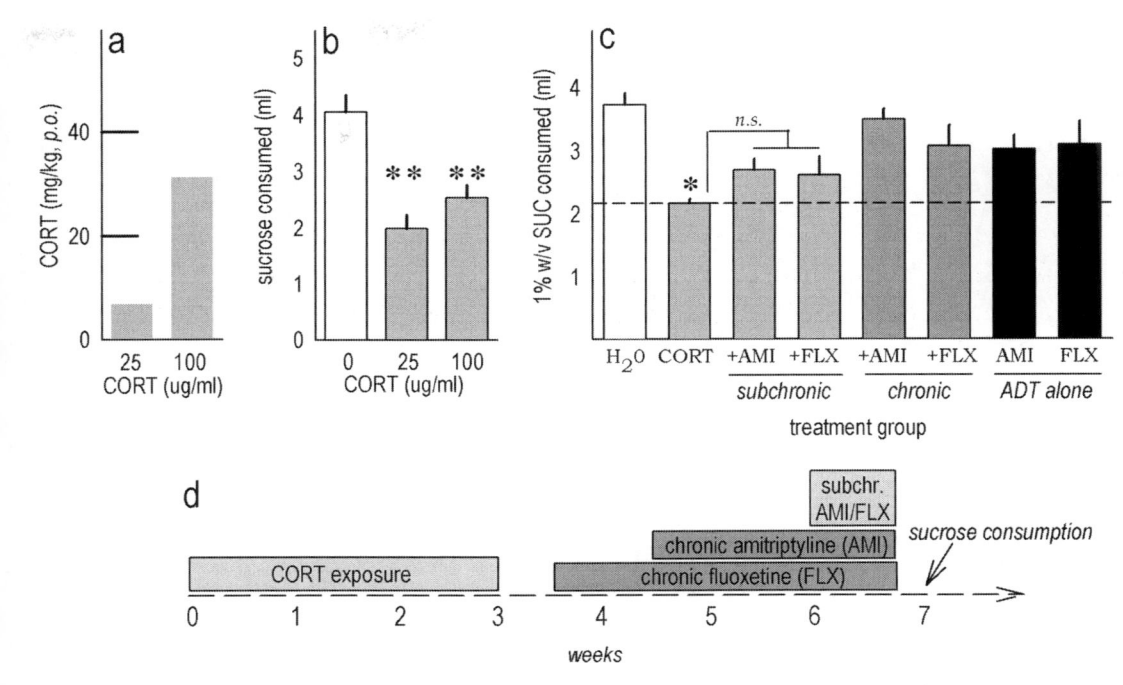

Fig. 1. Sucrose consumption is persistently diminished after CORT exposure. (a) Mice were exposed to either 25 or 100 µg/mL CORT in the drinking water, translating to daily doses of approximately 6 and 32 mg/kg/day, prior to weaning. (b) In sucrose consumption tests 2 weeks after weaning, previously CORT-exposed mice of the same body weight consumed significantly less 1% w/v sucrose solution that nonexposed animals ($F_{(2,22)} = 18.3$, $p < 0.001$; post hoc $ps < 0.001$). (c) In an effort to reverse the anhedonic-like response to sucrose with previously well-characterized antidepressants (ADT), mice in a separate cohort were exposed to 25 µg/mL CORT followed by chronic oral amitriptyline (AMI) or fluoxetine (FLX) administration. Subchronic administration regimens (4 days, p.o.) and antidepressant-only groups served as controls. One-factor ANOVA with post hoc tests against the CORT group revealed that both chronic amitriptyline and fluoxetine increased sucrose consumption above CORT-alone levels in mice of equivalent body weights ($F_{(7,77)} = 7.7$, $p < 0.001$, post hoc $ps < 0.05$), while subchronic antidepressants exerted no statistically significant effects, consistent with clinical observations that antidepressants require weeks to exert therapeutic benefits. Animals that had consumed antidepressants in the absence of prior CORT also consumed more sucrose than CORT-exposed mice, but did not significantly differ from control mice. (d) Experimental timeline indicating the time of antidepressant treatment relative to the completion of CORT exposure and initiation of habituation to sucrose. *Bars* represent group means (+SEM). *$p < 0.05$; **$p < 0.001$.

5. After 2 weeks of CORT administration, wean as follows: 3 days with 50% of the original CORT concentration, and 3 days with 25%, to allow for recovery of endogenous CORT secretion. Weaning is followed by a return to regular drinking water.

The entire exposure period is 20 days.

Consumption values and body weights can be monitored across the CORT consumption period and during washout using 2-factor (CORT×time bin) analysis of variance with repeated measures. Data are expected to be distributed normally. Significant interactions can be resolved with post hoc tests, with alpha levels corrected by Bonferoni correction for multiple comparisons as appropriate.

4. Experimental Variables

4.1. Selecting a Dose

Despite evidence for morphological and neurochemical alterations in the hippocampus after a high dose of oral CORT in rats (400 μg/mL in (14, 15)), the highest CORT doses used in our initial unpublished behavioral studies (300 and 400 μg/mL) were ineffective in producing a persistent depressive-like phenotype in mice, as measured by sucrose consumption, mobility in the forced swim test, and assays of primary motivation. We did not pursue these negative findings because the health of the mice also suffered with these CORT concentrations. By contrast, behavioral and biochemical findings after "low-dose" CORT (25–100 μg/mL) are robust and replicable without obvious threat to the health of the animal, particularly at the lower, 25 μg/mL, concentration, which is now used exclusively in our laboratories.

4.2. Use of a Washout Period Prior to Behavioral Testing

In our typical experiments, we employ a washout period of 2–5 weeks after the cessation of CORT exposure before behavioral testing because of our interest in the long-term effects of CORT exposure. In doing so, we have confirmed that many behavioral effects are quite persistent; these include anhedonic-like insensitivity to a palatable sucrose solution that persists up to 6 weeks after CORT. In our experience, however, a washout period is not necessary for the detection of depressive-like behaviors including anhedonic-like insensitivity to sucrose and diminished motivation to acquire food reward in an operant conditioning paradigm. One caveat to this observation, however, is that the precise nature of the behavioral phenotype may evolve as a function of washout time. One example of this phenomenon is disrupted dark cycle locomotor activity – in this case, mice appear to show malaise-like diminished activity immediately after washout, but this phenotype evolves into an early waking-like phenotype approximately 1 week after washout, and locomotor patterns appear to recover approximately 4 weeks after washout. See Fig. 2 for greater detail.

At a biochemical level, many immediate vs. long-term consequences of prior chronic CORT (25 μg/mL) exposure are similar, though not identical: For example, expression of trkB, the receptor for brain-derived neurotrophic factor (BDNF), is decreased in the dentate gyrus of the hippocampus immediately after CORT exposure, while phosphorylation of the receptor is attenuated 2 weeks and 1 month after exposure (17). Both decreased receptor expression and phosphorylation are suggestive of decreased activity of intracellular cascades associated with receptor binding. Nonetheless, we feel the use of a washout period is advantageous because it allows for the identification of molecular correlates of a specifically depressive-like state without the confound of anxiety-like behaviors

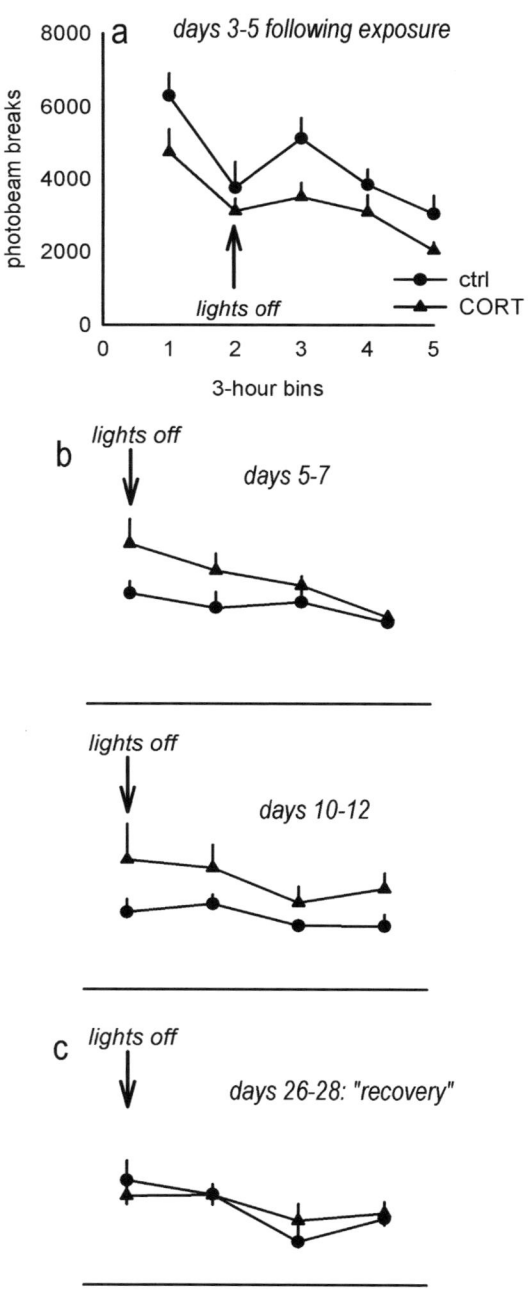

Fig. 2. Evolution of ambulatory activity patterns in prior chronic CORT-exposed mice. (a) Locomotor activity in C57bl/6 mice was continuously monitored in the weeks following CORT exposure (25 μg/mL). Photobeam breaks are represented in 3-h time bins with an *arrow* indicating lights out. Early after CORT, exposed mice broke fewer photobeams, indicative of diminished locomotor activity ($F_{(1,9)} = 7$, $p = 0.03$). (b) Five to 7 days after exposure, an "early waking"-like profile emerged, in that mice broke more photobeams during the dark cycle, particularly during the earliest time bins ($F_{(1,9)} = 5.9$, $p = 0.04$). (c) Approximately 1 month after exposure, locomotor cycles appeared to have recovered, as photobeam breaks were indistinguishable between groups ($F < 1$). *Symbols* represent group means (+SEM).

that transiently emerge during and immediately after CORT exposure (Gourley and Taylor, unpublished observations; (13)). This topic is addressed in greater detail in Sect. 6.

5. Troubleshooting

5.1. CORT Degradation

CORT solutions must be changed within 72 h of dissolution, as CORT will begin to degrade once in solution. Dissolved CORT should be stored at 4°C; stirring at 4°C will also slow degradation. In solid form, CORT hemisuccinate is stable and can be used until the manufacturer's expiration date. Solid CORT can be stored at room temperature or at 4°C.

5.2. My Mouse Isn't Drinking

In the authors' hands, mice readily consume CORT-infused water and do not show signs of avoidance. Strains other than C57bl/6 may, however, show different sensitivities. If animals appear to avoid the CORT solution, the authors suggest adding 2% w/v saccharin, as this approach has been previously used to make other agents palatable in mice ((23); Fig. 1c).

5.3. Health Concerns

We recommend the experimenters monitor body weights periodically during and after the CORT exposure period, first, to confirm weight loss is not excessive or injurious to the animal. (Animals are not expected to lose weight during the CORT exposure period, but note that exposure to these concentrations stalls normal body weight gain that will otherwise be observed in control animals.) Second, careful monitoring of consumption volumes and body weights will allow for conversion of CORT intake to CORT dose (i.e., in mg/kg terms). Although these values will vary slightly between experiments, these data are often requested during the peer review process.

Finally, the authors recommend being vigilant regarding conspecific aggression during the CORT exposure period. Although injury is relatively uncommon, CORT exposure does at times appear to increase aggressive encounters; thus, it is critical to separate animals at the first sign of heightened aggression to prevent injury.

5.4. Time Considerations

Because the CORT solution needs to be changed every 72 h, experimenters must be able to prepare CORT every 3 days. The exposure period is substantial – 20 days – and a wash-out period after CORT exposure is recommended, resulting in the addition of multiple weeks to the experiment duration. Pilot experiments will be required to determine an appropriate timeline for each experiment, depending on the experimental aims. One additional factor to consider is the possibility of testing animals before and after

CORT exposure, allowing for within-subjects comparisons, which may be statistically advantageous and potentially minimize the number of animals used for a given project. For example, the sucrose consumption protocol outlined below could easily be adapted to test animals both before and after CORT to investigate sucrose intake as a function of CORT exposure within subjects. This approach would add several days to the experimental timeline.

6. Anticipated Results

6.1. Materials and Methods for Experiments Testing the Persistent Effects of Prior CORT Exposure

Sucrose consumption (Fig. 1). Sucrose consumption was quantified in a model of anhedonia, a hallmark symptom of depression. The protocol we favor, which allows mice to remain group-housed for most of the experiment, first requires 48 h habituation to a 1% w/v sucrose (Sigma) solution, during which the sucrose solution replaces regular drinking water and provides the only source of hydration. Then, mice are habituated to modest water deprivation with 4, 14, and then 19 h of water deprivation over 3 consecutive days, followed by 1-h access to the sucrose solution on each day. Finally, each animal is allowed 1-h access to the sucrose solution in its home cage while its cage mates are temporarily group-housed in a clean cage in the colony room. Each successive individual animal has 1 h of access to the solution in such a way that the *average* water deprivation period per cage is 14 h. Habituation begins 1 week after CORT (25 and 100 μg/mL) exposure; as such, the intake measures reported here in Fig. 1b were collected 2 weeks after exposure.

To confirm the CORT-induced depressive-like phenotype is sensitive to reversal by chronic antidepressant treatment, we also report an experiment in which a separate group of mice was exposed to CORT (25 μg/mL), then provided amitriptyline hydrochloride (200 μg/mL in accordance with (23); Sigma) or fluoxetine hydrochloride (160 μg/mL; a generous donation from Dr. Ronald Duman, can be purchased from Sigma) in the drinking water for 2 or 3 weeks, respectively. Four-day treatment groups served as additional comparison groups, and all compounds were dissolved in 2% w/v saccharin and tap water. All mice consumed saccharin for 3 weeks to match the chronic fluoxetine-treated mice. See the timeline in Fig. 1 for information regarding when antidepressants were administered relative to weaning from CORT. Sucrose consumption testing proceeded as described above with the exception that the average deprivation period per cage was 19, rather than 14 h in this experiment.

Locomotor activity (Fig. 2). Ambulatory activity was monitored across the 24-h light/dark cycle in clean cages using the automated

Omnitech (Columbus, OH) Digiscan Micromonitor system equipped with 16 photocells. Data are represented as photobeams broken in 3-h time bins immediately preceding and during the dark cycle, i.e., when mice are expected to be most active. Mice were monitored for 4 weeks after CORT (25 µg/mL) exposure to assess effects on activity cycles.

Blood serum CORT (Fig. 3). Blood serum CORT was collected at multiple time points during and after CORT (25 µg/mL) exposure as indicated in Fig. 3 (x axes). Trunk blood was collected in chilled microcentrifuge tubes and centrifuged for 1 h at 5°C. Serum was extracted, and CORT content was assayed in duplicate using a CORT enzyme immunoassay kit in accordance with the manufacturer's instructions but excluding the extraction steps (Assay Designs, Inc., Ann Arbor, MI). Samples were diluted 10–13-fold with assay buffer to fall within the detection limits of the assay.

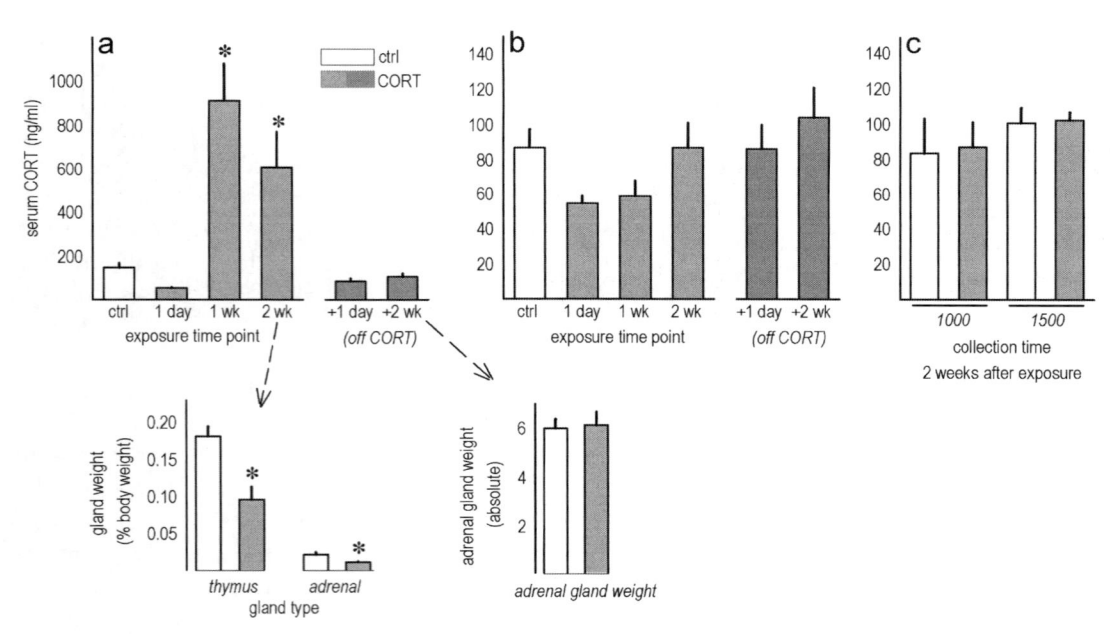

Fig. 3. Oral CORT (25 µg/mL) exposure impacts blood serum CORT at select time points. (a) Blood serum CORT was monitored at several time points during and after the oral CORT exposure period, revealing increased levels 1 and 2 weeks into the exposure period Kruskal-Wallis ANOVA-on-Ranks ($H_{(5,60)} = 38.4$, $p < 0.001$; post hoc $ps < 0.05$), but blood serum levels had normalized by 1 day after weaning. Blood was collected during the dark cycle (2 h after lights off). *Below: Arrows* point to plots representing diminished thymus and adrenal gland weights collected at the 2-week exposure time point (*left*), but normalized adrenal gland weights by the 2-week recovery time point (*right*). Thymus weights *after* CORT exposure have not to date been measured. (b) Blood serum measures were also collected during the light cycle (1 h after lights on). No statistically significant differences between the groups were identified either during or after the CORT exposure period. In other words, during the exposure period, blood serum CORT appears to be impacted only during the dark cycle, likely because mice are awake and consume CORT-infused water. (c) We also confirmed that blood serum CORT levels 2 weeks after exposure did not differ at two other time points during the light cycle – 3 h after lights on (1,000) and 8 h after lights on (1,500). Again, no differences were identified. *Bars* represent group means (+SEM). *$p < 0.05$.

The CORT exposure protocol described here induces a persistent depression-like state sensitive to antidepressant treatment in rodents. Specifically, prior chronic CORT exposure has behavioral consequences that include decreased sucrose consumption – because anhedonic-like insensitivity to sucrose is considered a gold-standard assay for depressive-like behavior in the rodent, a representative experiment is shown in Fig. 1b verifying diminished sensitivity after exposure to either the 25 or 100 μg/mL concentrations (reprinted with permission from (18)). We have also previously confirmed that basic gustatory discrimination processes – as measured by quinine:water discrimination – are intact, indicating the effects of prior CORT exposure on sucrose consumption are unlikely to be due to nonspecific effects on sensory discrimination skills (18). Prior CORT exposure also decreases instrumental responding for food reinforcement, particularly under progressive ratio conditions thought to reflect primary motivation for the food reinforcer (17–19); increases vulnerability to developing behavioral habits (24); and increases immobility in the forced swim and tail suspension tests of behavioral despair (17–19). These and other findings suggest this procedure results in a comprehensive depressive-like phenotype in mice.

Although a Major Depressive Disorder diagnosis in humans requires multiple symptoms, few established depression models or assays in rodents incorporate several behavioral features of depression, particularly symptom chronicity and sensitivity to chronic but not acute antidepressants. By contrast, CORT-exposed mice show a persistently decreased hedonic response to sucrose – which can be measured multiple times in the same animals – and sensitivity that can be restored only after chronic, and not subchronic, amitriptyline or fluoxetine (Fig. 1c), consistent with the typical time course of antidepressant efficacy in humans. By contrast, *non*-stress-related depression models produce conflicting results in tests of hedonic/motivational sensitivity to palatable outcomes (25–28), and the efficacy with which CMS decreases motivated behaviors has also been questioned (29).

We also show evidence of disrupted locomotor cycles in mice previously exposed to CORT (25 μg/mL) (Fig. 2). Although the C57bl/6 mouse strain is melatonin-deficient (30), this strain does show entrained locomotor cycling (e.g., (31)), and our data suggest that prior CORT exposure produces an initial transient decline in activity preceding and during the dark cycle when mice are most active (Fig. 2a); this phenotype may reflect a neurovegetative-like state. With further washout, CORT-exposed mice are more active than control mice during the early hours of the dark cycle, suggestive of "early waking," another common characteristic of depression in humans (Fig. 2b). Approximately 1 month after exposure, locomotor patterns appear to normalize, suggesting circadian cycles have recovered at this time point (Fig. 2c). One can thus imagine utilizing the "prior CORT exposure" method and targeting specific time points after exposure to study the biological bases of depressive-like neurovegetative behavior, early waking, and symptom recovery.

One concern regarding the use of CORT administration or CMS procedures to study the development and maintenance of depressive-like behaviors in rodents is the comorbid emergence of anxiety-like behaviors during and immediately after CORT or stressor exposure (e.g., (13)). Using the protocol described here, we have confirmed that exploration of an elevated plus maze, basal daytime endogenous CORT, and adrenal gland weights are unchanged 2 weeks after CORT (25 µg/mL) exposure ((22); Fig. 3, reprinted with permissions from (18)). This is despite diminished adrenal and thymus weights during the exposure period (inset Fig. 3a) and the persistence of depressive-like behavior after washout. In other words, current evidence indicates the use of a washout period allows for the emergence of a selectively depressive-like behavioral phenotype. Our evidence in rats indicates that this includes cognitive deficits also relevant to Post-traumatic Stress Disorder – specifically, prior chronic CORT retards fear extinction after contextual fear conditioning and disrupts the hypothalamic-pituitary-adrenal feedback response systems after foot shock but does not increase anxiety-related behavior in a neutral environment or blood serum CORT in unshocked rats (20). Although these experiments have not been replicated in mice, we would expect similar findings, based on comparable effects on other measures between these species (22).

From a biochemical standpoint, the emergence and persistence of depressive-like behaviors provides utility in identifying molecular and physiological correlates of depressive-like, rather than anxiety-like, behaviors. For example, the Neurotrophic Hypothesis of Depression and Antidepressant Efficacy (32), which posits that a decline in neurotrophin expression and/or activity, particularly in the hippocampus, contributes to the emergence and persistence of depression, was based in part on observations that active stressor exposure decreases activity of hippocampal CREB, a transcriptional regulator of BDNF, or BDNF itself (33–37). Recently, we showed that CORT (25 and 100 µg/mL) was sufficient to persistently decrease hippocampal CREB phosphorylation up to 1 month after CORT (18). These findings echoed a report by Laifenfeld et al. (38) showing decreased hippocampal CREB phosphorylation 4 months after chronic environmental stress and suggest that diminished CREB phosphorylation is related to persistent depressive-like behavior, as opposed to transient stress-related anxiety-like behaviors. These findings were novel because much of what is believed to be known about neurobiological factors of depression, including the potential importance of CREB-related signaling, is based upon immediate effects of stressor exposure and our understanding of rapid actions of standard antidepressants in otherwise naïve rodents (e.g., (36, 39, 40)). Whether identical neurobiological mechanisms are responsible for the persistence and reversal of a chronic depression-like phenotype is largely not known. Particularly given the increasing availability and utility of knockout and transgenic mice, we believe the CORT model described in this chapter is ideally suited to addressing these and other hypotheses regarding the basic molecular mechanisms of depression.

Acknowledgments

The authors thank Dr. Florence Wu and Ms. Jacqueline Barker for their assistance with the locomotor analyses reported here. These experiments were supported by grants from the National Institutes of Health.

References

1. Kendler KS, Karkowski LM, Prescott CA (1999) Causal relationship between stressful life events and the onset of major depression. Am J Psychiatry 156:837–841

2. de Kloet RE (2004) Hormones and the stressed brain. Ann N Y Acad Sci 1018:1–15

3. Willner P, Towell A, Sampson D, Sophokleous S, Muscat R (1987) Reduction of a sucrose preference by chronic unpredictable mild stress, and its restoration by a tricyclic antidepressant. Psychopharmacology 95:358–364

4. Willner P (2005) Chronic mild stress (CMS) revisited: Consistency and behavioural-neurobiological concordance in the effects of CMS. Neuropsychobiology 52:90–110

5. Matthews K, Forbes N, Reid IC (1995) Sucrose consumption as an hedonic measure following chronic unpredictable mild stress. Physiol Behav 57:241–248

6. Forbes NF, Stewart CA, Matthews K, Reid I (1996) Chronic mild stress and sucrose consumption: Validity as a model of depression. Physiol Behav 60:1481–1484

7. Vollmayr B, Henn FA (2001) Learned helplessness in the rat: Improvements in validity and reliability. Brain Res Protoc 8:1–7

8. Nestler EJ, Barrot M, DiLeone RJ, Eisch AJ, Gold SJ, Monteggia LM (2002) Neurobiology of depression. Neuron 34:13–25

9. Catalani A, Casolini P, Scaccianoce S, Patacchioli FR, Spinozzi P, Angelucci L (2000) Maternal corticosterone during lactation permanently affects brain corticosteroid receptors, stress response and behaviour in rat progeny. Neuroscience 100:319–325

10. Cinque C, Zuena AR, Casolini P, Ngomba RT, Melchiorri D, Maccari S, Nicoletti F, Gerevini D, Catalina A (2003) Reduced activity of hippocampal group 1 metabotropic glutamate receptors in learning-prone rats. Neuroscience 122:277–284

11. Deroche V, Piazza PV, Deminiere J-M, Le Moal M, Simon H (1993) Rats orally self-administer corticosterone. Brain Research 622:315–320

12. Piazza PV, Maccari S, Deminiere J-M, le Moal M, Mormede P, Simon H (1991) Corticosterone levels determine individual vulnerability to amphetamine self-administration. Proc Natl Acad Sci USA 88:2088–2092

13. Ardayfio P, Kim K-S (2006) Anxiogenic-like effects of chronic corticosterone in the light-dark emergence task in mice. Behav Neurosci 120:249–256

14. Magariños AM, Orchinik M, McEwen BS (1998) Morphological changes in hippocampal CA3 regions induced by non-invasive glucocorticoid administration: a paradox. Brain Res 809:314–318

15. Nacher J, Pham K, Gil-Fernandex V, McEwen BS (2004) Chronic restraint stress and chronic corticosterone treatment modulate differentially the expression of molecules related to structural plasticity in the adult rat piriform cortex. Neuroscience 126:503–509

16. Pekary AE, Sattin A, Blood J, Furst S (2008) TRH and TRH-like peptide expression in rat following episodic or continuous corticosterone. Psychoneuroendocrinology 33:1183–1197

17. Gourley SL, Wu FJ, Kiraly DD, Ploski JE, Kedves AT, Duman RS, Taylor JR (2008a) Regionally specific regulation of ERK MAP kinase in a model of antidepressant-sensitive chronic depression. Biol Psychiatry 63:353–359

18. Gourley SL, Kiraly DD, Howell JL, Olausson P, Taylor JR (2008b) Acute hippocampal BDNF restores motivational and forced swim performance after corticosterone. Biol Psychiatry 64:884–890

19. Gourley SL, Wu FJ, Taylor JR (2008c) Corticosterone regulates pERK1/2 in a chronic depression model. Ann N Y Acad Sci 1148:509–514

20. Gourley SL, Kedves AT, Olausson P, Taylor JR (2009) A history of corticosterone exposure regulates fear extinction and cortical NR2B, GluR2/3, and BDNF. Neuropsychopharmacology 34:707–716

21. David DJ, Samuels BA, Rainer Q, Wang JW, Marsteller D, Mendez I, Drew M, Craig DA, Guiard BP, Guilloux JP, Artymyshyn RP, Gardier AM, Gerald C, Antonijevic IA, Leonardo ED, Hen R (2009) Neurogenesis-dependent and -independent effects of fluoxetine in an animal model of anxiety/depression. Neuron 62:479–493

22. Gourley SL, Taylor JR (2009) Recapitulation and reversal of a persistent depression-likesyndrome in rodent. Current Protocols in Neuroscience. Chapter 9, Unit 9.32

23. Caldarone BJ, Karthigeyan K, Harrist A, Hunsberger JG, Witmack E, King SL, Jatlow P, Picciotto MR (2003) Sex differences in response to oral amitriptyline in three animal models of depression in C57BL/6J mice. Psychopharmacology 170:94–101

24. Gourley SL, Jacobs AM, Howell JL, Mo M, DiLeone RJ, Koleske AJ, Taylor JR (in revision) Action control is mediated by prefrontal BDNF and glucocorticoid receptor binding

25. Shalev U, Kafkafi N (2002) Repeated maternal separation does not alter sucrose-reinforced

26. Rüedi-Bettschen D, Pedersen E-M, Feldon J, Pryce CR (2005) Early deprivation under specific conditions leads to reduced interest in reward in adulthood in Wistar rats. Behav Brain Res 156:297–310

27. Barr AM, Phillips AG (1999) Withdrawal following repeated exposure to d-amphetamine decreases responding for a sucrose solution as measured by a progressive ratio schedule of reinforcement. Psychopharmacology 141:99–106

28. Russig H, Pezze M-A, Nanz-Bahr NI, Pryce CR, Feldon J, Murphy CA (2003) Amphetamine withdrawal does not produce a depressive-like state in rats as measured by three behavioral tests. Behav Pharmacol 14:1–18

29. Barr AM, Phillips AG (1998) Chronic mild stress has no effect on responding by rats for sucrose under a progressive ratio schedule. Physiol Behav 64:591–597

30. Goto M, Oshima I, Tomita T, Ebihara S (1989) Melatonin content of the pineal gland in different mouse strains. J Pineal Res 7:195–204

31. Olivero A, Malorni W (1979) Wheel running and sleep in two strains of mice: Plasticity and rigidity in the expression of circadian rhythmicity. Brain Res 163:121–133

32. Duman RS, Heninger GR, Nestler EJ (1997) A molecular and cellular theory of depression. Arch Gen Psychiatry 54:597–606

33. Smith MA, Makino S, Kvetnansky R, Post RM (1995) Stress and glucocorticoids affect the expression of brain-derived neurotrophic factor and neurotrophin-3 mRNAs in the hippocampus. J Neurosci 15:1767–1777

34. Schaaf MJM, Hoetelmans E, de Kloet R, Vreugdenhil E (1997) Corticosterone regulates expression of BDNF and trkB but not NT-3 and trkC mRNA in the rat hippocampus. J Neurosci Res 48:334–341

35. Schaaf MJM, de Jong J, de Kloet ER, Vreugdenhil E (1998) Down-regulation of BDNF mRNA and protein in the rat hippocampus by corticosterone. Brain Res 813:112–120

36. Nibuya M, Nestler EJ, Duman RS (1996) Chronic antidepressant administration increases the expression of cAMP response element binding protein (CREB) in rat hippocampus. J Neurosci 16:2365–2372

37. Prickaerts J, van den Hove DLA, Fieren FLP, Kia HK, Lenaerts I, Steckler T (2006) Chronic corticosterone manipulations in mice affect brain cell proliferation rates, but only partly affect BDNF protein levels. Neurosci Lett 396:12–16

38. Laifenfeld D, Kerry R, Grauer E, Klein E, Ben-Shacher D (2005) Antidepressant and prolonged stress in rats modulate CAM-L1, laminin, and pCREB, implicated in neuronal plasticity. Neurobiol Dis 20:432–441

39. Thome J, Sakai N, Shin K, Steffen C, Zhang YJ, Impey S, Storm D, Duman RS (2000) cAMP response element-mediated gene transcription is upregulated by chronic antidepressant treatment. J Neurosci 20:4030–4036

40. Tiraboschi E, Tardito D, Kasahara J, Moraschi S, Pruneri P, Gennarelli M, Racagni G, Popoli M (2004) Selective phosphorylation of nuclear CREB by fluoxetine is linked to activation of CaM kinase IV and MAP kinase cascades. Neuropsychopharmacology 29:1831–1840

Chapter 17

The Olfactory Bulbectomised Mouse

Michelle Roche

Abstract

Removal of the olfactory bulbs from the rodent induces neuronal reorganisation and the expression of behavioural, neurochemical, neuroendocrine and immune changes that resemble those observed in major depressive disorder. As such this model is widely used to examine the neurobiological substrates that may underlie the pathophysiology of depression and screen antidepressant agents. One of the most consistent changes observed in the olfactory bulbectomised (OB) mouse model is hyperactivity on exposure to a novel stressful environment. This behavioural response is attenuated selectively by chronic, but not acute, antidepressant treatment. This chapter provides a detailed protocol on the establishment of the OB mouse model and assessment of OB-related increase in locomotor activity in the open field test. Experimental variables which may impact on the results will be presented in addition to a short troubleshooting guide.

Key words: Olfactory bulbectomy, Depression, Antidepressants, Mouse, Mouse model of depression

1. Background and Historical Overview

Numerous attempts and approaches have been employed to develop animal models of major depressive disorder; however, due to the complex nature of the condition, no one model encompasses all of the hallmarks of the disorder or is without shortcomings. Models that most closely resemble the human condition attempt to fulfil criteria of construct, face and predictive validity. Evaluating models in terms of these criteria has revealed that the OB rodent possesses the highest degree of validity when compared against developmental and genetic models of predisposition to depression (1). Developed over 35 year ago by Cairncross and colleagues (2, 3), the OB rodent is a well-recognised and reproducible model of depression and antidepressant activity. Removal of the olfactory bulbs induces behavioural, neurotransmitter, neuroendocrine and immune changes resembling those reported in depressed patients (for reviews see (4, 5)). In addition to the strong

Todd D. Gould (ed.), *Mood and Anxiety Related Phenotypes in Mice: Characterization Using Behavioral Tests, Volume II,* Neuromethods, vol. 63, DOI 10.1007/978-1-61779-313-4_17, © Springer Science+Business Media, LLC 2011

face validity for this model, the OB model displays one of the best portfolios in terms of predictive validity of antidepressant activity following chronic administration. The behavioural and physiological alterations displayed cannot be explained due to the loss of smell (anosmia) alone (6, 7) and are believed to result from compensatory neuronal reorganisation in cortical-hippocampal-amygdaloid circuits following removal of the bulbs (5). The olfactory bulbs are integrally connected with the limbic system, particularly the amygdala, structural and functional alteration has been reported in both the depressed patient (8–10) and in the OB model (11–13). Neuronal degeneration and remodelling occurs in the cortex, amygdala, hippocampus, raphe nuclei and locus coeruleus following bulbectomy, effects reversed by chronic antidepressant treatment (14–18). The resultant neurochemical changes and dysinhibition of the amygdala have been proposed to underlie many of the behavioural changes in the model (12, 19–22). Furthermore, cognitive impairment in depression has been correlated with reduced hippocampal volume and cell density (23–25), alterations in which have also been observed in the OB model (11, 15). Hence, removal of the olfactory bulbs in the rodent adversely affects the homeostatic regulation of impulse traffic within cortical and limbic system structures, mimicking functional alterations that occur in the depressed state.

Correlating with symptoms observed in the human situation, behavioural changes reported in the OB rodent include anhedonia (6, 26–29), decreased social behaviour (30–32), deficits in learning and memory (7, 15, 21, 33–35), reduced sexual behaviour (36, 37) and impaired reactivity to stressful environments (38–42). Although predominantly assessed in the rat, an increasing number of studies have examined behavioural changes following bulbectomy in mice. It should be cautioned that data obtained in rats should not be over extrapolated to mice (43); however, to date similar behavioural changes have been demonstrated in both species following bulbectomy. Table 1 presents the current behavioural changes reported in OB mice. An overview of the more commonly used tests and paradigms employed to evaluate behavioural changes in the OB mouse model is presented forthwith. Particular emphasis is placed on assessment of activity in a novel open field area, the most widely evaluated behaviour in the model.

Alterations in learning and memory following bulbectomy have been evaluated in several behavioural paradigms (passive avoidance, novel object recognition, T-maze, Y-maze and morris water maze) (Table 1). Overall, OB mice exhibit cognitive impairments and deficits in spatial memory (34, 35, 44–48). Both acute and chronic antidepressant administration reverse the behavioural deficit in passive avoidance in rats (4, 5); however, considerably less studies have examined the response to antidepressants in OB mice. Chronic administration of the tricyclic antidepressants amitriptyline

Table 1
Behavioural changes observed in the OB mouse model

Behavioural test	Strain	OB-induced effect	References
Open field	C57BL/6	Increase in locomotor activity	(40–42, 48, 49, 52–54)
	DBA	Increase in locomotor activity	(49)
	Swiss	No effect	(47)
Saccharine/sucrose preference	C57BL/6	Reduction in saccharine/sucrose preference (anhedonia)	(28, 29, 64)
Elevated plus maze	ddY	No effect	(15)
Forced swim test	Swiss	Decreased duration of immobility	(47)
Hole-board test	ICR	Increased frequency of head dips	(57, 58)
		No effect on locomotor activity	
Interspecies aggression	Swiss	Reduced aggressive behaviour	(32, 47)
Passive avoidance	C57BL/6	Impaired passive avoidance	(42, 49)
	ddY	Impaired passive avoidance	(15, 35, 46)
	DBA	No effect	(49)
Active avoidance	C57BL/6	Impaired active avoidance	(49)
	DBA	No effect	(49)
Novel object test	C57BL/6	Increased time exploring novel object	(41)
	ddY	Fail to discriminate between new and old object	(45, 50)
T-maze	C57BL/6	No preference for novel arm	(41)
Y-maze	ddY	Reduced spontaneous alternation	(45, 50)
Morris water maze	C57BL/6	Impaired spatial memory	(51)
	Swiss	Impaired spatial memory	(47)
	NMRI	Impaired spatial memory	(34, 65)
Maternal behaviour	ddY	Impaired maternal behaviour	(64, 66)

and imipramine, and the atypical antidepressant trazodone, attenuates OB-induced impairments in passive avoidance in C57BL/6 mice (49); however, a subsequent study failed to demonstrate an effect of amitriptyline (42) on this behavioural response. Similarly, reports of both the effectiveness (42) and ineffectiveness (48) of chronic treatment with the selective serotonin re-uptake inhibitor citalopram, in reversing the passive avoidance deficits in OB mice, have been presented. However, passive avoidance deficits in DBA mice following bulbectomy were not attenuated by antidepressant treatment (49), highlighting that background strain may influence behavioural responding of OB mice. Impaired learning and memory in the model are accompanied by a loss of cholinergic neurons (15, 45, 46, 50) and increased brain β-amyloid (34, 51); therefore, the OB mouse also provides a useful experimental model for examining dementia associated with Alzheimer's disease. Acute and chronic treatment with cholinesterase inhibitors, muscarinic agonists (15) and potential cognitive enhancers (34, 45, 46, 50) ameliorates the memory impairments exhibited by OB mice.

On exposure to a novel, open field environment, OB mice exhibit behavioural hyperactivity (40–42, 48, 49, 52–54). This characteristic behavioural hallmark of the model has been proposed to exemplify psychomotor agitated depression (36). OB-induced hyperactivity is dependent on a number of factors, including the shape, size and aversiveness of the open field area (see experimental variables) and is associated with stress-induced behaviours such as thigmotaxis (time spent and activity along the walls of the arena) and defecation. A detailed protocol for evaluating OB-related hyperactivity in the open field test has been provided below. Based on behavioural observations in OB rats, the nature of this hyperactivity has been attributed to an inability to mount appropriate stress or defensive responses (38, 39, 55). The inability to habituate and inhibit behaviour in novel stressful environments or situations has also demonstrated in other acute stress regimes such as the elevated plus maze, T-maze, passive avoidance and Vogel's conflict test (38, 41, 56). Similarly, increases in emotionality and impulsivity have been demonstrated in OB mice exposed to the hole-board test (57, 58). OB-induced hyperactivity in the open field is the only behaviour currently known to selectively respond to chronic, but not acute, treatment with antidepressants. The necessity for repeated administration to correct this behavioural aberration mimics the clinical time-course of antidepressant action and distinguishes the OB model from many other simulations of depression and tests of antidepressant action. As such, attenuation of this behavioural response in OB rats is widely used as a means to screen potential new antidepressant agents. In comparison, there has not been extensive validation of pharmacologically diverse types of antidepressant drugs in mice using the OB model. Chronic treatment of OB mice with the antidepressants amitriptyline, imipramine, trazodone and citalopram reverses the hyperactivity observed on exposure to the open test (40, 42, 48, 49). Extending the characterisation of the model in mice has enabled genetically modified animals to be examined and therefore the study of neural mechanisms that subserve emotional responses. OB mice with targeted deletion of the tac 1 gene, which encodes for the neuropeptide substance P, do not exhibit behavioural hyperactivity in the open field test (54) or decreased preference for saccharine solution (measure of anhedonia) (28). Thus, substance P neurotransmission mediates, at least in part, the behavioural responses observed in the OB mouse model.

2. Experimental Procedures

2.1. Animals and Housing

Mice may be housed in groups or individually, usually in plastic bottom cages containing wood shavings as bedding. The animals should be maintained in standard housing conditions of constant temperature ($20 \pm 2°C$) and standard lighting (e.g. 12:12 h light–dark). Food and water should be available ad libitum. At least 7 days should be allowed for animals to acclimatise following any changes to environmental conditions prior to behavioural testing or surgery. The most widely assessed behavioural parameter in the OB model is hyperactivity on exposure to a novel environment. As such it is advisable that animals are tested in the open field test or equivalent prior to surgery to ensure comparable baseline activity between mice assigned to sham and OB groups.

2.2. OB and Sham Surgery

All procedures should be carried out under appropriate ethical approval and comply with national and international regulations regarding the use of animals for research purposes.

2.2.1. Equipment, Materials and Setup

1. Appropriate surgical facilities and equipment are required. The surgical area should be cleaned and disinfected (e.g. 1–3% Milton disinfectant (Procter and Gamble, Ireland) or 10–70% alcohol). Surgical instruments (scalpel and retractors (Fine Science Tools, Germany)) should be sterilised prior to surgery. Setup should be in close proximity to a sink area equipped with an aspiration pump (also known as water aspirator or venturi pump, e.g. Vacu/Trol vacuum water aspirator, Spectrum Europe, the Netherlands), which efficiently generates a vacuum when connected to a standard laboratory tap. This is the most commonly used method to remove the olfactory bulbs from the olfactory cavity. This process may be facilitated by attachment of a blunt 16G hypodermic needle or fine glass pipette to the vacuum hose on the aspirator. Additional equipment required include mouse stereotactic frame (Harvard Apparatus, UK), homoeothermic blanket with temperature controller or equivalent (CMA, Sweden or Harvard Apparatus, UK), high speed micro drill with 2.1 mm burr (Fine Science Tools, Germany) and hair clippers. Surgical area may be set up similar to that depicted in Fig. 1.

2. The most common means of anaesthetising animals is the use of a combination of Xylazin (80 mg/kg; Bayer, Germany) and

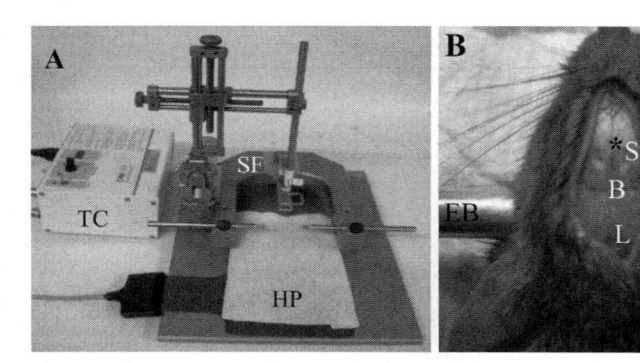

Fig. 1. (a) Example of a surgical setup area. *SF* stereotactic frame; *HP* heating pad; *TC* temperature controller. (b) Representative landmarks on the mouse skull. *Asterisk* represents proposed site for burr hole required to remove the olfactory bulbs. *B* bregma; *L* lambda; *SS* sagittal suture; *EB* ear bars.

Ketamine (100 mg/kg; Aventis Pharma, Germany), although other forms of injectable anaesthesia such as pentobarbital sodium (50 mg/kg; Sigma, Ireland) have also been used. OB surgery may be performed under inhalation anaesthesia such as Isoflurane (Isoflo (1–3% in O_2; 0.5 L/min), Abbott, Ireland) which allows for excellent control of anaesthesia and rapid recovery following surgery.

3. Additional reagents required: Betadine or Povidone (iodine) solution, sterile alcohol swabs, local anaesthetic (e.g. provocain), cotton gauze/swabs, haemostatic sponges (cut to 2 mm pieces; Dental supply companies) and/or bone wax (Fine Science Tools, Germany), antibiotic powder (e.g. Neomycin, Sigma, Ireland), sutures (3-0 Vicryl) or equivalent (e.g. Histoacryl, Aesculap Germany), sterile injectable saline (0.89% NaCl), analgesia (e.g. the nonsteroidal anti-inflammatory agent, carprofen (1.25 mg/25 µL s.c., Rimadyl, Pfizer, UK)), and hypodermic needles (25–30G).

2.2.2. Procedure

1. Remove the mouse from the cage and record body weight.

2. Anaesthetise the mouse and monitor depth of anaesthesia by toe and tail pinch.

3. Once unconscious, secure the head in the stereotactic frame using the ear bars and nose clamp. Take care not the damage the skull when mounting on the ear bars. Ensure that the head is horizontal to the stereotactic frame.

4. Place temperature probe into animal's rectum and note body temperature. The heating blanket/pad and temperature controller will act to maintain body temperature (~37°C) during surgery. Extra fabric can be placed over animal to prevent excess heat loss.

5. Apply saline to animal's eyes to prevent corneal damage.

6. Shave the superior surface of the head. Clean the area to be incised using iodine solution followed by an alcohol swab. Apply local anaesthetic to the area.

7. Analgesia may be administered prior to surgery to manage post-operative pain.

8. Make a midline longitudinal incision (2–3 cm) with a sterile scalpel blade in the scalp over the junction of the frontal and nasal bone of the skull (midway between the eyes). Retract the skin, scrape back the periosteum and dry the skull using a cotton swab.

9. Identify the sagittal suture and bregma (Fig. 1). Again ensure that the skull is parallel to the horizontal plane of the stereotactic apparatus by adjusting the nose clamp.

10. Mark the coordinates of the entry point on the surface of the skull. Coordinates should be chosen prior to surgery. For mice one hole (~2 mm diameter), 4 mm rostral to bregma (or 1 mm rostral to the sagittal suture) on the midline, is sufficient. Alternatively, mark two 1 mm holes directly above each bulb, 1 mm rostral to the sagittal suture and 1 mm lateral to midline (ML±1 mm). Drill the burr hole(s) at the marked entry point(s) and pierce the dura with a fine hypodermic needle.

11. For sham surgery, stop any bleeding that may have resulted, dry the skull with a cotton swab, fill burr hole with bone wax, apply antibiotic power to the area and close the skin with interrupted sutures or equivalent (proceed to point 16). Should damage to the olfactory bulbs during the procedure be suspected, it is advisable that the bulbs are removed.

12. For bulbectomy surgery, run water through the aspirator pump to create a vacuum.

13. In order to remove the olfactory bulbs, pierce the bulbs with a sterile hypodermic needle and carefully aspirate the bulbs from the individual cavities with the aid of the blunt needle or glass pipette attached to the vacuum hose. Ensure that the needle is inserted vertically, not at an angle, to avoid damage to the frontal cortex or olfactory tubercles.

14. Fill the burr hole with haemostatic sponge to stop bleeding. Bone wax may be used to seal over the burr hole(s). Ensure all bleeding has stopped prior to wound closure.

15. Apply antibiotic power to the wound and close the skin using interrupted sutures or equivalent.

16. Remove the mouse from the frame and administer 1 mL of sterile saline (i.p.) to maintain hydration.

17. Transfer the mouse to a clean warm cage until recovery from anaesthesia.

18. Re-house in new cage with fresh bedding and free access to food and water. As OB animals are rendered anosmic, it is advisable that food be placed in the base of the animals' home cage for the first 24 h in order to encourage food consumption.

19. Post-operative care: Body weight should be carefully monitored throughout the experiment as OB mice exhibit a more profound reduction in body weight following surgery when compared to sham-operated mice. Animals that lose more than 20% of their body weight in the first 24 h should be excluded. Animals should be checked for post-operative wound infection and handled regularly post surgery in order to reduce aggression that is known to develop following surgery.

2.2.3. Anticipated Results

Following OB surgery, the olfactory bulbs should be completely removed and the animal should be rendered anosmic. The most common means of verifying olfactory bulb removal is gross inspection following completion of the study. Careful examination of olfactory cavity for incomplete removal of one or both bulbs or damage to the frontal cortex following OB surgery necessitates removal of the animal from analysis. Verification that neuronal damage to the frontal cortex was not induced following removal of the olfactory bulbs may also be confirmed using Nissl (47) or thionin (32) staining and histological examination. A criterion of removal of at least two-thirds of the olfactory bulbs and lesioning of part of the olfactory nuclei has also been used as indications of successful bulbectomy. However, olfactory function can remain with relatively small bulb remnants (36) and therefore should an investigator employ such criteria it is important that anosmia is confirmed. Animals should be eliminated if damage to the olfactory bulb(s) is observed following sham surgery.

Although not routinely examined, anosmia may be confirmed following removal of the olfactory bulbs by replacing water with a bitter, scented (0.1% amyl acetate + 0.4% quinine) (32) or lithium chloride (0.12 M) (29) solution. Sham mice learn to associate the smell with the bitter unpleasant taste and avoid the solution, drinking only when plain water is presented. By contrast, if mice have been rendered anosmic, they lick both solutions in order to differentiate between water and the unpleasant solution. Anosmia may also be determined by examining the latency to approach a novel odour (e.g. vanilla extract) or find hidden food in the home cage.

2.3. OB-Induced Hyperactivity

Enhanced locomotor activity is the most commonly examined behavioural change observed in the OB mouse model, generally assessed using the open field test, although also observed in other behavioural paradigms including locomotor activity monitors (actometers), the T-maze and upon exposure to a novel home cage (41, 42) (Table 1). A comprehensive protocol relating to the open field test has been provided by Gould and colleagues in Volume I of the Neuromethods series on Mood and Anxiety Related Phenotypes in Mice (59). As such the protocol presented below pays particular attention to amendments in the procedure required to detect OB-induced increases in locomotor activity in this test. Activity in the open field is most commonly assessed 14 days following surgery.

2.3.1. Equipment, Materials and Setup

1. The room used to conduct the open field test should be isolated from sound; however, if this is not possible a while noise generator (San Diego Instruments, US or equivalent) can be used. Ensure that noise generator is turned on prior to introducing animal into the test room.

2. Open field arenas have been constructed of various materials (wood, metal, plastic), in numerous shapes (circular, square, rectangle) and sizes. Our studies have found that OB-induced hyperactivity is reliably detected when assessed in a large aversive arena (e.g. circular arena (75–90 cm diameter) with aluminium walls (60 cm high) and white floor) (see experimental variables) (Fig. 2). Large arenas also allow for behaviours such as thigmotaxis (distance moved or time in outer perimeter) and anxiety-related behaviours (time in the centre zone) to be assessed. Although no consensus on whether arenas should be cleaned between test subjects has been reached, the majority of studies employ a regime of cleaning with mild detergent or disinfectant (e.g. 1% Milton disinfectant solution or 10–50% alcohol).

3. Various lighting conditions have been reported in the open field test; however, OB-induced hyperactivity is most consistently observed when testing occurs in brightly lit open field arena (Lux > 100) (see experimental variables). In general, the open field should be evenly illuminated, the light intensity recorded using a lux metre and reported in all publications.

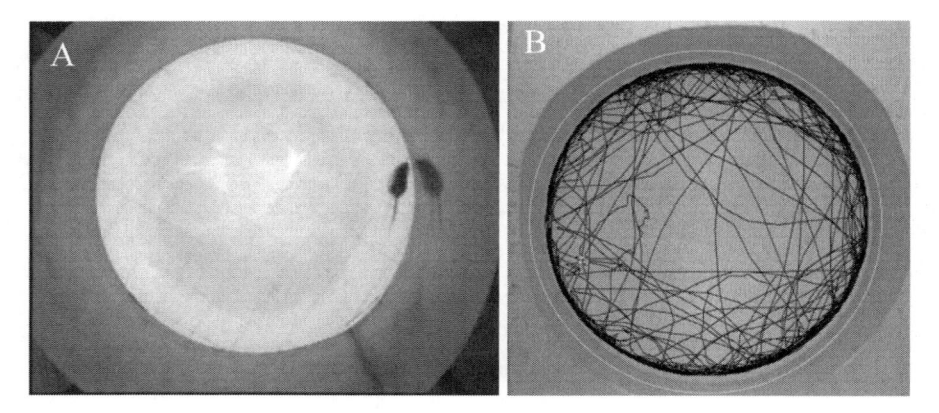

Fig. 2. Representative images of (**a**) OB mouse in the open field test and (**b**) automated video tracking of the open field test using EthoVision software (Noldus, the Netherlands).

4. Behaviour in the open field test is most commonly assessed over a period of 3–5 min although longer periods have also been examined. Manual rating of open field activity generally involves determining the number of line crosses on the base of the arena (e.g. 10×10 cm squares painted on base of arena) and frequency of rearing. An increasing number of studies now employ either photocell automated open field (Opto M3, Columbus, US) or video tracking software (EthoVision, Noldus, the Netherlands; Fig. 2b) to assess locomotor activity. Irrespective of the means of assessing behaviour in the open field, it is advisable that the open field test be recorded by video camera onto DVD to enable reassessment of behaviour at a later date if required.

5. Drug treatment generally commences following the establishment and confirmation of OB-induced hyperactivity (14 days post surgery) and continues for a minimum of 14 days. Chronic administration of a number of antidepressants (amitriptyline, imipramine and citalopram) has been demonstrated to attenuate OB-related increase in locomotor activity in the open field (40, 42, 48, 49). When evaluating antidepressant-like activity of novel compounds in the OB model, it is advisable that a reference antidepressant be included in the experimental design.

2.3.2. Procedure

Prior to behavioural testing, animals within an experiment should be handled in the same manner. Due to the potential for the development of OB-induced aggression (particularly in individually housed animals), our experimental protocol calls for daily handling of mice post surgery. The animals' home cage should not be changed within the last 24 h prior to open field testing. Overall these procedures act to minimise background stress that may impact on behavioural assessments.

1. As the open field test is used to determine not only locomotor activity but also behavioural responding of OB mice in a novel, stressful environment, animals need not be acclimatised to the test room, particularly if located close to the holding room and there is minimal change in lighting, temperature and humidity difference between the rooms. In instances where animals must be transported and habituated to the test room, mice should acclimatised to the room but not to the high illumination associated with the open field test, for at least an hour prior to testing.

2. Clean open field arena thoroughly with mild detergent or disinfectant prior to testing.

3. Turn on video camera, photocell and/or video tracking software. In order to easily rescore behaviour of individual animals from DVD at a later time point, it is recommended that an

identity number be placed beside the arena (out of sight of the animal) that can be seen by the video camera.

4. Place the mouse in the open field. Although the position which investigators place the mouse in the open field varies between laboratories (centre vs. perimeter), positioning along the perimeter allows for the latency to enter the centre zone to be assessed, a useful measure of anxiety-related behaviour.

5. Set the timer.

6. During the test the investigator should either leave the room or position themselves as far away from the arena and remaining as still as possible during the trial.

7. At the end of the trial the mouse should be removed from the arena and placed into a new cage.

8. The number of faecal boli should be counted (as a measure of anxiety). Other behaviours may be assessed manually from video or automatically recorded using photobeam or tracking software.

9. The arena should be thoroughly cleaned and dried before the next mouse is introduced.

Data collected: Behaviours recorded manually are primarily the number of line crosses, frequency of rearing and number of faecal boli. Although, it is possible to determine the time spent, latency to enter and number of line crosses in the centre zone manually, these events are rarely reported in studies employing this method of evaluation. In comparison, large amounts of data can be easily and quickly generated with the use of automated systems, particular video tracking software (Fig. 2). This technology also eliminates much of the subjectivity associated with manual recording. Behavioural output from these systems generally includes distance moved (in the entire arena and in different zones), duration of time spent in and latency to enter the centre zone and frequency of rearing. In addition, data can be assessed over the entire duration of the trial or in individual time bins.

2.3.3. Data Analysis and Anticipated Results

Examining the effects of bulbectomy alone on behavioural responding (e.g. distance moved) is generally assessed using t-test (sham vs. OB) or repeated measures ANOVA (sham vs. OB effect over time). Two-way ANOVA are routinely used to determine effects of drugs/treatments in the model (with the factors of surgery and treatment).

Prior to surgery, no difference should be noted between groups assigned to sham or OB for any of the behavioural parameters examined, e.g. distance moved or frequency of rearing. However following bulbectomy, mice exhibit enhanced locomotor activity (number of line crosses or distance moved) in the open field

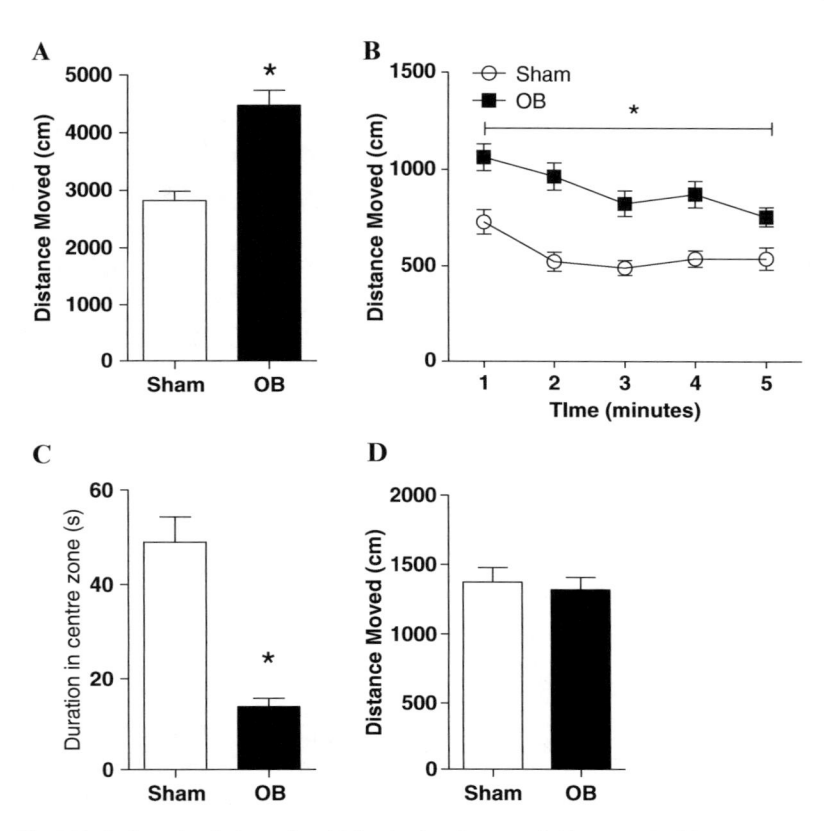

Fig. 3. Typical results that may be obtained using the open field test. (**a**, **b**) OB mice demonstrate an increase in distance moved in the open field over 5 min when compared to sham controls. (**c**) OB mice spend less time during the trial in the centre of the open field when compared to sham controls. (**d**) No significant difference in distance moved between sham and OB mice when tested in a small, dimly lit arena. *$P < 0.05$ sham vs. OB.

(Fig. 3a, b). Although this behavioural change is most commonly assessed 14 days post surgery, development of OB-induced hyperactivity may be observed at an earlier time points (e.g. 7 days) post surgery. The majority of studies assess open field activity over 3–5-min period during which time OB mice do not habituate to the test arena (Fig. 3b). The OB-related increase in distance moved is primarily associated with increased thigmotactic but not anxiolytic behaviour, as OB mice spend less time in the anxiety-related centre of the arena (Fig. 3c).

The antidepressant-like effect of compounds may be determined by their ability to attenuate OB-related increases in locomotor activity in the open field, an effect observed by traditional antidepressants (e.g. tricyclics and SSRIs) following chronic, but not acute, administration.

3. Experimental Variables

A number of experimental variables have been identified which may affect the successful outcome of both the surgery and subsequent behavioural assessment in this particular model and are discussed below.

3.1. Surgical Procedure

As with any surgical procedure, the experience of the surgeon, aseptic technique, the ability to maintain body temperature during surgery and appropriate post-operative care will affect the success of the surgery and the recovery of the animal. OB surgery is a relatively simple procedure and often successful even for those with minimal prior surgical experience. In order to allow for full recovery prior to the nocturnal phase, during which time mice are naturally active and consume most food, it is recommended that surgery is performed during the early light phase. Although not examined directly, it is possible that mice that have not recovered completely from surgery will consume less food during this period and as such loose greater body weight. This may be particularly relevant for the OB mice as they will be rendered anosmic and possibly hypophagic as a consequence, resulting in greater body weight loss post surgery when compared to sham-operated controls. In accordance, anaesthetics that result in rapid recovery from surgery (such an inhalation anaesthetics) should be considered for this procedure. If choosing to use an injectable form of anaesthesia, nonbarbiturates are recommended due to shorter recovery period and reduced complications associated with prolonged anaesthesia. Consideration should also be paid to the appropriate coordinates for surgery and the aspirator vacuum pressure, inaccuracy of which may result in either an inability to remove the olfactory bulbs completely or damage to the frontal cortex.

3.2. Genetic Background

Bilateral olfactory bulbectomy has been performed in several mouse strains although OB-induced hyperactivity has not been assessed in all studies. The most commonly used mouse strain is that of the inbred C57BL/6, where it has been consistently demonstrated that removal of the olfactory bulbs induces hyperactivity in the open field test under several different experimental conditions, an effect reversed by repeated antidepressant treatment. (40, 41, 48, 49, 52–54). Deficits in passive and active avoidance (42, 49) and sucrose and saccharine preference (28, 29) have also been demonstrated following bulbectomy in this strain (Table 1). By contrast, although DBA mice exhibit hyperactivity in the open field test following bulbectomy that is attenuated by chronic antidepressant treatment, a deficit in passive or active avoidance was not observed in this strain of mouse (49). Removal of the olfactory bulbs from Swiss and ICR mice did not result in increased locomotor activity in the open field (47) or hole-board test (57) respectively, although factors other than genetic background may participate in the inability to detect OB-induced hyperactivity in these strains. The effects of bulbectomy on memory, emotionality and circadian rhythm, but not locomotor activity, have been assessed in ddY, CF-1 and NMRI mice (Table 1). Careful consideration should be paid to the background strain of the mouse, particularly when working with genetically modified mice, when examining OB-induced behavioural alterations and the ability of pharmacological treatments to attenuate such alterations.

3.3. Gender

Male and female rats exhibit differential effects related to sucrose preference but not open field activity following bulbectomy (27, 39). However, no such study has been conducted in mice. It has been demonstrated that both male and female mice exhibit anhedonia following bulbectomy (28, 29); however, it remains to be determined if the magnitude of this response is comparable between the genders. OB-induced behavioural changes have predominantly been conducted in male mice and further studies are required in order to determine if gender differences exist in the OB mouse model. If examining OB-induced effects in both male and female mice, it is important that gender be recognised as a separate factor during analysis.

3.4. Age

The majority of studies using the OB model have been conducted in mice between 2 and 4 months of age (20–30 g body weight), although mice ranging from 18 days to 30 weeks old have also been used. The effect of age on behavioural responses in OB mice is largely unknown; however, age had been demonstrated to induce significant changes on OB behaviour in the rat. Older (19 month) OB rats exhibit greater locomotor activity when compared to young (9 week) rats and although young OB rats demonstrate a loss of passive avoidance and startle reflex when compared to sham-operated animals, this was not observed in older rats (60). Based on these observations, it has been proposed that examining the effects of bulbectomy in aged rodents may provide a model of geriatric depression. When comparing across and between studies, it is important to be aware that the age of the mouse may impact on the behavioural response observed in the model.

3.5. Housing Conditions

Saitoh and colleagues demonstrated that singly housed OB mice demonstrate enhanced emotional behaviour as assessed using the hole-board test but not locomotor activity when compared to group-housed counterparts (57). In addition, our studies indicate that housing conditions do not alter the development of OB-induced hyperactivity in the open field test, although the magnitude of the response was reduced in group-housed (10 per cage; 5 sham + 5 OB) when compared to singly housed mice. As such housing conditions (single vs. group) may alter OB-induced behavioural responses. Mice are predominantly housed in groups per cage as they are social animals and exhibit stress-like behaviour when individually housed. There is a lack of detail from several studies on whether all sham, all OB or equal amounts of both are housed together following surgery. Group housing configurations may have a profound impact on behavioural assessments in the model particularly as OB mice display impaired social and aggressive behaviour (32, 47). Several studies report housing mice individually post surgery (40, 41, 52). In an attempt to reduce stress associated with singly housing, environmental enrichment may be provided and animals may be handled on a regular basis. It should be noted that environmental enrichment may alter emotional behaviour of mice in the open field test (61). Irrespective of the housing conditions employed, similar conditions should be maintained between studies and reported on research papers.

3.6. Handling

It is well recognised that OB rats are irritable and aggressive following surgery and that frequent handling prior to and post surgery reduces these behavioural traits (22). Similarly, unless irritability and aggressive behaviour are the behavioural outcomes of interest in the experiment, most investigators employ a regime of regular handling of mice following bulbectomy. It is unknown what direct effect handling may have on OB-behaviour other than aggression, although in naïve mice handling does not modify open field behaviour (62). Damage to the frontal cortex during surgery often results in animals remaining irritable despite extensive handling, and these animals should be removed from the analysis. The time of day, means and duration of handling may differ between laboratories; however, it is important that the handling procedure remains consistent between studies.

3.7. Open Field Size, Shape and Lighting

Behavioural hyperactivity in OB mice is most commonly assessed in highly illuminated (100–320 lux) large (50–90 cm diameter) arenas (40–42, 52, 53). Although it has been demonstrated in certain studies that OB mice exhibit hyperactivity in a dimly lit small open field areas (54), observations from both our own laboratory (Fig. 3d) and that of others (47) failed to demonstrate an effect of bulbectomy on locomotor activity when assessed under similar conditions (arena $(22 \times 40 \times 25)$ with low level illumination (lux 30–35)). Large arenas also allow for anxiety-related behaviour (time spent in and latency to enter centre area) to be concurrently assessed. The open field arenas used are primarily either black or white and constructed of plastic to allow ease of cleaning. In order to increase the aversive nature of the test, several laboratories have included reflective walls into the area. Illumination is generally provided by light(s) positioned directly above the arena. It is important that the arena is evenly illuminated; otherwise, the mouse will most likely spend a greater proportion of the test time in the darker parts of the arena. In general, testing is conducted without bedding in the arena. Therefore, as previously demonstrated for the rat (4, 63), the design and aversiveness of open field apparatus appears to affect behavioural responses of OB mice.

4. Troubleshooting

4.1. Incomplete Removal of the Olfactory Bulbs or Damage to the Frontal Cortex During OB Surgery

Effects such as these may result from inappropriate vacuum pressure created by the aspirator. Placing a pressure gauge indicator on the pump will allow for the recording and maintenance of an appropriate vacuum pressure to aspirate the bulbs. Damage to the frontal cortex may also occur due to incorrect surgical coordinates. It is recommended that the OB surgery is carried out on a test mouse in order to confirm appropriate coordinates and vacuum pressure required for bulb removal prior to embarking on a large scale study.

4.2. Irritability and Aggression in OB Mice

This is most commonly due to a lack of handling post surgery or damage to the frontal cortex during removal of the bulbs. As such it is recommended that animals are regularly handled prior to and following surgery. Although it remains to be examined, it is also possible that removal of the olfactory bulbs in certain strains of mice may result in enhancement of aggressive behaviour.

4.3. Inability to Detect OB-Induced Hyperactivity

This is most commonly attributed to the design of the testing apparatus. OB-induced hyperactivity is most pronounced when assessed in a large aversive (high illuminated) arena when compared to a small, low illuminated area (see experimental variables). Alternatively, a lack of effect may be due to small sample size. Although the number of animals per group varies drastically between studies, based on our observations, we recommend at least 8–10 mice per group when assessing behavioural changes in the OB model. Ensure all mice are of similar age and have been housed and handled under similar conditions. Consideration should also be paid to the strain of mice used. Olfactory bulbectomy in Swiss mice results in the characteristic impairment in spatial memory but not hyperactivity in the open field (47).

4.4. Inability to Detect Reversal of OB-Induced Hyperactivity Following Treatment

Careful consideration should be given to the strain of mouse (e.g. C57BL/6 vs. DBA), handling, housing (single vs. group) and dosing regimen (dose, frequency of administration, time tested post administration, etc.) employed. Confirm that the studies have sufficient power – a minimum of 8–10 mice per group. Effects of novel compounds should be compared to those of a reference antidepressant (e.g. imipramine or citalopram) known to attenuate OB-induced hyperactivity. Longer treatment regimes (e.g. 4–6 weeks) may be required when examining effects of novel compounds/treatments, particularly if not monoamine based.

5. Conclusion

In conclusion, the OB mouse model is a robust, reproducible and valid animal model of depression and antidepressant activity. Taking into consideration the experimental variables, investigators with minimal prior surgical and behavioural experience can develop this model with relative ease. Characterisation of the OB model in mice continues to be evaluated in a host of behavioural paradigms, and may extend the models' utility in evaluating novel conditions and treatments. Furthermore, development of the model in mice has afforded the opportunity to examine the involvement of novel neural mechanisms involved in mediating OB-induced behavioural changes through the use of genetically modified mice.

References

1. Willner P, Mitchell PJ (2002) The validity of animal models of predisposition to depression. Behav Pharmacol 13(3):169–188.

2. Cairncross KD, Cox B, Forster C, Wren AF (1977) The olfactory bulbectomized rat: a simple model for detecting drugs with antidepressant potential [proceedings]. Br J Pharmacol 61(3):497P.

3. Cairncross KD, King MG, Schofield SP (1975) Effect of amitriptyline on avoidance learning in rats following olfactory bulb ablation. Pharmacol Biochem Behav 3(6):1063–1067.

4. Kelly JP, Wrynn AS, Leonard BE (1997) The olfactory bulbectomized rat as a model of depression: an update. Pharmacol Ther 74(3):299–316.

5. Song C, Leonard BE (2005) The olfactory bulbectomised rat as a model of depression. Neurosci Biobehav Rev 29(4–5):627–647.

6. Calcagnetti DJ, Quatrella LA, Schechter MD (1996) Olfactory bulbectomy disrupts the expression of cocaine-induced conditioned place preference. Physiol Behav 59(4–5):597–604.

7. van Rijzingen IM, Gispen WH, Spruijt BM (1995) Olfactory bulbectomy temporarily impairs Morris maze performance: an ACTH (4–9) analog accelerates return of function. Physiol Behav 58(1):147–152.

8. Drevets WC (2000) Neuroimaging studies of mood disorders. Biol Psychiatry 48(8):813–829.

9. Sheline YI (2000) 3D MRI studies of neuroanatomic changes in unipolar major depression: the role of stress and medical comorbidity. Biol Psychiatry 48(8):791–800.

10. Townsend JD, Eberhart NK, Bookheimer SY, Eisenberger NI, Foland-Ross LC, Cook IA, Sugar CA, Altshuler LL (2010) fMRI activation in the amygdala and the orbitofrontal cortex in unmedicated subjects with major depressive disorder. Psychiatry Res 183(3):209–217.

11. Wrynn AS, Mac Sweeney CP, Franconi F, Lemaire L, Pouliquen D, Herlidou S, Leonard BE, Gandon J, de Certaines JD (2000) An invivo magnetic resonance imaging study of the olfactory bulbectomized rat model of depression. Brain Res 879(1–2):193–199.

12. Shibata S, Watanabe S (1994) Facilitatory effect of olfactory bulbectomy on 2-deoxyglucose uptake in rat amygdala slices. Brain Res 665(1):147–150.

13. Mucignat-Caretta C, Bondi M, Caretta A (2004) Animal models of depression: olfactory lesions affect amygdala, subventricular zone, and aggression. Neurobiol Dis 16(2):386–395.

14. Nesterova IV, Gurevich EV, Nesterov VI, Otmakhova NA, Bobkova NV (1997) Bulbectomy-induced loss of raphe neurons is counteracted by antidepressant treatment. Prog Neuropsychopharmacol Biol Psychiatry 21(1):127–140.

15. Hozumi S, Nakagawasai O, Tan-No K, Niijima F, Yamadera F, Murata A, Arai Y, Yasuhara H, Tadano T (2003) Characteristics of changes in cholinergic function and impairment of learning and memory-related behavior induced by olfactory bulbectomy. Behav Brain Res 138(1):9–15.

16. Tadano T, Hozumi S, Yamadera F, Murata A, Niijima F, Tan-No K, Nakagawasai O, Kisara K (2004) Effects of NMDA receptor-related agonists on learning and memory impairment in olfactory bulbectomized mice. Methods Find Exp Clin Pharmacol 26(2):93–97.

17. Carlsen J, De Olmos J, Heimer L (1982) Tracing of two-neuron pathways in the olfactory system by the aid of transneuronal degeneration: projections to the amygdaloid body and hippocampal formation. J Comp Neurol 208(2):196–208.

18. Norrholm SD, Ouimet CC (2001) Altered dendritic spine density in animal models of depression and in response to antidepressant treatment. Synapse 42(3):151–163.

19. McNish KA, Davis M (1997) Olfactory bulbectomy enhances sensitization of the acoustic startle reflex produced by acute or repeated stress. Behav Neurosci 111(1):80–91.

20. Watanabe A, Tohyama Y, Nguyen KQ, Hasegawa S, Debonnel G, Diksic M (2003) Regional brain serotonin synthesis is increased in the olfactory bulbectomy rat model of depression: an autoradiographic study. J Neurochem 85(2):469–475.

21. Grecksch G, Zhou D, Franke C, Schroder U, Sabel B, Becker A, Huether G (1997) Influence of olfactory bulbectomy and subsequent imipramine treatment on 5-hydroxytryptaminergic presynapses in the rat frontal cortex: behavioural correlates. Br J Pharmacol 122(8):1725–1731.

22. Leonard BE, Tuite M (1981) Anatomical, physiological, and behavioral aspects of olfactory bulbectomy in the rat. Int Rev Neurobiol 22:251–286.

23. Bremner JD, Narayan M, Anderson ER, Staib LH, Miller HL, Charney DS (2000) Hippocampal volume reduction in major depression. Am J Psychiatry 157(1):115–118.

24. Ballmaier M, Narr KL, Toga AW, Elderkin-Thompson V, Thompson PM, Hamilton L, Haroon E, Pham D, Heinz A, Kumar A (2008) Hippocampal morphology and distinguishing late-onset from early-onset elderly depression. Am J Psychiatry 165(2):229–237.

25. Campbell S, Macqueen G (2004) The role of the hippocampus in the pathophysiology of major depression. J Psychiatry Neurosci 29(6):417–426.

26. Romeas T, Morissette MC, Mnie-Filali O, Pineyro G, Boye SM (2009) Simultaneous anhedonia and exaggerated locomotor activation in an animal model of depression. Psychopharmacology (Berl) 205(2):293–303.

27. Stock HS, Ford K, Wilson MA (2000) Gender and gonadal hormone effects in the olfactory bulbectomy animal model of depression. Pharmacol Biochem Behav 67(1):183–191.

28. Frisch P, Bilkei-Gorzo A, Racz I, Zimmer A (2010) Modulation of the CRH system by substance P/NKA in an animal model of depression. Behav Brain Res 213(1):103–108.

29. Zukerman S, Touzani K, Margolskee RF, Sclafani A (2009) Role of olfaction in the conditioned sucrose preference of sweet-ageusic T1R3 knockout mice. Chem Senses 34(8):685–694.

30. Cain DP (1974) Olfactory bulbectomy: neural structures involved in irritability and aggression in the male rat. J Comp Physiol Psychol 86(2):213–220.

31. Kolunie JM, Stern JM (1995) Maternal aggression in rats: effects of olfactory bulbectomy, ZnSO4-induced anosmia, and vomeronasal organ removal. Horm Behav 29(4):492–518.

32. Liebenauer LL, Slotnick BM (1996) Social organization and aggression in a group of olfactory bulbectomized male mice. Physiol Behav 60(2):403–409.

33. Jaako-Movits K, Zharkovsky A (2005) Impaired fear memory and decreased hippocampal neurogenesis following olfactory bulbectomy in rats. Eur J Neurosci 22(11):2871–2878.

34. Ostrovskaya RU, Gruden MA, Bobkova NA, Sewell RD, Gudasheva TA, Samokhin AN, Seredinin SB, Noppe W, Sherstnev VV, Morozova-Roche LA (2007) The nootropic and neuroprotective proline-containing dipeptide noopept restores spatial memory and increases immunoreactivity to amyloid in an Alzheimer's disease model. J Psychopharmacol 21(6):611–619.

35. Nakagawasai O, Hozumi S, Tan-No K, Niijima F, Arai Y, Yasuhara H, Tadano T (2003) Immunohistochemical fluorescence intensity reduction of brain somatostatin in the impairment of learning and memory-related behaviour induced by olfactory bulbectomy. Behav Brain Res 142(1–2):63–67.

36. Lumia AR, Teicher MH, Salchli F, Ayers E, Possidente B (1992) Olfactory bulbectomy as a model for agitated hyposerotonergic depression. Brain Res 587(2):181–185.

37. Chambliss HO, Van Hoomissen JD, Holmes PV, Bunnell BN, Dishman RK (2004) Effects of chronic activity wheel running and imipramine

on masculine copulatory behavior after olfactory bulbectomy. Physiol Behav 82(4): 593–600.

38. Primeaux SD, Holmes PV (1999) Role of aversively motivated behavior in the olfactory bulbectomy syndrome. Physiol Behav 67(1):41–47.

39. Stock HS, Hand GA, Ford K, Wilson MA (2001) Changes in defensive behaviors following olfactory bulbectomy in male and female rats. Brain Res 903(1–2):242–246.

40. Roche M, Shanahan E, Harkin A, Kelly JP (2008) Trans-species assessment of antidepressant activity in a rodent model of depression. Pharmacol Rep 60(3):404–408.

41. Zueger M, Urani A, Chourbaji S, Zacher C, Roche M, Harkin A, Gass P (2005) Olfactory bulbectomy in mice induces alterations in exploratory behavior. Neurosci Lett 374(2):142–146.

42. Jarosik J, Legutko B, Unsicker K, von Bohlen Und Halbach O (2007) Antidepressant-mediated reversal of abnormal behavior and neurodegeneration in mice following olfactory bulbectomy. Exp Neurol 204(1):20–28.

43. Cryan JF, Mombereau C (2004) In search of a depressed mouse: utility of models for studying depression-related behavior in genetically modified mice. Mol Psychiatry 9(4):326–357.

44. Bobkova NV, Nesteroval IV, Dana R, Dana E, Nesterov VI, Aleksandrova Y, Medvinskaya NI, Samokhin AN (2004) Morphofunctional changes in neurons in the temporal cortex of the brain in relation to spatial memory in bulbectomized mice after treatment with mineral ascorbates. Neurosci Behav Physiol 34(7):671–676.

45. Han F, Shioda N, Moriguchi S, Yamamoto Y, Raie AY, Yamaguchi Y, Hino M, Fukunaga K (2008) Spiro[imidazo[1,2-a]pyridine-3,2-indan]-2(3H)-one (ZSET1446/ST101) treatment rescues olfactory bulbectomy-induced memory impairment by activating Ca2+/calmodulin kinase II and protein kinase C in mouse hippocampus. J Pharmacol Exp Ther 326(1):127–134.

46. Nakajima A, Yamakuni T, Haraguchi M, Omae N, Song SY, Kato C, Nakagawasai O, Tadano T, Yokosuka A, Mimaki Y, Sashida Y, Ohizumi Y (2007) Nobiletin, a citrus flavonoid that improves memory impairment, rescues bulbectomy-induced cholinergic neurodegeneration in mice. J Pharmacol Sci 105(1):122–126.

47. Mucignat-Caretta C, Bondi M, Caretta A (2006) Time course of alterations after olfactory bulbectomy in mice. Physiol Behav 89(5): 637–643.

48. Legutko B, Dudys D, Branski P, Znojek P, Pilc A (2006) Olfactory bulbectomy in C57BL/6J mice: behavioural deficits and effects of chronic citalopram treatment. FENS Forum Abstracts 3:A161:115.

49. Otmakhova NA, Gurevich EV, Katkov YA, Nesterova IV, Bobkova NV (1992) Dissociation of multiple behavioral effects between olfactory bulbectomized C57Bl/6J and DBA/2J mice. Physiol Behav 52(3):441–448.

50. Han F, Shioda N, Moriguchi S, Qin ZH, Fukunaga K (2008) The vanadium (IV) compound rescues septo-hippocampal cholinergic neurons from neurodegeneration in olfactory bulbectomized mice. Neuroscience 151(3):671–679.

51. Aleksandrova IY, Kuvichkin VV, Kashparov IA, Medvinskaya NI, Nesterova IV, Lunin SM, Samokhin AN, Bobkova NV (2004) Increased level of beta-amyloid in the brain of bulbectomized mice. Biochemistry (Mosc) 69(2): 176–180.

52. Hellweg R, Zueger M, Fink K, Hortnagl H, Gass P (2007) Olfactory bulbectomy in mice leads to increased BDNF levels and decreased serotonin turnover in depression-related brain areas. Neurobiol Dis 25(1):1–7.

53. Licht CL, Kirkegaard L, Zueger M, Chourbaji S, Gass P, Aznar S, Knudsen GM (2010) Changes in 5-HT4 receptor and 5-HT transporter binding in olfactory bulbectomized and glucocorticoid receptor heterozygous mice. Neurochem Int 56(4):603–610.

54. Bilkei-Gorzo A, Racz I, Michel K, Zimmer A (2002) Diminished anxiety- and depression-related behaviors in mice with selective deletion of the Tac1 gene. J Neurosci 22(22): 10046–10052.

55. Mar A, Spreekmeester E, Rochford J (2000) Antidepressants preferentially enhance habituation to novelty in the olfactory bulbectomized rat. Psychopharmacology (Berl) 150(1):52–60.

56. Wieronska JM, Papp M, Pilc A (2001) Effects of anxiolytic drugs on some behavioral consequences in olfactory bulbectomized rats. Pol J Pharmacol 53(5):517–525.

57. Saitoh A, Hirose N, Yamada M, Nozaki C, Oka T, Kamei J (2006) Changes in emotional behavior of mice in the hole-board test after olfactory bulbectomy. J Pharmacol Sci 102(4): 377–386.

58. Kamei J, Hirose N, Oka T, Miyata S, Saitoh A, Yamada M (2007) Effects of methylphenidate on the hyperemotional behavior in olfactory bulbectomized mice by using the hole-board test. J Pharmacol Sci 103(2):175–180.

59. Gould TD, Doa DT, Kovacsis CE. (2009). The open field test. In Gould TD, editor Mood and Anxiety Related Phenotypes in Mice: characterization using behavioral tests. Humana Press, New York. p 1–21.

60. Slotkin TA, Miller DB, Fumagalli F, McCook EC, Zhang J, Bissette G, Seidler FJ (1999) Modeling geriatric depression in animals: biochemical and behavioral effects of olfactory bulbectomy in young versus aged rats. J Pharmacol Exp Ther 289(1):334–345.

61. Lin EJ, Choi E, Liu X, Martin A, During MJ (2010) Environmental enrichment exerts sex-specific effects on emotionality in C57BL/6J mice. Behav Brain Res 10.1016/j.bbr.2010.08.019.

62. Gariepy JL, Rodriguiz RM, Jones BC (2002) Handling, genetic and housing effects on the mouse stress system, dopamine function, and behavior. Pharmacol Biochem Behav 73(1):7–17.

63. Mar A, Spreekmeester E, Rochford J (2002) Fluoxetine-induced increases in open-field habituation in the olfactory bulbectomized rat depend on test aversiveness but not on anxiety. Pharmacol Biochem Behav 73(3):703–712.

64. Sato A, Nakagawasai O, Tan-No K, Onogi H, Niijima F, Tadano T (2010) Effect of non-selective dopaminergic receptor agonist on disrupted maternal behavior in olfactory bulbectomized mice. Behav Brain Res 210(2):251–256.

65. Ostrovskaya RU, Retyunskaya MV, Bondarenko NA, Gudasheva TA, Bobkova NV, Samokhin AN (2005) Cholinopositive effect of dilept (neurotensin peptidomimetic) as the basis of its mnemotropic effect. Bull Exp Biol Med 139(3):340–344.

66. Sato A, Nakagawasai O, Tan-No K, Onogi H, Niijima F, Tadano T (2010) Influence of olfactory bulbectomy on maternal behavior and dopaminergic function in nucleus accumbens in mice. Behav Brain Res 215(1):141–145.

Chapter 18

Differential-Reinforcement-of-Low-Rate Behavior in Rodents as a Screen for Antidepressant Efficacy

Lindsay M. Lueptow and James M. O'Donnell

Abstract

Differential-reinforcement-of-low-rate (DRL) is an operant conditioning schedule that requires behavioral inhibition for a given length of time, known as the interresponse time (IRT), in order to receive reinforcement. For rodents under a DRL 36 or 72-s schedule, antidepressant treatment results in an increased reinforcement rate, decreased response rate, and a rightward shift in the IRT histogram distribution. While these characteristic changes are seen following a wide variety of clinically proven antidepressant treatments, including selective serotonin and norepinephrine reuptake inhibitors, tricyclic antidepressants, monoamine oxidase inhibitors, atypical antidepressants, electroconvulsive shock, and numerous putative antidepressants, these changes are not seen with other psychotherapeutic drug classes including anxiolytics, stimulants, antihistamines, anticholinergics, opioids, or sedative-hypnotics. Due to the increasing popularity and utility of genetic engineering, mouse DRL is an essential tool for not only the identification of antidepressant efficacy of new compounds but also for the identification of underlying neurochemical and neuroanatomical mechanisms involved in the onset of antidepressant effects. When properly implemented, the DRL protocol has high predictive validity and reliability as a screen for antidepressant efficacy.

Key words: Differential-reinforcement-of-low-rate behavior, Mouse, Depression, Antidepressant, Operant conditioning

1. Introduction

Depression is a multi-symptomatic, heterogeneous mood disorder experienced by millions of individuals each year. According to the DSM-IV, a depressive episode is characterized by any combination of symptoms that involve changes in sleep, energy level, cognitive abilities, or weight, as well as a loss of interest and pleasure in daily activities, a markedly depressed mood, and/or recurrent thoughts of death or suicide (1). Although many pharmacological treatments are available, few diverge from the mechanism of action of monoamine reuptake inhibition (i.e., inhibition of the serotonin

Todd D. Gould (ed.), *Mood and Anxiety Related Phenotypes in Mice: Characterization Using Behavioral Tests, Volume II*, Neuromethods, vol. 63, DOI 10.1007/978-1-61779-313-4_18, © Springer Science+Business Media, LLC 2011

transporter, SERT, the norepinephrine transporter, NET, and to a lesser extent the dopamine transporter, DAT), and it typically takes weeks of repeated treatment for global symptom remission (2). This, in combination with the heterogeneous nature of the disorder, makes studying the pathogenesis and treatment of depression in animals very difficult. In order to assess the likelihood of clinical efficacy of new antidepressants for a patient population, researchers must have reliable screens to identify antidepressant-like effects in animals.

2. Background

In 1980, McGuire and Seiden identified an operant conditioning task with rats that appeared to predict antidepressant efficacy in humans. Under a differential-reinforcement-of-low-rate 18-second (DRL 18-s) schedule, reinforcement comes only for responses with an interresponse time (IRT) greater than 18-s. Responses that occur before the 18-s has elapsed reset the timer, and no reinforcer is provided. In order to receive the reward, animals must learn to wait at least 18-s between each response. Seiden's group demonstrated that treatment with a number of different tricyclic antidepressants (TCAs) results in a decreased response rate (number of lever presses) and an increased reinforcement rate (number of lever presses that result in a reinforcer), shifting responses to longer IRTs (3–5). Plotting IRT values in a frequency histogram results in a bimodal distribution, with an initial peak at low IRT values, known as "burst" IRTs, and an inverted-U distribution at longer IRT values (pause IRTs), with a second peak near or below the IRT criterion level. Following treatment with TCAs, longer IRT values result in an orderly shift of this U-distribution to the right.

Seiden's group also showed that this antidepressant profile is dependent on the criterion level. When using a criterion level of 9-s, they did not find specific shifts in response rate, reinforcement rate, or IRT distribution following treatment with the TCA imipramine (4). By contrast, using a DRL 72-s schedule, not only did they show consistent changes in behavior following antidepressant treatment (e.g., decreased response rate, increased reinforcement rate, and a shift in the IRT distribution; Figs. 1 and 2), but they also found that this effect was specific to drugs that were known to have antidepressant activity, including monoamine oxidase inhibitors (MAOIs) and in some cases, atypical antidepressants. However, similar outcomes were not seen with other psychoactive drugs such as alcohol or morphine, or drugs that have stimulant in addition to antidepressant effects, such as dopamine reuptake inhibitors (see Table 1 for a summary of data).

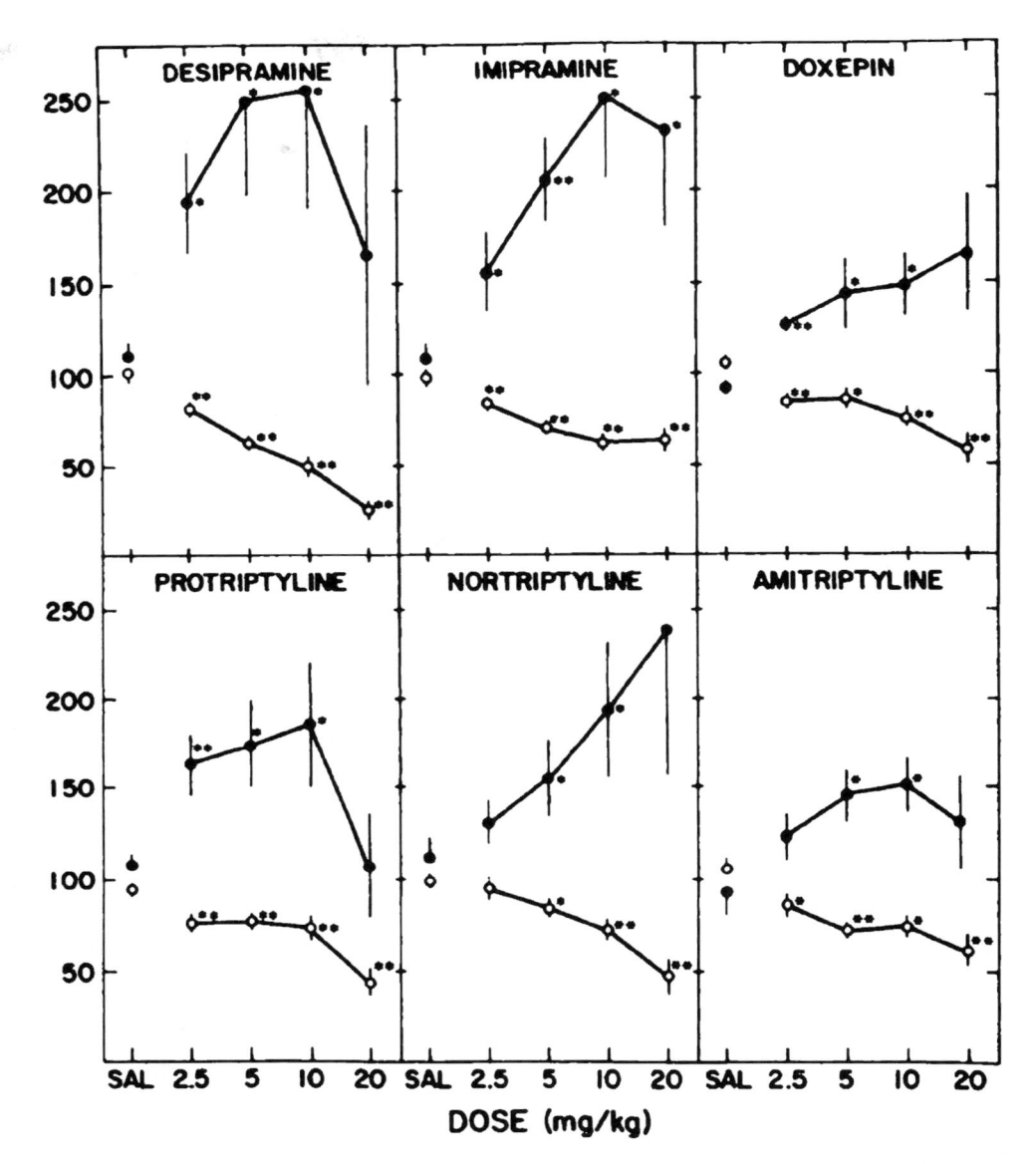

Fig. 1. Administration of tricyclic antidepressants causes a decrease in response rate (*open circles*) and an increase in reinforcement rate (*closed circles*) in rats under a DRL 72-s schedule. Values shown are mean ± SE for 8–10 rats expressed as a percentage of control. *$p < 0.05$, **$p < 0.01$ vs. control (from O'Donnell and Seiden (6) and is reprinted with the copyright permission of the American Society for Pharmacology and Experimental Therapeutics).

3. DRL Behavior: Effects of Antidepressants

3.1. Tricyclic Antidepressants

Early studies investigating the effects of imipramine on rat DRL performance indicated that behavioral changes were greatest when using schedules requiring longer IRTs (4, 6). Subsequently, most researchers have chosen to use a DRL 72-s schedule when

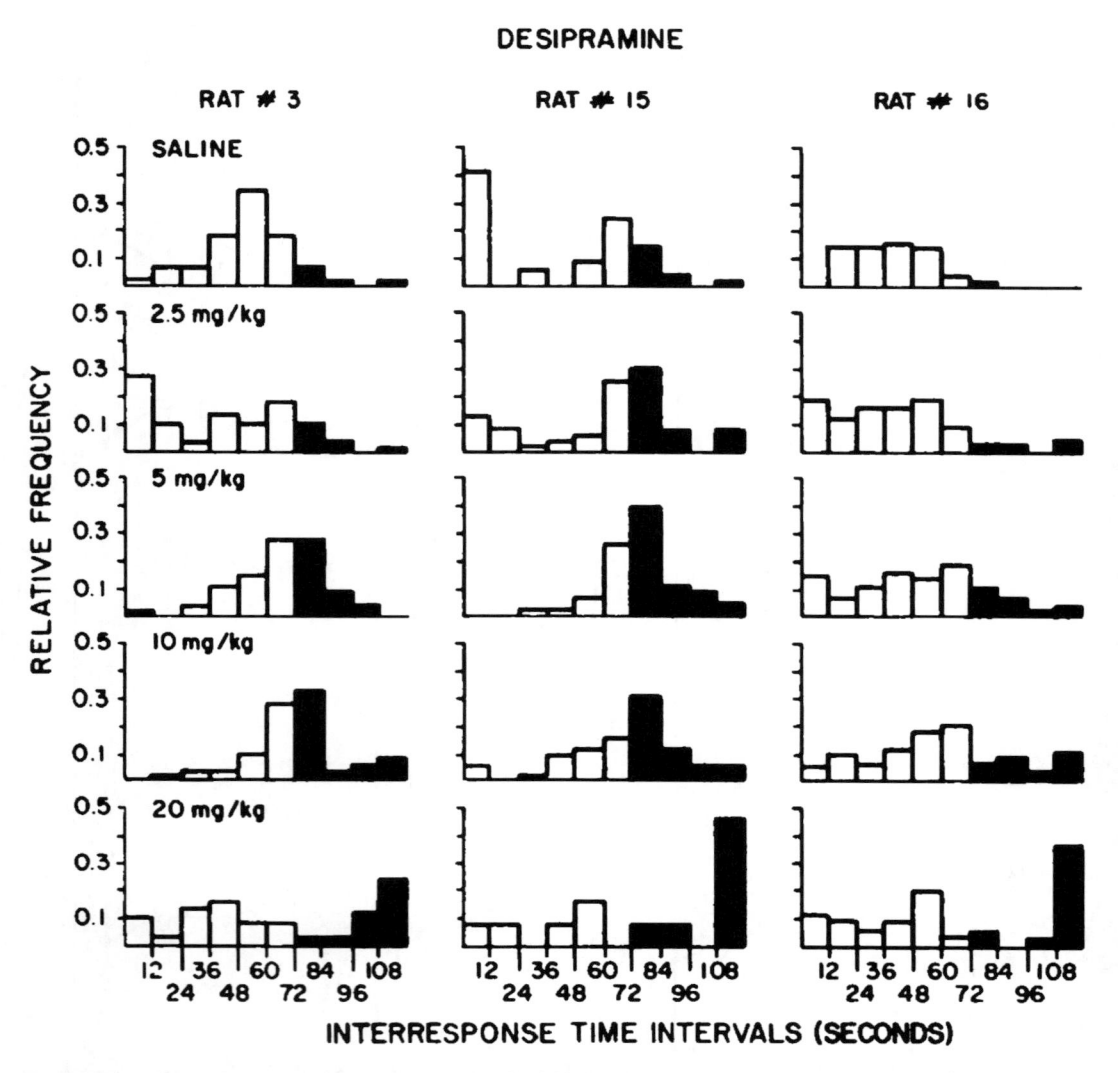

Fig. 2. Desipramine treatment results in a rightward shift of IRT duration to longer IRTs without disrupting the overall temporal pattern of responding, except at the highest doses, in rats under a DRL 72-s schedule. Data from three representative rats is similar to effects seen with antidepressants from other therapeutic classes. Data are plotted as relative frequency with each bar representing a 12-s interval. *Shaded bars* represent reinforced responses (from O'Donnell and Seiden (6) and is reprinted with copyright permission of the American Society for Pharmacology and Experimental Therapeutics).

assessing the antidepressant efficacy of a compound in rats, and a 36-s schedule when using mice. With the exception of one study by Jackson et al. (22), which failed to show a change in reinforcement rate for desipramine, research involving TCAs such as imipramine, desipramine, protriptyline, nortriptyline, doxepin, and amitriptyline all show dose-dependent decreases in response rate and increases in reinforcement rate (Fig. 1). When looking at the IRT distribution histograms, TCA treatment also results in an orderly shift of the IRT distribution toward longer IRTs (Fig. 2).

Table 1
Changes in DRL behavior following antidepressant treatment, by drug class

Treatment	Reinforcement rate	Response rate	IRT shift
Tricyclic ADs	↑ (3, 4, 6–8, 17, 19, 23, 36, 37, 41, 45, 47, 51, 54, 56–62) ↔ (58)	↓ (3, 4, 6, 7, 17 19, 23, 36, 37, 45, 47, 51, 54, 56, 58–60)	R (3, 6, 23, 45, 47, 51, 58–60) O (19)
SSRIs	↑ (7–11, 13, 14, 19, 63, 64) ↔ (17, 62, 65)	↓ (7–11, 13, 14, 17, 19, 63, 64) ↔ (62, 65)	O (13, 19, 62–65)
NRIs	↑ (15–17)	↓ (15–17)	R (15)
SNRIs	↔ (17)	↔ (17)	
DRIs	↔ (10, 17) ↓ (6, 19)	↔ (17) ↑ (6, 10, 19)	L (19) O (6)
TRIs	↑ (19)	↔ (19)	L (19)
Monoamine oxidase inhibitors	↑ (5, 6, 23, 24) ↔ (22)	↓ (5, 6, 22–24)	R (5, 6, 22–24)
ECS	↑ (10)	↓ (10)	
Atypical ADs	↑ (6, 10, 24, 26, 66) ↔ (17, 22, 25)	↓ (6, 10, 22, 24, 66) ↔ (17, 25, 26)	R (6)
5HT2A antagonist	↑ (23, 25, 45, 59, 65) ↔ (59)	↓ (23, 25, 45, 59, 65) ↔ (59)	R (23, 59, 65)
+classical AD	↑↑ (23, 65)	↓↓ (23, 65)	R (65)
Nonselective 5HT2A antagonist	↔ (11, 25, 59)	↔ (11, 25, 59)	O (59)
5HT1A agonists	↑ (8, 11, 45, 67) ↔ (49)	↓ (8, 11, 45, 49, 67)	O (45, 67)

(continued)

Table 1
(continued)

Treatment	Reinforcement rate	Response rate	IRT shift
5-Hydroxy-tryptophan	↑ (11, 49, 67)	↓ (49, 67) ↔ (11)	R (67)
5HT-1A KO	↔ (62, 68)	↔ (62, 68)	R (62)
+SSRI	↑ (62)	↓ (62)	
5HT-1B KO	↓ (68)	↑ (68)	L (68)
α1 Adrenergic antagonist	↔ (25, 59)	↓ (25, 59)	O (59)
β2 Adrenergic agonists	↑ (69–72)	↓ (69–72)	R (69, 70, 72)
β-KO + tricyclic AD	↑ (57)	↓ (57)	
Neurokinin-2 antagonist	↑ (36)	↓ (36)	R (36)
Neurosteroids	↓ (73)	↑ (73)	L (73)
Neuropeptide Y	↑ (37)	↓ (37)	R (37)
PDE4 inhibitors	↑ (29, 30)	↓ (29, 30)	R (29, 30)
+β adrenergic agonists	↑↑ (74)	↓↓ (74)	
mGluR 5 antagonist	↑ (40)	↓ (40)	R (40)
mGluR 2 positive allosteric modulators	↑ (41, 75)	↓ (41, 75)	R (75)
Corticotropin releasing factor 1 antagonist	↑ (35)	↓ (35)	R (35)
Vasopressin 1b antagonist	↑ (35)	↓ (35)	R (35)
Antipsychotics	↓ (10, 22) ↔ (6, 8) ↑ (19, 54, 56, 61)	↓ (6, 8, 10, 19, 22, 54, 56, 61)	O (6, 19, 22, 54)
Anticholinergic/histamine 1 antagonist	↔ (6)	↔ (6)	

Benzodiazepines	↔ (8, 19, 47, 67, 76) ↓ (5, 73) ↑ (53)	↓ (53, 76) ↑ (5, 73, 77) ↔ (8, 19, 47, 67)	O (19, 47)
Amphetamines	↓ (8, 47, 52, 78)	↑ (8, 47, 52, 76, 78)	L (47, 52, 78)
Caffeine	↓ (79, 80)	↑ (79, 80)	L (79, 80)
Barbiturates	↓ (5)	↔ (5)	
Alcohol	↑ (81) ↔ (5, 73)	↓ (5, 73)	
Opioid agonists	↔ (5)	↓ (5)	

Response rate/reinforcement rate: Numbers indicate papers reporting data (see References)

↑ increase; ↓ decrease; ↔ no consistent change; double arrows indicate a synergistic or additive effect of coadministration

Treatment: *AD* antidepressant; *SSRI* selective serotonin reuptake inhibitor; *NRI* norepinephrine reuptake inhibitor; *SNRI* serotonin and norepinephrine reuptake inhibitor; *DRI* dopamine reuptake inhibitor; *TRI* triple reuptake inhibitor; *ECS* electroconvulsive shock; *KO* knockout; *PDE4* phosphodiesterase-4; *mGluR* metabotropic glutamate receptor IRT shift: *R* rightward shift; *L* leftward shift; *O* effect other than R or L shift

3.2. Serotonin and Norepinephrine Reuptake Inhibitors

Treatment with selective serotonin reuptake inhibitors (SSRIs) results in altered DRL performance, similar to that seen with TCA treatment. Studies using fluoxetine, fluvoxamine, paroxetine, sertraline, and zimelidine show increases in reinforcement rate and decreases in response rate under a DRL 72-s schedule (7–14). However, individual SSRIs vary somewhat in their DRL response profile, and the IRT distribution analysis is often more variable than with other classes of antidepressants. For example, fluoxetine, paroxetine, and sertraline produce increases in reinforcement rate, but only fluoxetine and sertraline show consistent decreases in response rate (13). In this same study, fluoxetine and sertraline shift the peak IRT without affecting the overall temporal pattern of responding. However, the IRT distribution for paroxetine is flattened to resemble more of a random pattern of responding. Overall, quantitative analysis of IRT distributions for many SSRIs reveals only a modest effect on the IRT distribution, with some drugs actually shifting the distribution toward a more random pattern of responding; most SSRIs do cause an overall increase in response efficiency.

Treatment with norepinephrine reuptake inhibitors (NRIs) results in a classical antidepressant DRL profile. Reboxetine and (+)-oxaprotiline administration results in increased reinforcement rate and a decreased response rate (15–17). Rats treated with oxaprotiline also display a rightward shift of the IRT distribution. Desipramine, a TCA that exhibits marked selectivity for inhibition of norepinephrine reuptake, reliably shows an antidepressant profile in DRL schedules for both mice and rats (Fig. 2; Table 1). Although certain dopamine reuptake inhibitors are known to have some degree of antidepressant efficacy, the increase in dopaminergic neurotransmission has a psychomotor stimulant effect that interferes with DRL responding, and often results in a decrease in reinforcement efficiency. For example, treatment with nomifensine and high doses of bupropion, both of which interact with the DAT, results in an increase in response rate and decrease in reinforcement rate under a DRL 72-s schedule (6, 10, 17).

More recently developed serotonin and norepinephrine reuptake inhibitors (SNRIs) have yet to be fully evaluated using DRL behavior. One study by Dekeyne et al. (17) showed no change in response efficiency following treatment with venlafaxine or S33005. However, their rat strain, training protocol, and experimental conditions differed from the previously established protocols. Triple reuptake inhibitors that block DAT, NET, and SERT are currently being developed and have shown antidepressant efficacy in other tests of antidepressant-like behavior (18, 19). Early tests with the triple reuptake inhibitor DOV216,303 have shown a limited antidepressant-like profile, with an increase in reinforcement rate but no change in response rate (19). As the authors point out, this is in agreement with previous studies using other dopamine enhancing drugs, such as bupropion, that tend to increase locomotor activity, and therefore interfere with DRL performance.

3.3. Monoamine Oxidase Inhibitors

One additional therapeutic class is the MAOIs. These drugs have reliably shown an antidepressant profile in rats under a DRL 72-s schedule. Much like TCAs, MAOIs also cause a rightward shift in the IRT distribution without a major disruption in the temporal pattern of responding; these effects do not persist beyond the day of treatment, even though the drugs inhibit the enzyme irreversibly. Clinically, subtype-selective MAO-A inhibitors, but not MAO-B selective inhibitors, have antidepressant efficacy (20), although an MAO-B selective inhibitor administered transdermally has been shown to be efficacious in treating depression (21). Similarly, researchers have shown increases in reinforcement rate, decreases in response rate, and a rightward shift in IRT distributions for nonselective MAOIs isocarboxazid, phenelzine, and tranylcypromine as well as MAO-A selective inhibitors clorgyline and CGP11305A (5, 6, 22–24). The MAO-B selective inhibitor selegiline does not produce an antidepressant profile, but pargyline, which is somewhat MAO-B selective, begins to take on an antidepressant-like profile on DRL 72-s behavior when its dose is high enough to inhibit at least 90% of MAO-A (24).

3.4. Atypical and Alternative Antidepressants

Although monoamine reuptake inhibitors are currently the most widely used drugs for the treatment of depression, a number of other classes of drugs have shown antidepressant efficacy. Some of these have been developed for antidepressant use, but may have unknown or mixed mechanisms of action. Drugs such as mianserin, mirtazapine, trazodone, nefazodone, iprindole, pizotifen, and AHR 9377, which are fairly nonselective and interact with numerous biogenic amine receptors and transporters, have shown an antidepressant profile on DRL 72-s behavior (see Table 1; (6, 17, 25–28)). Selective inhibitors of phosphodiesterase-4 have demonstrated antidepressant activity in clinical trials and have also shown an antidepressant-like profile under a DRL 72-s schedule (29–31). Other alternative treatments with clinical antidepressant efficacy that have also been shown to increase reinforcement rate for DRL include treatment with electroconvulsive shock or an extended photoperiod, as well as administration with extracts from *Mimosa pudica* or *Morinda officinalis* (10, 32–34).

The focus of research and development of antidepressant drugs has widened from a more traditional monoamine-centered approach to include other signaling systems that may be involved in the therapeutic response. Researchers have begun looking at the role of neuropeptides in depression, such as corticotropin releasing factor and vasopressin receptor antagonists (35), as well as tachykinin signaling, including neurokinin 2 receptor antagonists (36), and neuropeptide Y (37). All of these studies have shown antidepressant-like effects on DRL, and others have shown efficacy in preclinical and early clinical trials (38, 39). Researchers have also investigated the therapeutic effects of metabotropic glutamate receptor antagonists under a DRL 72-s schedule, finding antidepressant-like effects (40, 41), with the exception of one study, which only reported antidepressant-like behavior in the mouse forced swim test (42).

4. Materials, Equipment, and Setup

Rodents are placed in an operant conditioning chamber, which is enclosed in a sound-attenuated box containing a fan for ventilation and masking noise. Modular chambers and sound-attenuating boxes are available from MedAssociates (Model ENV-307A; St. Albans, VT) or Colbourn Instruments (Model E10-10; Lehigh Valley, PA). The chamber is equipped with a house light, two nose-poke holes for mice (ENV-313M; MedAssociates), or levers for rats (ENV-112CM; MedAssociates) on either side of a water access port (2.2 cm in diameter), when water is used as a reinforcer. Water reinforcement is supplied via a dipper (ENV-302M; MedAssociates) submerged in water that rises up through a small hole in the floor of the access port, providing approximately 0.01 mL (for mice) or 0.02 mL (for rats) of water for 4 s (see Fig. 3 for standard mouse setup). When using food pellets as a reward, a food pellet receptacle is placed next to a lever, with water available throughout the experiment.

4.1. Protocol: Reaching a Stable Baseline of Responding

Following a 1-week acclimation to the animal facility, animals can begin operant training. Sessions are run once daily, 5–7 days per week for a period of approximately 6–8 weeks to reach stable baseline behavior. Animals are maintained on a regimen of restricted access to water or food, depending on the reinforcer used. Typically, this involves providing postsession access to water or food in an amount that results in body weights of about 80–90% of animals allowed to drink or eat freely. For the first few sessions, use an alternate fixed-ratio 1 (FR1), fixed-time 1-minute (FT1) schedule for 1 h each day to establish reliable lever-pressing or nose-poking behavior. The rodent receives a reinforcer for every nose poke/lever press (i.e., FR1), as well as a reward every minute without responding (i.e., FT1). After 1 or 2 days, when the animal exhibits steady responding, the rodent should be switched to a simple FR1 schedule, without the FT1 component. The FR1 schedule should be maintained for 1–2 weeks, depending on the animal's ability to acquire the desired response behavior (nose-poke for mice or lever-pressing for rats). Once the rodents have learned to respond at a high rate for the reinforcer, they should progress to the DRL 18-s schedule. At any point during the training, if the rodents are

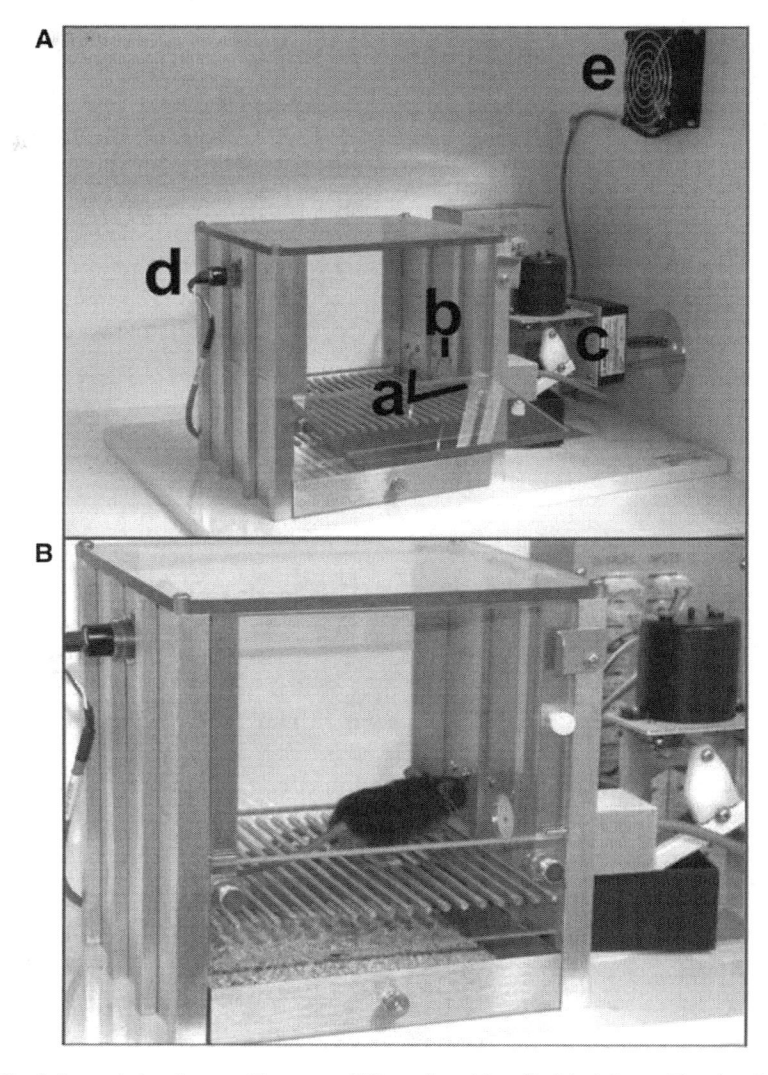

Fig. 3. Operant chamber used for mouse DRL, equipped for a liquid reinforcer. The chamber and test apparatus (**a**) contain the following: *a* response apparatus; *b* water access port; *c* liquid dipper; *d* house lights; *e* fan/masking noise. (**b**) Mouse activating the liquid dipper via the nose poke hole.

performing sub-optimally, postsession water or food rations should be reduced temporarily. Following DRL 18-s, animals should be put on DRL 36-s (mice) or DRL 72-s (rats) schedule. Mice can also progress to a DRL 72-s schedule; however, DRL 36 s is sufficient to show effects of antidepressant treatment. Since mice are more variable in their DRL performance, a 36-s criterion level may provide a more stable baseline in a shorter amount of training time.

4.2. Protocol:
Experimental Design

One major benefit of using DRL as a screen for antidepressant-like behavior is the ability to use a within-subjects design. Under this protocol, the researcher can use an individual animal's baseline behavior as a means of comparison in addition to comparison with between-subject treatments. Because baseline behavior, in addition to genetic and environmental variables, can greatly affect rodent behavior (43–45), specifically for DRL, a within-subjects design minimizes confounding variables and error introduced by using only a between-subjects design. When animals have reached the baseline performance criteria (e.g., for rats, more than 70 responses, fewer than 12 reinforcements maintained for 5 days), drug testing can begin. For animals tested 5 days per week, it is best to inject animals with drug or vehicle on Tuesdays and Fridays and use Thursdays as noninjection control days, unless pharmacokinetic or other factors limit twice-weekly drug administration. Remember to expose all animals to vehicle injections prior to data collection in order to acclimate them to the injection procedure, which can affect behavior.

5. Data Analysis

The data for the effects of drugs on DRL behavior are best normalized to each animal's control response and reinforcement values. When assessing the effects of acute administration of a drug, the noninjected control values are those from Thursdays during the period when the dose-response function is determined; this assumes drug or vehicle is administered on Tuesdays and Fridays. Simply plotting the data as a percent of control provides a fast and effective way for visually analyzing trends in the data. For daily drug treatments, control values for each animal are usually the average of five vehicle-injected days prior to the start of daily drug treatments. Plot the response and reinforcement means ± S.E.M. as a percent of control for each dose of drug injected (Fig. 4). Data are collected using MedAssociates software and subsequently analyzed by any number of available software programs. MedAssociates does offer data transfer software that can export raw data files into Excel spreadsheets, where the data are more manageable. Once the raw data are exported and processed, statistical significance can be determined using repeated measures ANOVA followed by post hoc tests between individual treatment conditions.

5.1. IRT Histograms

IRTs can be grouped into specific intervals, or bins, and graphed in histogram format, providing an overall temporal IRT distribution. For these graphs, plot the relative frequency, or the proportion of responses, per bin. The IRT distribution should be graphed to the temporal resolution used during data collection. For rats performing DRL 72-s, 12-s bins are typically used (Fig. 2). For mice performing DRL 36 s, a smaller bin time (such as 3 or 6 s) may be desired (Fig. 5).

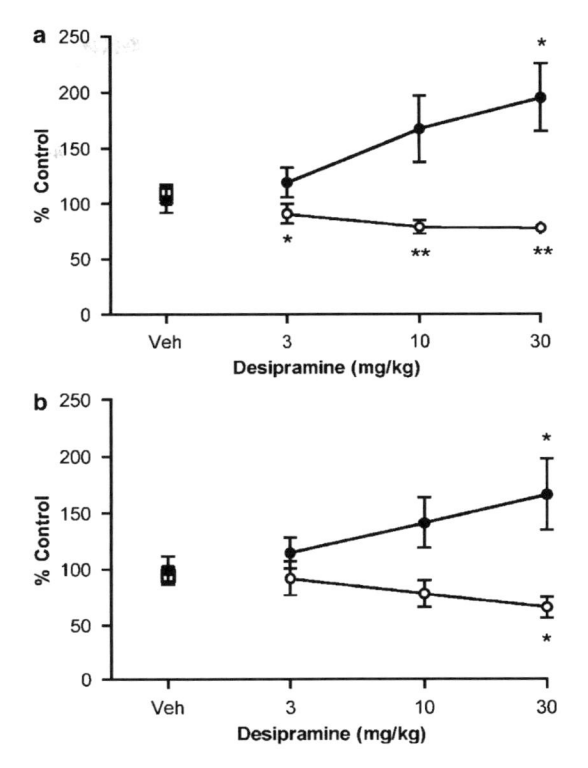

Fig. 4. Treatment with desipramine results in a dose-dependent change in DRL 36-s behavior in mice. Wild-type (**a**) and β-1 and β-2 adrenergic receptor knockout mice (**b**) show increased reinforcement rates (*filled circles*) and decreased response rates (*open circles*) following desipramine treatment 30 min prior to testing. Data represent mean ± SEM as a percentage of control. *$p < 0.05$, **$p < 0.01$ vs. corresponding vehicle (Veh) treatment (reprinted from Zhang et al. (57)).

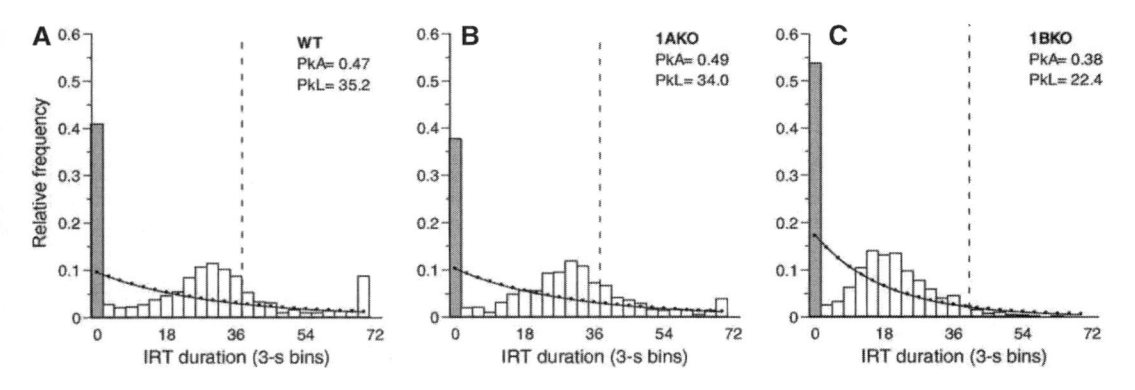

Fig. 5. Peak deviation metrics graphed on an interresponse time (IRT) histogram for wild-type (WT; **a**), 5HT-1A (**b**), and 5HT-1B (**c**) receptor knockout mice under a DRL 36-s schedule. The graphs show a bimodal distribution, with a burst peak at <3 s (*gray bar*), and a second peak (quantified as PkL) located left of the criterion level (*dashed line*) represents pause responding (*white bars*). Mice lacking the 5HT1B receptor (1BKO) have a significantly lower PkL than either 5HT1A knockout mice (1AKO) or WT mice. The area above the corresponding negative exponential (*solid dotted line*), which represents random performance, is quantified as the PkA (reprinted from Pattij et al. (68), with permission from Elsevier).

5.2. Peak Deviation Analysis

Although IRT distributions provide a graphical means for comparison, quantitatively they are difficult to compare without further statistical analysis. In 1991, Richards and Seiden described a statistical approach for quantitatively comparing IRT distributions that is based on a principle that states the IRT distribution would be a negative exponential function for an animal that is responding randomly in time, but at a constant overall rate (46). Richards and Seiden extended this principle to their own data analysis by comparing individually obtained IRT distributions to their corresponding negative exponential (CNE) distributions. The CNE distribution is the negative exponential distribution that corresponds to the obtained IRT distribution. The CNE distributions are based upon each animal's mean IRT duration and therefore are unique for each individual animal. Every CNE then serves as an adjusted baseline for comparison of each animal.

In 1993, Richards et al. extended this analysis by providing three specific metrics for comparison: the peak area (PkA), peak location (PkL), and the burst ratio (BR). The BR is the ratio of obtained to predicted responses and represents the burst component of responding, which is not factored into the peak analysis. The BR is not an absolute or relative frequency, and therefore corrects for the increasing chance of an IRT occurring in the burst category as the IRT distribution shrinks. The pause component of responding is described by the PkA and the PkL. The PkA is the peak area of the obtained IRT distribution that is above the CNE distribution. The PkL is the median IRT duration in the peak area. IRT distributions often are bimodal, with an initial burst phase peak and a later pause phase peak (Fig. 5). It is therefore important to choose an appropriate burst cutoff point, so as to not affect integrity of the PkA and PkL values. For rat DRL 72-s, it has been reported that a cutoff of IRTs < 6 s is appropriate (for detailed instructions on peak deviation analysis, see (47)).

6. Experimental Variables

6.1. Alternate Training Schedules

When beginning the training schedule, animals may have a difficult time making an association between the response behavior (nose poke or lever press) and the reinforcer (e.g., water or food). Autoshaping programs may be used to encourage the acquisition of the response behavior. This involves the addition of cue lights to encourage lever pressing (32, 40, 48). If, after one or two sessions, animals are not learning the response–reinforcer association, manual shaping may be employed by the experimenter. When the animal approaches the response apparatus, engage the reward. As the animal begins to learn the association, one can place a drop of water on the lever to encourage lever pressing. For the acquisition of DRL responding, some groups employ overnight training sessions (17, 49–52).

6.2. Setting Criterion Levels and Establishing Baseline Behavior

The training progression for animals can vary somewhat, and criterion levels to advance to the next stage of training may depend on the animal's intrinsic level of responding. For example, Balcells-Olivero et al. (45) show that two separate stocks of Sprague-Dawley rats not only have different baseline behavior but also show different responses to lesions and drug treatment. For rat DRL, our lab sets a specific criterion level that must be obtained in order to qualify for the next phase of training. Under the FR1 schedule, rats must make more than 100 responses in less than 20 min. For the DRL 18-s phase, rats must obtain more than 200 responses per 60-min session for at least 1–2 weeks straight. Finally, baseline behavior for rat DRL 72-s is defined as greater than 70 responses and less than 12 reinforcements per 60-min session for 5 consecutive days. Mouse criterion levels are not as stringent, as mouse baseline behavior has greater variability. We train our mice for 2 weeks on an FR1 schedule, followed by DRL 18-s for 3 weeks, and finally DRL 36 s until a stable baseline is reached. For mice, a stable baseline is defined as less than 10% within-subjects variability across 5 days.

6.3. Factors Affecting Baseline Responding and Antidepressant Sensitivity

While some more recent studies have made use of Wistar or Long Evans rats, historically, Sprague-Dawley rats have been the most commonly used strain of rat in DRL 72-s schedules. Studies have shown that gender (53), previous training schedules (43), as well as different suppliers (45) (e.g., Harlan vs. Holtzman) can affect the baseline rate of responding, even within the same strain. Additionally, the baseline rate of responding will affect behavioral outcomes following drug treatment (6, 54), so it is therefore important to consider these variables when designing an experiment. With animals that have a lower baseline of responding, it may be advantageous to use a lower criterion level, e.g., 36 s, in order to detect more subtle changes following drug treatment (54). For mouse DRL, shorter criterion levels may be used as well for this reason.

Traditionally, water deprivation has been used to keep animals at 80–90% of their normal body weight. However, the choice of water vs. food deprivation may be limited to the lab's available apparatus. For animals on a water deprivation schedule, 0.1% saccharin or sucrose may be used to encourage higher baseline responding. For food deprivation, condensed milk is an alternative to food pellets. As with any behavior test, be sure to slowly introduce the novel food or liquid to the animal over a few days, prior to testing. There is limited research regarding the effects of food vs. water deprivation on animal behavior. One recent study demonstrated that food-deprived mice have larger fluctuations in body weight, as well as decreases in exploration of novel environments and alterations in timing behavior in a conditional learning task (55). While further research is needed, this is an indication that water restriction protocols may be better suited for DRL tasks.

References

1. American Psychiatric Association Task Force On DSM-IV (2000) Diagnostic and statistical manual of mental disorders: DSM-IV-TR. American Psychiatric Association, Washington.

2. Katz MM, Tekell JL, Bowden CL, Brannan S, Houston JP, Berman N, Frazer A (2004) Onset and early behavioral effects of pharmacologically different antidepressants and placebo in depression. Neuropsychopharmacology. 29: 566–579.

3. Mcguire PS, Seiden LS (1980) The effects of tricyclic antidepressants on performance under a differential-reinforcement-of-low-rates schedule in rats. J Pharmacol Exp Ther. 214: 635–641.

4. Mcguire PS, Seiden LS (1980) Differential effects of imipramine in rats as a function of DRL schedule value. Pharmacol Biochem Behav. 13: 691–694.

5. O'Donnell JM, Seiden LS (1982) Effects of monoamine oxidase inhibitors on performance during differential reinforcement of low response rate. Psychopharmacology. 78: 214–8.

6. O'Donnell JM, Seiden LS (1983) Differential-reinforcement-of-low-rate 72-second schedule: selective effects of antidepressant drugs. J Pharmacol Exp Ther. 224: 80–88.

7. Danysz W, Plaznik A, Kostowski W, Malatynska E, Järbe TU, Hiltunen A, Archer T (1988) Comparison of desipramine, amitriptyline, zimeldine and alaproclate in six animal models used to investigate antidepressant drugs. Pharmacol Toxicol. 62: 42–50.

8. van Hest A, van Drimmelen M, Olivier B (1992) Flesinoxan shows antidepressant activity in a DRL 72-s screen. Psychopharmacology. 107: 474–9.

9. Olivier B, Bosch L, van Hest A, van der Heyden J, Mos J, van der Poel G, Schipper J, Tulp M (1993) Preclinical evidence on the psychotropic profile of fluvoxamine. Pharmacopsychiatry. 26 (Suppl 1): 2–9.

10. Seiden LS, Dahms JL, Shaughnessy RA (1985) Behavioral screen for antidepressants: the effects of drugs and electroconvulsive shock on performance under a differential-reinforcement-of-low-rate schedule. Psychopharmacology. 86: 55–60.

11. Marek GJ, Li AA, Seiden LS (1989) Evidence for involvement of 5-hydroxytryptamine1 receptors in antidepressant-like drug effects on differential-reinforcement-of-low-rate 72-second behavior. J Pharmacol Exp Ther. 250: 60–71.

12. Marek GJ, McDougle CJ, Price LH, Seiden LS (1992) A comparison of trazodone and fluoxetine: implications for a serotonergic mechanism of antidepressant action. Psychopharmacology. 109: 2–11.

13. Sokolowski JD, Seiden LS (1999) The behavioral effects of sertraline, fluoxetine, and paroxetine differ on the differential-reinforcement-of-low-rate 72-second operant schedule in the rat. Psychopharmacology. 147: 153–61.

14. Cousins MS, Seiden LS (2000) The serotonin-1A receptor antagonist WAY-100635 modifies fluoxetine's antidepressant-like profile on the differential reinforcement of low rates 72-s schedule in rats. Psychopharmacology. 148: 438–442.

15. Marek GJ, Li A, Seiden LS (1988) Antidepressant-like effects of (+)-oxaprotiline on a behavioral screen. Eur J Pharmacol. 157: 183–8.

16. Wong EH, Sonders MS, Amara SG, Tinholt PM, Piercey MF, Hoffmann WP, Hyslop DK, Franklin S, Porsolt RD, Bonsignori A, Carfagna N, McArthur RA (2000) Reboxetine: a pharmacologically potent, selective, and specific norepinephrine reuptake inhibitor. Biol Psychiatry. 47: 818–29.

17. Dekeyne A, Gobert A, Auclair A, Girardon S, Millan MJ (2002) Differential modulation of efficiency in a food-rewarded "differential reinforcement of low-rate" 72-s schedule in rats by norepinephrine and serotonin reuptake inhibitors. Psychopharmacology. 162: 156–67.

18. Caldarone BJ, Paterson NE, Zhou J, Brunner D, Kozikowski AP, Westphal KGC, Korte-bouws GAH, Prins J, Korte SM, Olivier B, Ghavami A (2010) The novel triple reuptake inhibitor JZAD-IV-22 exhibits an antidepressant pharmacological profile without locomotor stimulant or sensitization properties. J Pharmacol Exp Ther. 335: 762–770.

19. Paterson NE, Balci F, Campbell U, Olivier BE, Hanania T (2010) Triple reuptake inhibitor DOV216,303 exhibits limited antidepressant-like properties in the differential reinforcement of

low-rate 72-second responding assay, likely due to dopamine reuptake inhibition. J Psychopharmacol. doi: 10.1177/0269881110364272

20. Rudorfer MV (1992) Monoamine oxidase inhibitors: reversible and irreversible. Psychopharmacol Bull. 28: 45–57.

21. Feiger AD, Rickels K, Rynn MA, Zimbroff DL, Robinson DS (2006) Selegiline transdermal system for the treatment of major depressive disorder: an 8-week, double-blind, placebo-controlled, flexible-dose titration trial. J Clin Psychiatry. 67: 1354–61.

22. Jackson A, Koek W, Colpaert FC (1995) Can the DRL 72-s schedule selectively reveal antidepressant drug activity? Psychopharmacology. 117: 154–61.

23. Ardayfio PA, Benvenga MJ, Chaney SF, Love PL, Catlow J, Swanson SP, Marek GJ (2008) The 5-Hydroxytryptamine2A receptor antagonist (M100907) attenuates impulsivity after both drug-induced disruption (dizocilpine) and enhancement (antidepressant drugs) of differential-reinforcement-of-low-rate 72-s in the rat. J Pharmacol Exp Ther. 327: 891–897.

24. Marek GJ, Seiden LS (1988) Selective inhibition of MAO-A, not MAO-B, results in antidepressant-like effects on DRL 72-s behavior. Psychopharmacology. 96: 153–60.

25. Marek GJ, Li AA, Seiden LS (1989) Selective 5-hydroxytryptamine2 antagonists have antidepressant-like effects on differential-reinforcement-of-low-rate 72-second schedule. J Pharmacol Exp Ther. 250: 52–9.

26. Hand TH, Marek GJ, Seiden LS (1991) Comparison of the effects of mianserin and its enantiomers and metabolites on a behavioral screen for antidepressant activity. Psychopharmacology. 105: 453–8.

27. Andrews JS, Jansen JHM, Linders S, Princen A, Drinkenburg WHIM, Coenders CJH, Vossen JHM (1994) Effects of imipramine and mirtazapine on operant performance in rats. Drug Dev Res. 32: 58–66.

28. Eison AS, Eison MS, Torrente JR, Wright RN, Yocca FD (1990) Nefazodone: preclinical pharmacology of a new antidepressant. Psychopharmacol Bull. 26: 311–5.

29. O'Donnell JM (1993) Antidepressant-like effects of rolipram and other inhibitors of cyclic adenosine monophosphate phosphodiesterase on behavior maintained by differential reinforcement of low response rate. J Pharmacol Exp Ther. 264: 1168–78.

30. O'Donnell J, Zhang H (2004) Antidepressant effects of inhibitors of cAMP phosphodiesterase (PDE4). Trends Pharmacol Sci. 25: 158–63.

31. Zhang H, Zhao Y, Huang Y, Deng C, Hopper A, De Vivo M, Rose G, O'Donnell J (2006) Antidepressant-like effects of PDE4 inhibitors mediated by the high-affinity rolipram binding state (HARBS) of the phosphodiesterase-4 enzyme (PDE4) in rats. Psychopharmacology. 186: 209–17.

32. Molina-Hernández M, Téllez-Alcántara N (2000) Long photoperiod regimen may produce antidepressant actions in the male rat. Prog Neuropsychopharmacol Biol Psychiatry. 24: 105–16.

33. Molina M, Contreras CM, Téllez-Alcántara N (1999) Mimosa pudica may possess antidepressant actions in the rat. Phytomedicine. 6: 319–23.

34. Zhang Z, Yuan L, Yang M, Luo Z, Zhao Y (2002) The effect of Morinda officinalis How, a Chinese traditional medicinal plant, on the DRL 72-s schedule in rats and the forced swimming test in mice. Pharmacol Biochem Behav. 72: 39–43.

35. Louis C, Cohen C, Depoorte R, Griebel G (2006) Antidepressant-like effects of the corticotropin-releasing factor 1 receptor antagonist, SSR125543, and the vasopressin 1b receptor antagonist, SSR149415, in a DRL-72 s schedule in the rat. Neuropsychopharmacology. 31: 2180–2187.

36. Louis C, Stemmelin J, Boulay D, Bergis O, Cohen C, Griebel G (2008) Additional evidence for anxiolytic- and antidepressant-like activities of saredutant (SR48968), an antagonist at the neurokinin-2 receptor in various rodent-models. Pharmacol Biochem Behav. 89: 36–45.

37. Molina-Hernández M, Téllez-Alcántara N, Olivera-Lopez JI, Jaramillo MT (2010) Antidepressant-like or anxiolytic-like actions of topiramate alone or co-administered with intralateral septal infusions of neuropeptide Y in male Wistar rats. Peptides. 31: 1184–9.

38. Palucha A, Pilc A (2007) Metabotropic glutamate receptor ligands as possible anxiolytic and antidepressant drugs. Pharmacol Ther. 115: 116–47.

39. Zorrilla EP, Koob GF (2010) Progress in corticotropin-releasing factor-1 antagonist development. Drug Discov Today. 15: 371–83.

40. Molina-Hernández M, Téllez-Alcántara N, Pérez-García J, Olivera-Lopez JI, Jaramillo MT (2006) Antidepressant-like and anxiolytic-like actions of the mGlu5 receptor antagonist MTEP, microinjected into lateral septal nuclei of male Wistar rats. Prog Neuropsychopharmacol Biol Psychiatry. 30: 1129–35.

41. Fell MJ, Witkin JM, Falcone JF, Katner JS, Perry KW, Hart J, Rorick-Kehn LM, Overshiner CD, Rasmussen K, Chaney SF, Benvenga MJ, Li X, Marlow DL, Thompson LK, Luecke SK,

Wafford KA, Seidel WF, Edgar DM, Quets AT, Felder CC, Wang X, Heinz BA, Nikolayev A, Kuo M-S, Mayhugh D, Khilevich A, Zhang D, Ebert PJ, Eckstein JE, Ackermann BL, Swanson SP, Catlow JT, Dean RA, Jackson K, Tauscher-Wisniewski S, Marek GJ, Schkeryantz JM, Svensson KA (2011) THIIC, a novel mGlu2 potentiator with potential anxiolytic/antidepressant properties: In vivo profiling suggests a link between behavioral and CNS neurochemical changes. J Pharmacol Exp Ther. 336: 165–177.

42. Bespalov AY, van Gaalen MM, Sukhotina IA, Wicke K, Mezler M, Schoemaker H, Gross G (2008) Behavioral characterization of the mGlu group II/III receptor antagonist, LY-341495, in animal models of anxiety and depression. Eur J Pharmacol. 592: 96–102.

43. Pizzo MJ, Kirkpatrick K, Blundell PJ (2009) The effect of changes in criterion value on differential reinforcement of low rate schedule performance. J Exp Anal Behav. 92: 181–98.

44. Pattij T, Broersen LM, Peter S, Olivier B (2004) Impulsive-like behavior in differential-reinforcement-of-low-rate 36 s responding in mice depends on training history. Neurosci Lett. 354: 169–71.

45. Balcells-Olivero M, Cousins MS, Seiden LS (1998) Holtzman and harlan Sprague-Dawley rats: differences in DRL 72-sec performance and 8-hydroxy-di-propylamino tetralin-induced hypothermia. J Pharmacol Exp Ther. 286: 742–52.

46. Blough D (1966) Reinforcement of least-frequent sequences of choices. J Exp Anal Behav. 9: 581–591.

47. Richards JB, Sabol KE, Seiden LS (1993) DRL interresponse-time distributions: quantification by peak deviation analysis. J Exp Anal Behav. 60: 361–85.

48. van der Stelt HM, Broersen LM, Olivier B, Westenberg HGM (2004) Effects of dietary tryptophan variations on extracellular serotonin in the dorsal hippocampus of rats. Psychopharmacology. 172: 137–44.

49. Jolly DC, Richards JB, Seiden LS (1999) Serotonergic mediation of DRL 72-s behavior: receptor subtype involvement in a behavioral screen for antidepressant drugs. Biol Psychiatry. 45: 1151–62.

50. Sabol KE, Richards JB, Layton K, Seiden LS (1995) Amphetamine analogs have differential effects on DRL 36-s schedule performance. Psychopharmacology. 121: 57–65.

51. Borsini F, Cesana R, Kelly J, Leonard BE, McNamara M, Richards J, Seiden LS (1997) BIMT 17: a putative antidepressant with a fast onset of action? Psychopharmacology. 134: 378–86.

52. Balcells-Olivero M, Richards JB, Seiden LS (1997) Sensitization to amphetamine on the differential-reinforcement-of-low-rate 72-s schedule. Psychopharmacology. 133: 207–13.

53. van Haaren F, Katon E, Anderson KG (1997) The effects of chlordiazepoxide on low-rate behavior are gender dependent. Pharmacol Biochem Behav. 58: 1037–43.

54. Howard JL, Pollard GT (1984) Effects of imipramine, bupropion, chlorpromazine, and clozapine on differential reinforcement of low rate (DRL)>72 Sec and>36 sec schedules in rat. Drug Dev Res. 4: 607–616.

55. Tucci V, Hardy A, Nolan PM (2006) A comparison of physiological and behavioural parameters in C57BL/6J mice undergoing food or water restriction regimes. Behav Brain Res. 173: 22–9.

56. Britton K, Koob G (1989) Effects of corticotropin releasing factor, desipramine and haloperidol on a DRL schedule of reinforcement. Pharmacol Biochem Behav. 32: 967–70.

57. Zhang H, Whisler L, Huang Y, Xiang Y, O'Donnell J (2009) Postsynaptic alpha-2 adrenergic receptors are critical for the antidepressant-like effects of desipramine on behavior. Neuropsychopharmacology. 34: 1067–77.

58. Richards JB, Seiden LS (1991) A quantitative interresponse-time analysis of DRL performance differentiates similar effects of the antidepressant desipramine and the novel anxiolytic gepirone. J Exp Anal Behav. 56: 173–92.

59. Marek GJ, Seiden LS (1988) Effects of selective 5-hydroxytryptamine-2 and nonselective 5-hydroxytryptamine antagonists on the differential-reinforcement-of-low-rate 72-second schedule. J Pharmacol Exp Ther. 244: 650–8.

60. O'Donnell JM, Seiden LS (1984) Altered effects of desipramine on operant performance after 6-hydroxydopamine-induced depletion of brain dopamine or norepinephrine. J Pharmacol Exp Ther. 229: 629–35.

61. Pollard G, Howard J (1986) Similar effects of antidepressant and non-antidepressant drugs on behavior under an interresponse-time greater than 72-s schedule. Psychopharmacology. 89: 253–8.

62. Scott-McKean JJ, Wenger GR, Tecott LH, Costa ACS (2010) 5-HT1A receptor null mutant mice responding under a differential-reinforcement-of-low-rate 72-second schedule of reinforcement. Neuropsychopharmacology. 1: 24–32.

63. Richards JB, Sabol KE, Seiden LS (1993) Fluoxetine prevents the disruptive effects of fen-

fluramine on differential-reinforcement-of-low-rate 72-second schedule performance. J Pharmacol Exp Ther. 267: 1256–63.

64. Cousins MS, Vosmer G, Overstreet DH, Seiden LS (2000) Rats selectively bred for responsiveness to 5-hydroxytryptamine(1A) receptor stimulation: differences in differential reinforcement of low rate 72-second performance and response to serotonergic drugs. J Pharmacol Exp Ther. 292: 104–13.

65. Marek GJ, Martin-Ruiz R, Abo A, Artigas F (2005) The selective 5-HT2A receptor antagonist M100907 enhances antidepressant-like behavioral effects of the SSRI fluoxetine. Neuropsychopharmacology. 30: 2205–15.

66. Li A, Marek GJ, Hand TH, Seiden LS (1990) Antidepressant-like effects of trazodone on a behavioral screen are mediated by trazodone, not the metabolite m-chlorophenylpiperazine. Eur J Pharmacol. 177: 137–44.

67. Richards JB, Sabol KE, Hand TH, Jolly DC, Marek GJ, Seiden LS (1994) Buspirone, gepirone, ipsapirone, and zalospirone have distinct effects on the differential-reinforcement-of-low-rate 72-s schedule when compared with 5-HTP and diazepam. Psychopharmacology. 114: 39–46.

68. Pattij T, Broersen LM, van der Linde J, Groenink L, van der Gugten J, Maes RAA, Olivier B (2003) Operant learning and differential-reinforcement-of-low-rate 36-s responding in 5-HT1A and 5-HT1B receptor knockout mice. Behav Brain Res. 141: 137–145.

69. O'Donnell JM (1987) Effects of clenbuterol and prenalterol on performance during differential reinforcement of low response rate in the rat. J Pharmacol Exp Ther. 241: 68–75.

70. O'Donnell JM (1988) Psychopharmacological consequences of activation of beta adrenergic receptors by SOM-1122. J Pharmacol Exp Ther. 246: 38–46.

71. O'Donnell JM (1993) Effects of the beta-2 adrenergic agonist zinterol on DRL behavior and locomotor activity. Psychopharmacology. 113: 89–94.

72. Dunn R, Richards JB, Seiden LS (1993) Effects of salbutamol upon performance on an operant screen for antidepressants. Psychopharmacology. 113: 1–10.

73. Amato RJ, Lewis PB, He H, Winsauer PJ (2010) Effects of positive and negative modulators of the γ-aminobutyric acid A receptor complex on responding under a differential-reinforcement-of-low-rate schedule of reinforcement in rats. Behav Pharmacol. doi: 10.1097/FBP.0b013e32833fa7c7.

74. Zhang H, Huang Y, Mishler K, Roerig S, O'Donnell J (2005) Interaction between the antidepressant-like behavioral effects of beta adrenergic agonists and the cyclic AMP PDE inhibitor rolipram in rats. Psychopharmacology. 182: 104–115.

75. Nikiforuk A, Popik P, Drescher KU, van Gaalen M, Relo A-lucia, Mezler M, Marek G, Schoemaker H, Gross G, Bespalov A (2010) Effects of a positive allosteric modulator of group II metabotropic glutamate receptors, LY487379, on cognitive flexibility and impulsive-like responding in rats. J Pharmacol Exp Ther. 335: 665–673.

76. Bayley PJ, Bentley GD, Dawson GR (1998) The effects of selected antidepressant drugs on timing behaviour in rats. Psychopharmacology. 136: 114–22.

77. Burke TF, Miller LG, Moerschbaecher JM (1994) Acute effects of benzodiazepines on operant behavior and in vivo receptor binding in mice. Pharmacol Biochem Behav. 48: 69–76.

78. Fowler SC, Pinkston J, Vorontsova E (2009) Timing and space usage are disrupted by amphetamine in rats maintained on DRL 24-s and DRL 72-s schedules of reinforcement. Psychopharmacology. 204: 213–225.

79. Marek GJ, Heffner TG, Richards JB, Shaughnessy RA, Li AA, Seiden LS (1993) Effects of caffeine and PD 116,600 on the differential-reinforcement-of-low rate 72-S (DRL 72-S) schedule of reinforcement. Pharmacol Biochem Behav. 45: 987–90.

80. Webb D, Levine TE (1978) Effects of caffeine on DRL performance in the mouse. Pharmacol Biochem Behav. 9: 7–10.

81. Järbe TU, Hiltunen AJ (1988) Ethanol and Ro 15-4513: behaviour maintained by operant procedures (DRL-72s and PTZ-drug discrimination) in rats. Drug Alcohol Depend. 22: 83–90.

Chapter 19

The Intracranial Self-Stimulation Procedure Provides Quantitative Measures of Brain Reward Function

Astrid K. Stoker and Athina Markou

Abstract

Since the discovery of the intracranial self-stimulation (ICSS) procedure in the 1950s, studies using this method have greatly expanded our knowledge of the neurobiology of motivation and reward. ICSS is an operant behavioral procedure in which laboratory rodents prepared with stimulating electrodes learn to deliver brief electrical pulses into brain structures that are part of the brain reward pathway. The ICSS procedure is unique because it enables researchers to quantitatively assess brain reward function in laboratory animals. This procedure has predominantly been used in rats until recently and is now also used in mice. With the recent advances in genetic engineering in this species, the mouse serves as an excellent subject for investigating the neurobiology of reward and motivation. The ICSS procedure, however, is often perceived as too difficult and elaborate to perform in mice, despite the advantages of this technique and the unique research opportunities that mice offer. This chapter describes the two most commonly used ICSS procedures in mice – the discrete-trial current-intensity and rate-frequency curve-shift procedures – and provides suggestions for the successful implementation of ICSS in mice.

Key words: Intracranial self-stimulation, Brain reward thresholds, Motivation, Mouse, Discrete-trial current-intensity, Rate-frequency curve-shift

1. Background and Historical Significance

1.1. Historical Perspective

Intracranial self-stimulation (ICSS) behavior was discovered in the 1950s, when James Olds and Peter Milner serendipitously misplaced an electrode in one of their experiments and found that the rat showed conditioned place preference for the corner of an operant chamber in which it had previously received electrical brain stimulation. In his writings in *Scientific American*, Olds describes how the electrical stimulus was first assumed to provoke curiosity of some sort and how only subsequent testing revealed that the electrical brain stimulation was experienced by the rat as rewarding (1).

Todd D. Gould (ed.), *Mood and Anxiety Related Phenotypes in Mice: Characterization Using Behavioral Tests, Volume II*, Neuromethods, vol. 63, DOI 10.1007/978-1-61779-313-4_19, © Springer Science+Business Media, LLC 2011

The exact location of the electrode in this first self-stimulating rat is unknown, but the finding motivated Olds and Milner to begin exploring the brain sites in which electrical stimulation is experienced as rewarding in an attempt to identify brain reward circuits (2).

1.2. Brain Stimulation

Electrical brain stimulation can be perceived as rewarding (positive reinforcement) or aversive. ICSS, by definition as the term self-stimulation implies, is positive reinforcement that supports operant behavior. ICSS is supported by activation of brain reward pathways, using stimulation parameters that induce action potentials. That is, the electrical stimulus consists of a train of brief, repeated, identical electrical pulses interspaced with interstimulus intervals that are longer than the refractory periods of neurons (3).

1.3. Brain Reward Thresholds

By varying various parameters of the rewarding electrical stimulus, researchers can determine the value of the stimulation parameter above which laboratory animals will perform operant responses to self-administer this electrical stimulus (or, in other words, below which the subjects will no longer respond for the stimulus). This value represents a "threshold" value, termed the brain reward threshold. At a given set of stimulation parameters, this brain reward threshold of an animal is stable across repeated tests and days under baseline conditions. Deviations from an animal's baseline reward threshold value are considered to reflect changes in the animal's brain reward sensitivity and thus its "mood" state. An animal with increased reward sensitivity will respond to less intense electrical stimuli and vice versa. Therefore, lowered brain reward thresholds are considered to reflect a hedonic mood state (increased brain reward sensitivity), whereas elevated ICSS thresholds are considered to reflect anhedonia, which is the loss of the ability to experience pleasure (decreased brain reward sensitivity). The quantifiable value of the reward threshold allows the quantifiable assessment of brain reward states.

2. Overview of the ICSS Procedure

2.1. General Procedure

An ICSS experiment consists of several phases: habituation to the animal facility, surgery, recovery, training, threshold stabilization, and testing. Upon arrival at the animal facility, animals are allowed several days of acclimatization and handling by the experimenter. The habituation period is followed by a surgical procedure during which an electrode is implanted into one of the brain areas of the reward pathway (most commonly in the lateral hypothalamus or ventral tegmental area and along the medial forebrain bundle). Animals are allowed at least 7 days of recovery time, which is followed by the training phase of the experiment. During training, animals are introduced to the operant chambers and learn to operate a manipulandum, such as a wheel, to obtain electrical stimulation into the brain structure in which the electrode was implanted. After successful acquisition of self-stimulation behavior, one of the parameters of the stimulus is varied (see below) and a brain reward threshold is obtained. The threshold value is subsequently stabilized by repeated daily threshold assessments until it is stable over several days. Successful establishment of stable thresholds is followed by the testing phase, during which experimental manipulations are introduced.

2.2. Parameters

The stimulation train is defined by four parameters:

1. Current intensity: the magnitude of the electrical stimulation in μA.
2. Pulse duration: the duration of the pulse in microseconds.
3. Number of pulses: the total number of pulses that are delivered within one stimulation train.
4. Interpulse interval: the time interval between electrical pulses within the stimulation train measured in microseconds.

Two other parameters that are often used to describe the electrical stimulation are frequency and train duration. Frequency is measured in hertz and is defined by the pulse duration and interpulse interval. Train duration is measured in milliseconds and reflects the total time span of the stimulation train. Train duration is defined by the pulse duration, interpulse interval, and number of pulses. Today, because of the sophisticated stimulators available to investigators, cathodal stimulation, which optimally activates brain reward pathways (3), is used almost exclusively, whereas alternating current (both cathodal and anodal pulses) was used in the past.

Brain reward thresholds are obtained by varying one parameter while maintaining all other parameters at a constant level. The brain reward threshold is defined as the magnitude of one of the above parameters that sustains a predetermined performance level (e.g., operant responses in two of three stimulus presentations or 50% of maximal response rate). The reward threshold can be manipulated by varying any one of the different parameters, thereby varying the intensity of the stimulus and thus the rewarding value of the stimulus. Decreasing the value of a parameter elevates brain reward thresholds measured in another stimulation parameter and vice versa. For example, when the number of pulses within a stimulation train is decreased, and thus the stimulation frequency is decreased, then the rewarding value of the stimulus will be decreased, and the current-intensity threshold will be elevated. An example of how variations in the value of one stimulation parameter alter reward thresholds, defined by another parameter of the stimulation, is demonstrated in Fig. 1.

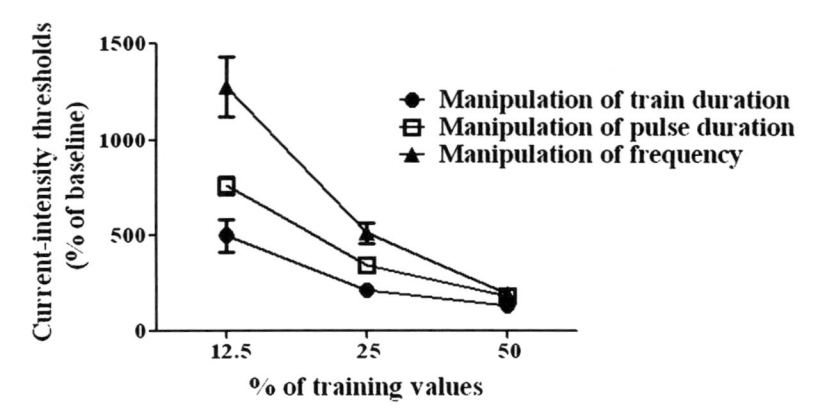

Fig. 1. Manipulations of stimulation parameters alter reward thresholds measured in another stimulation parameter in C57BL/6J mice and demonstrate the trade-off between stimulation parameters. Current-intensity threshold data are expressed as mean ± SEM. C57BL/6J mice ($n = 10$) were trained and subsequently allowed to establish stable self-stimulation performance with electrical stimuli that consisted of 100 µs pulses, a 10,000 µs interpulse interval, and 50 pulses administered per stimulation train (100% stimulation: 100 Hz). Upon stabilization of current-intensity thresholds, manipulations were introduced in six daily sessions per manipulation. First, the duration of the stimulation train was shortened by decreasing the number of pulses administered within one stimulation train (i.e., 50% for 25 pulses, 25% for 12 pulses, and 12.5% for six pulses), with all other stimulation parameters remaining constant, which corresponded to train durations of 242.5, 111.2, and 50.6 ms. After thresholds were restabilized on 100% stimulation, the duration of the pulses was decreased to 50% (50 µs), 25% (25 µs), and 12.5% (12.5 µs), with all other stimulation parameters remaining constant. Thresholds were again restabilized on 100% stimulation, after which the frequency was decreased to 50% (100 µs pulse duration, 20,000 µs interpulse interval, 25 pulses), 25% (100 µs pulse duration, 40,000 µs interpulse interval, 12 pulses), and 12.5% (100 µs pulse duration, 80,000 µs interpulse interval, six pulses), corresponding to frequencies of 50, 25, and 12.5 Hz.

3. Advantages and Disadvantages of the ICSS Procedure

In the past 60 years, the ICSS procedure has become a very important and valuable tool in the study of the neurobiology of motivation and reward. Implanting an electrode directly into a brain structure that is part of the brain reward circuit enables assessment of the involvement of this brain structure in reward and motivational processes. ICSS has additionally been very valuable in the field of psychopharmacology, where it has been widely used to study drug effects on reinforcement and reward processes. A major strength of the ICSS procedure is that it directly stimulates brain reward circuits, bypassing most of the input system and thus avoiding potential confounding effects related to sensory inputs. For example, food reward is communicated to the brain reward circuits originally through activation of taste buds and subsequently through communication of these sensory signals to brain reward sites. Malfunction of this sensory input may lead to the interpretation of reduced reward when in fact the taste buds may be damaged or the bitter taste of the food may have interfered with its rewarding nutritional properties. Additionally, the ICSS procedure provides a means of quantitatively measuring brain reward in rodents. Importantly, these quantitative measurements of reward are robust and remain stable over prolonged periods of time under baseline conditions. The stability of ICSS thresholds enables experiments that use within-subjects experimental designs or the study of the effects of manipulations over prolonged periods of time, such as the study of chronic administration and withdrawal from drugs of abuse.

Nevertheless, the ICSS procedure also has some disadvantages. Implantation of a stimulating electrode requires stereotaxic surgery techniques. Failure to accurately position the electrode in the intended brain site leads to unstable ICSS behavior. For some subjects, this is likely the primary reason that they never exhibit stable ICSS behavior, and these subjects will need to be excluded from subsequent experimentation. An additional disadvantage is that both the stereotaxic surgery and training, such as training in the discrete-trial current-intensity procedure (see below), are labor-intensive. The surgery typically requires approximately 30–45 min per animal by an experienced surgeon. The duration of the training period is approximately 1–2 weeks for a cohort of mice, and stabilization of the thresholds requires a minimum of 3 weeks. Nevertheless, as discussed above, stable reward thresholds maintained over several months allow for long-term repeated testing of each subject and thus the collection of much data from each subject.

4. ICSS in Laboratory Mice

Until recently, ICSS research has predominantly focused on rats, with only a handful of publications reporting ICSS testing in mice. Because of the great advances in genetic engineering techniques in the mouse, the demand for ICSS studies in mice has increased rapidly. However, the ICSS procedure is often considered too intricate and elaborate to perform in this species. Therefore, only few laboratories have undertaken ICSS studies in mice.

In the 1970s, Pierre Cazala and Bernard Cardo were pioneers in the study of ICSS in mice. They compared self-stimulation of the hypothalamus between DBA/2 and BALB/c mice (4). The same investigators subsequently studied differences between dorsal and ventral hypothalamic self-stimulation (5) and the effects of different isomers of amphetamine on dorsal and ventral hypothalamic self-stimulation in C57BL/6, DBA/2, and BALB/c mice (6). Few investigators used the ICSS procedure in mice during the first 3 decades after Cazala and Cardo first published their studies, but the popularity of the mouse ICSS model has increased during the past decade. The first study to report ICSS responding in genetically modified mice was performed in the laboratory of Taketoshi Ono in 2002 (7). In this study, dopamine D_2 receptor knockout mice were initially trained to self-stimulate the lateral hypothalamus by nosepoking. During the testing phase of the experiment, the mice were required to reach predetermined locomotor distance criteria or find a certain location in an open field setup to obtain stimulation (7, 8). Thus, rather than using ICSS as a tool to assess brain reward states, this study used brain stimulation as a reward for performing a particular task. A more conventional study of brain reward function using the rate-frequency curve shift method of ICSS (see below) also used dopamine D_2 receptor knockout mice and was published by the laboratory of Roy Wise in 2005 (9).

The mouse ICSS procedure has been demonstrated to be a powerful tool in studying the underlying mechanisms of mood states and disorders, such as depression and mania. Elevations in brain reward thresholds may be considered an operational measure of anhedonia, defined as diminished interest or pleasure in hedonic stimuli. Anhedonia is one of the core symptoms of depression, one of the negative symptoms of schizophrenia and a symptom of withdrawal from a variety of drugs of abuse (10). Elmer et al. (9) have demonstrated that mice null for the dopamine D_2 receptor required higher current intensities compared to their wildtype counterparts to maintain ICSS responding. DiNieri and colleagues showed that administration of the κ-opioid receptor agonist U50488, a compound that induces a negative mood state (11), resulted in depressive-like effects reflected in elevations in ICSS frequency thresholds (12).

By contrast, the extreme euphoria characterizing mania (10) may be reflected in lowered brain reward thresholds. Using the mouse ICSS procedure, Roybal et al. (13) demonstrated that mice carrying a mutation in the *CLOCK* gene required lower current intensities to sustain ICSS response rates on a fixed-ratio 1 schedule of reinforcement compared with control mice. Additionally, *CLOCK* mutant mice required lower doses of cocaine to exhibit lowering of ICSS frequency thresholds compared with control mice. A similar behavioral profile has been demonstrated for mice in which cAMP response element-binding protein function was disrupted in the nucleus accumbens (12).

The mouse ICSS procedure additionally appears to be a promising assessment tool in the field of psychopharmacology. Increased brain reward sensitivity has been demonstrated in mice after acute administration of drugs of abuse, including amphetamine (6, 9, 14), cocaine (15–17), and morphine (9), reflecting the hedonic state and reward enhancement induced by these drugs. Similarly, after the discontinuation of prolonged administration of drugs of abuse, mice demonstrate a decrease in brain reward sensitivity, reflecting the anhedonic effects of drug withdrawal. This drug withdrawal-induced anhedonia has been shown for withdrawal from nicotine (18, 19) and cocaine (unpublished data from our laboratory).

Several laboratories have now established the mouse ICSS procedure and shown robust ICSS responding in numerous inbred mouse strains (for review, see Table 1). Although many ICSS procedures have been developed, only two reliable experimental procedures have been experimentally validated and are widely used in research today: the discrete-trial current-intensity procedure and rate-frequency curve-shift procedure. Both of these procedures have been established in the laboratory mouse and allow the

Table 1
Summary of ICSS studies that have been performed using mice as the subjects

References	Mouse strain	Summary of results
Cazala and Cardo (4)	DBA/2 BALB/c	BALB/c mice exhibited increased ICSS response rates compared with DBA/2 mice
Cazala et al. (5)	DBA/2 BALB/c C57BL/6	BALB/c mice exhibited the highest and C57BL/6 mice the lowest response rates for both dorsal and ventral hypothalamic self-stimulation. Seizure frequency during self-stimulation behavior was highest in the DBA/2 strain and lowest in the BALB/c strain
Cazala (6)	DBA/2 BALB/c C57BL/6	Differential effects of isomers of the psychomotor stimulant amphetamine were demonstrated between DBA/2, BALB/c, and C57BL/6 strains and between dorsal and ventral hypothalamic stimulation sites
Kokkinidis and Zacharko (14)	Swiss	Mice demonstrated reliable self-stimulation behavior in a modified hole-board task. Amphetamine dose-dependently decreased self-stimulation rates
Cazala (29)	C57BL/6	Low doses of the α-noradrenergic agonist clonidine did not affect ICSS responding, whereas intermediate doses of clonidine increased responding for dorsal lateral hypothalamic stimulation but decreased responding for ventral lateral hypothalamic stimulation. Higher doses of clonidine disrupted ICSS responding for both ventral and dorsal lateral hypothalamic stimulation. Phentolamine disrupted ICSS responding of the dorsal hypothalamus compared with the ventral hypothalamus

(continued)

Table 1
(continued)

References	Mouse strain	Summary of results
Cazala and Guenet (30)	Recombinant inbred strains DBA/2 × BALB/c	The distribution pattern of ICSS performance among recombinant inbred strains suggested that the difference between the low ICSS response rates of DBA/2 mice and high response rates of BALB/c mice is regulated by a single gene. The genetic determination of sensitivity to convulsions was suggested to be more complex and independent of the genetic determination of ICSS response rates. Mice with higher response rates exhibited lower current-intensity threshold values, suggesting that response rates and current-intensity thresholds could be under the same genetic control mechanism
Garrigues and Cazala (31)	BALB/c DBA/2	One week after mice were trained in the ICSS procedure, the lateral hypothalamus was experimenter-stimulated, and catecholamine levels were assessed post-mortem. Stimulation of the lateral hypothalamus did not affect catecholamine content but did affect catecholamine turnover
Zacharko et al. (32)	CD-1	Whereas escapable shock did not influence ICSS performance, inescapable shock reduced ICSS responding of the medial forebrain bundle and nucleus accumbens but not substantia nigra
Bowers et al. (33)	CD-1	The one-hole head-dipping task and two-hole discrimination procedure were used to assess ICSS behavior with electrodes implanted in the A9 cell group of mice treated with pimozide, a dopamine receptor blocker. ICSS response rates decreased upon drug treatment, but response accuracy remained unaffected
Kokkinidis et al. (34)	CD-1	Amphetamine withdrawal decreased ICSS response rates
Zacharko et al. (35)	DBA/2J BALB/cByJ C57BL/6J	Uncontrollable shock decreased rates of responding for nucleus accumbens ICSS in DBA/2J mice but did not affect response rates in C57BL/6J mice. Remarkably, uncontrollable footshock increased ICSS response rates in BALB/cByJ mice
Zacharko et al. (36)	DBA/2J BALB/cByJ C57BL/6J	ICSS response rates of the mesocortex decreased after uncontrollable footshock in BALB/cByJ and DBA/2J mice and remained unaffected in C57BL/6J mice
Zacharko et al. (37)	CD-1	Decreases in response rates for dorsal ventral tegmental area ICSS were observed after uncontrollable footshock, whereas ventral tegmental area ICSS remained unaffected by uncontrollable stressors

(continued)

Table 1
(continued)

References	Mouse strain	Summary of results
Zacharko et al. (38)	CD-1	The distribution of ICSS response rates was suggested to be heterogeneous along rostral-caudal or dorsal-ventral gradients of several brain regions within the mesolimbic and nigrostriatal systems and their connecting pathways. Variations in ICSS response rates were suggested to occur within different regions in specific brain areas
Wolfe and Zacharko (39)	CD-1	Decreases in response rates for ICSS of the prefrontal cortex induced by uncontrollable footshock were ameliorated by chronic administration of the tricyclic antidepressant and norepinephrine reuptake inhibitor desmethylimipramine
Hebb et al. (40)	CD-1	Intraventricular interleukin-2 administration decreased frequency thresholds in the dorsal VTA but did not affect ICSS of the VTA
Yavich and Tiihonen (41, 42)	BALB/c	Electrically evoked dopamine release was recorded in freely moving mice after ICSS of the medial forebrain bundle. Decreasing the rate of self-stimulation, by either increasing the fixed-ratio schedule of reinforcement or decreasing the rewarding value of the electrical stimulation, allowed for reliable recording of dopamine overflow peaks
Ikeda et al. (27)	C57BL/6J	The head-dipping and place-learning ICSS procedures were compared in C57BL/6J mice
Tran et al. (7)	Dopamine D_2 receptor knockout mice	Dopamine D_2 receptor knockout mice did not show disruptions in ICSS behavior of the lateral hypothalamus but exhibited delays in acquiring a spatial learning task in which brain stimulation was used as the reward
Gillis et al. (16)	Swiss-Webster	Cocaine and the dopamine D_1-like receptor agonist SKF-82958 dose-dependently potentiated the rewarding effects of the lateral hypothalamus in Swiss-Webster mice in the discrete-trial current-intensity ICSS procedure
Hebb et al. (43)	CD-1	Exposure to a predator odor did not affect ICSS of the ventral tegmental area in CD-1 mice
Hebb and Zacharko (44)	CD-1	Evidence was provided for an interface between cholecystokinin and opioids in modulating ICSS behavior
Tran et al. (8)	Dopamine D_1 receptor knockout mice	Dopamine D_1 receptor knockout mice required higher current intensities to sustain stable ICSS behavior. Learning was delayed in these mice during the acquisition of a spatial learning task in which brain stimulation was used as the reward compared with wildtype counterparts

(continued)

**Table 1
(continued)**

References	Mouse strain	Summary of results
Elmer et al. (9)	Dopamine D_2 receptor knockout mice	ICSS frequency thresholds in dopamine D_2 receptor knockout mice were lower at baseline, and 50% higher stimulation currents were required than for wildtype mice. Morphine potentiation of brain reward stimulation was abolished in D_2 receptor knockout mice, whereas amphetamine-potentiated ICSS frequency thresholds remained unchanged
Oksman et al. (45)	C57BL/6J	Spontaneous mutation resulting in the absence of α-synuclein in C57BL/6J mice increased ICSS response rates
Roybal et al. (13)	*CLOCK* knockout mice	Mania-like behavior resulted from mutation of the *CLOCK* gene. Mutant mice were demonstrated to require lower minimal current intensities to sustain reliable ICSS response rates on a fixed-ratio 1 schedule of reinforcement compared with control mice. Additionally, *CLOCK* mutant mice required lower doses of cocaine to decrease frequency reward thresholds
Stoker et al. (19)	C57BL/6J	Chronic nicotine administration lowered current-intensity ICSS thresholds, whereas spontaneous and precipitated nicotine withdrawal elevated ICSS thresholds in C57BL/6J mice
Johnson et al. (18)	C57BL/6J	Spontaneous nicotine withdrawal elevated current-intensity ICSS thresholds in C57BL/6J mice
Takahashi et al. (46)	CBA/J	ICSS of the medial forebrain bundle enhanced cell proliferation of the hippocampal dentate gyrus
DiNieri et al. (12)	Inducible bitrans-genic mice	mCREB-expressing mice were increasingly sensitive to the reward-potentiated effects of cocaine in the rate-frequency curve-shift ICSS procedure but were unaffected by the κ-opioid antagonist U50,488
Straub et al. (17)	C57BL/6	Using the rate-frequency curve-shift ICSS procedure, frequency thresholds were potentiated by the psychomotor stimulant cocaine and the benzodiazepine diazepam in C57BL/6J mice
Fish et al. (15)	C57BL/6J DBA/2J	Alcohol and cocaine lowered ICSS frequency thresholds in both C57BL/6J and DBA/2J strains. High drug doses that were ineffective in C57BL/6J mice continued to lower thresholds in DBA/2J mice
Elmer et al. (26)	DBA/2J C57BL/6J	Administration of the opiate morphine lowered ICSS frequency thresholds in C57BL/6J mice but elevated thresholds in DBA/2J mice

quantitative measurement of brain reward. This chapter describes the experimental setup for both the discrete-trial current intensity procedure (originally developed for rats by Kornetsky and Esposito [20]) and rate-frequency curve-shift procedure (originally developed by the laboratory of Gallistel [21]) as they have been described in the literature by the laboratories using these procedures. Notably, although laboratories practicing the ICSS procedure tend to be consistent in their preferred study design, many aspects of the methodology for both procedures are interchangeable. For example, instead of varying the current intensity when using the discrete-trial procedure, the frequency can be varied. Similarly, when applying the method described for the rate-frequency curve-shift procedure, the current intensity, rather than the frequency, can be manipulated to provide rate-intensity functions. These two ICSS procedures reflect different behavioral contingencies, but the stimulation parameters and characteristics are not unique to any one ICSS procedure.

5. Equipment, Materials, and Setup

5.1. Equipment for Surgical Procedure

1. Stereotaxic frame for mice (Kopf Instruments, Tujunga, CA, USA, or similar).

2. Bipolar or monopolar electrodes (Plastics One, Roanoke, VA, USA, or similar).

3. Surgical screws (3.2 mm length, Plastics One, Roanoke, VA, USA, or similar).

4. Anesthetics: isoflurane (requires vaporizer and oxygen supply) or ketamine and acepromazine.

5. Dental acrylic (Ortho-Jet, Lang Dental, Wheeling, IL, USA) and/ or resin-ionomer cement (Den-Mat, Santa Maria, CA, USA).

6. Surgical tools: stainless steel blades, forceps, surgical and/or electric hand drill, precision screwdriver, spatula for dental cement application, wound clips to hold skin away from the skull.

7. Postsurgical analgesics, according to institutional guidelines.

5.2. Surgery

Surgeries are generally similar for the discrete-trial current-intensity and rate-frequency procedures. Traditionally, however, bipolar electrodes have predominantly been used for the discrete-trial current-intensity procedure, whereas monopolar electrodes have mostly been used for the rate-frequency procedure. Nevertheless, as mentioned above, bipolar and monopolar electrodes can be used with either procedure. The main difference between these two electrodes is that monopolar electrodes must be grounded by wrapping a non-insulated stainless steel wire around a screw that is threaded into the skull.

1. Mice are anesthetized during surgery by 1.5% isoflurane inhalation in oxygen. Alternatively, a 1 mL/kg injection of a 100 mg/kg ketamine/2.5 mg/kg acepromazine cocktail may be administered intraperitoneally.

2. The hair on the head is shaved carefully to ensure that the whiskers are left intact.

3. The mouse is positioned in a stereotaxic frame. Protective ointment is applied to the eyes, and the skin is disinfected with iodine and alcohol.

4. A rostrocaudal incision approximately 1.5 cm long is made starting from the level of the eyes toward the back of the head. The skin is pulled down from the skull toward the surface of the table and secured with wound clips. The exposed skull is cleaned of meninges and blood. The skull is then very slightly roughened with a surgical blade to increase the adhesive surface on the skull for the later application of the dental cement.

5. A hand drill is used to drill holes for fixation of stainless steel screws to the skull to keep the electrode in place. Four screws are used to attach the electrode to the skull. The use of several screws increases the chances that the head mount will remain in position for the duration of the experiments. To ensure that the screws adhere well to the skull, the diameter of the drilled holes should be slightly smaller than the diameter of the skull screws so that some small pressure needs to be applied for the skull screws to be screwed into the skull.

6. Bregma and lambda measurements are subsequently taken and used to ensure that the skull is in a flat position (bregma and lambda are on an equal dorsoventral position, as are points equidistantly to the left and right of bregma).

7. A hole is drilled by an electric drill over the intended stimulation site. An electric drill is preferred over a hand drill when drilling the electrode insertion site because it provides more precision. A stainless steel bipolar electrode (0.20 mm diameter, 6 mm length for implantation into the lateral hypothalamus) is implanted through the hole into the intended stimulation site. Different electrode lengths may be used, depending on the intended stimulation site. A site commonly used for electrode placement is the medial forebrain bundle at the level of the lateral hypothalamus using the following coordinates as previously described for C57BL/6J mice (19, 22): anterior/posterior, 1.6 mm; medial/lateral, 1.0 mm; dorsal/ventral, 5.3 mm from flat skull (23).

8. A resin ionomer is applied around the electrode and screws and is supplemented with a top layer of dental acrylic.

9. The wound is covered with iodine or triple antibiotic ointment, and postsurgical analgesics are administered.

10. Mice are single-housed after surgery. Although protective caps are commercially available that will protect the electrode from biting and other damage caused by cagemates, these are relatively large in size and therefore more suitable for rats than mice.

11. Mice are allowed at least 7 postsurgery recovery days before the beginning of training.

5.3. Apparatus

1. Intracranial self-stimulation operant chambers are enclosed within light- and sound-attenuated chambers (Fig. 2). Preferably, these chambers should contain a wheel manipulandum (see Sect. 7 for details), but levers or nosepoke holes may also be used as manipulanda. The manipulandum should be mounted low on one of the walls of the operant chamber to ensure accessibility for the mice. Metal rod floors allow for easy cleaning of the operant chambers. There are no specific requirements for the size of the operant chambers or for materials used. A transparent door at the front of the operant chamber, however, allows the experimenter to observe the mouse behavior within the operant chamber.

2. Constant current stimulators (Stimtek, Acton, MA, USA; Med Associates, St. Albans, VT, USA) are used to deliver the electrical stimulation.

3. An oscilloscope is needed to ensure that the stimulators deliver the specified type of stimulation in terms of parameters and to troubleshoot equipment problems.

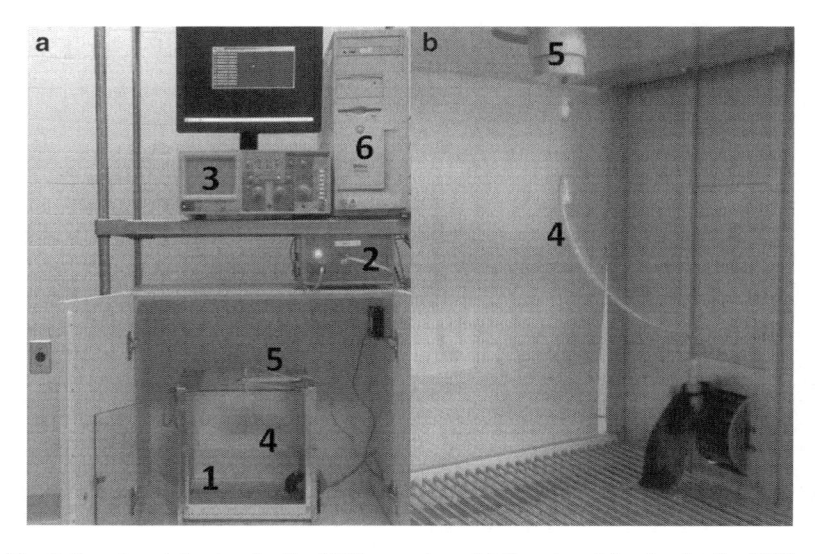

Fig. 2. Experimental setup for the ICSS procedure. (**a**) Experimental setup for the ICSS procedure: (*1*) intracranial self-stimulation operant chamber, (*2*) electrical stimulator, (*3*) oscilloscope, (*4*) flexible lead, (*5*) commutator, (*6*) computer. (**b**) A laboratory mouse in the ICSS testing chamber.

4. Flexible monopolar or bipolar leads (depending on the type of electrodes implanted during surgery) coated with plastic (Plastics One, Roanoke, VA, USA) are used to connect the stimulation electrodes to the stimulators.

5. Gold-contact swivel commutators (Plastics One, Roanoke, VA, USA) are used to connect the leads to the stimulator to allow the mice free movement without tangling the leads.

6. A standard computer is used to control the stimulation parameters, data collection, and all test session functions.

7. Software to program the computer is often provided by the manufacturer and customized upon request with the purchase of the stimulators.

5.4. Discrete-Trial Current-Intensity Procedure

5.4.1. Training

The following parameters are used for stimulation of the lateral hypothalamus during the training sessions. The electrical current is set to 180 µA. The pulse duration is set to 100 µs. The pulse interval is set to 10,000 µs, and 50 pulses are administered within one stimulation train. These parameters result in a stimulation train duration of 495 ms with a 100 Hz frequency. On the first day of training, mice are allowed to explore the operant chambers, and responding on the manipulandum is reinforced on a fixed-ratio 1 schedule of reinforcement. Two consecutive sessions are conducted; each is terminated when the mouse earns 200 reinforcements or when at least 60 min have elapsed. On the second day of training, mice are again allowed to respond on a fixed-ratio 1 schedule of reinforcement until successful acquisition of responding under this schedule, defined as two sessions during which the mouse receives 200 reinforcements in less than 10 min.

Mice are subsequently trained in the discrete-trial current-threshold procedure. Each trial begins with the delivery of a non-contingent electrical stimulus followed by a 7.5-s response window within which the subject is able to make a response to receive a second contingent stimulus identical in all parameters to the initial noncontingent stimulus. A response during this time window (i.e., limited hold period) is labeled a positive response (Fig. 3b), whereas the lack of a response is labeled a negative response (Fig. 3a). During a 2-s period immediately after a positive response, additional responses have no consequences. The intertrial interval that follows either a positive response or the end of the response window (in the case of a negative response) has an average duration of 10 s (7.5–12.5-s range). Responses that occur during the intertrial interval are labeled timeout responses and result in a further delay of the onset of the next trial (Fig. 3c). During training in the discrete-trial procedure, the duration of the intertrial interval and delay periods induced by timeout responses are gradually increased using four steps. The initial settings use a 3-s intertrial interval, and

The Discrete-Trial Current-Intensity Threshold Procedure

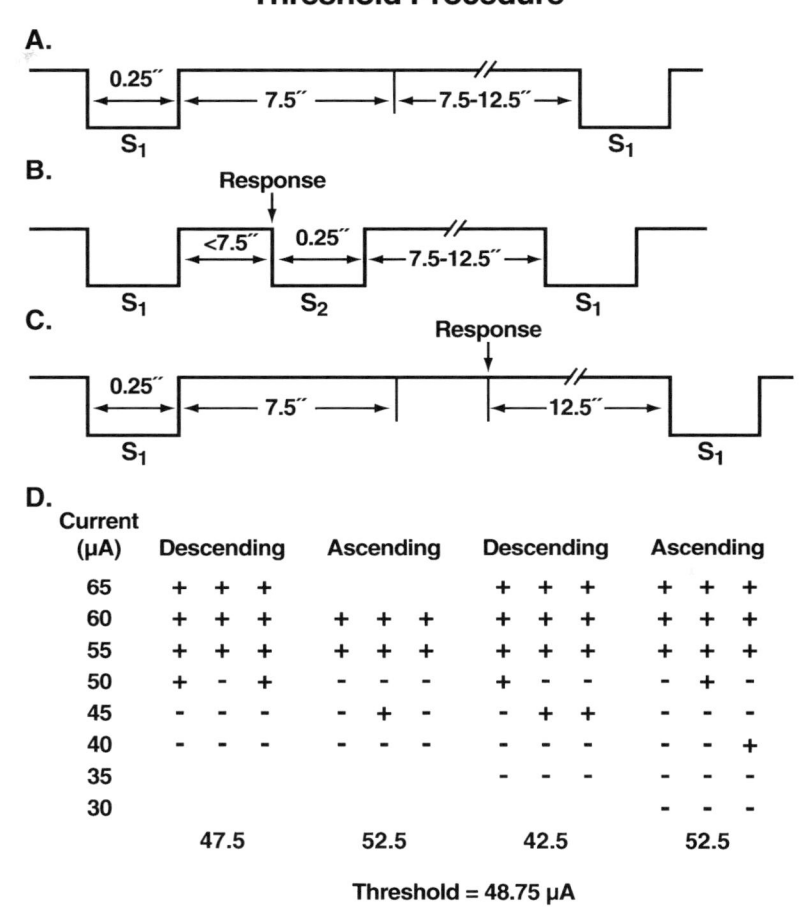

Fig. 3. The discrete-trial current-intensity procedure. (**a**) Negative response: noncontingent stimulus S$_1$ is followed by a 7.5-s response window and subsequently by a 7.5–12.5-s intertrial interval. (**b**) Positive response: noncontingent stimulus S$_1$ is followed by a positive response within the 7.5-s response window. Contingent stimulus S$_2$ is delivered and followed by the 7.5–12.5-s intertrial interval. (**c**) Timeout response: a response made during the intertrial interval results in a 12.5-s delay before delivery of the next noncontingent stimulus. (**d**) Schematic representation of threshold estimation: the threshold of an ICSS session is defined as the average of the thresholds of four individual series of descending and ascending current intensities. Adapted with permission from Markou and Koob (28).

1-s timeout "penalty" added to the intertrial interval, if the mouse responds during the intertrial interval. When a minimum of 70% positive responses is reached over at least 100 trials, mice are advanced to a 5-s intertrial interval period with a 3-s timeout window and subsequently to an 8-s intertrial interval and 5-s timeout, and finally a 10-s intertrial interval with a 10-s timeout "penalty." Mice are considered to have reached criterion performance if they complete 70% positive responses over at least 100 trials.

5.4.2. Testing

After successful completion of the training phase, mice are tested in the discrete-trial procedure in which stimulation current intensities are varied according to the classical psychophysical method of limits. A test session consists of four alternating series of descending and ascending current intensities starting with a descending series (Fig. 3d). Blocks of three trials are presented to the subject at a given stimulation intensity, and the intensity is changed by 5 μA steps between blocks of trials. The initial current intensity is set at approximately 40 μA above the baseline current threshold for each animal. Each test session lasts approximately 30–40 min. Importantly, the thresholds should be stable over days (standard deviation over 3 consecutive days should be less than 10% of the mean threshold during these 3 days) before any experimental manipulations are implemented. Although some mice may appear to exhibit stable thresholds within a few days, they should be tested under baseline conditions for at least 3 weeks (5 days per week) because thresholds typically tend to lower significantly within this time period.

5.4.3. Data Analysis

Four dependent variables are provided by the discrete-trial current-intensity procedure for behavioral assessment: threshold, response latency, extra responses, and timeout responses.

The threshold value of each series is defined as the midpoint in microamperes between the current intensity level at which the animal makes two or more positive responses out of the three stimulus presentations and the level at which the animal makes fewer than two positive responses (Fig. 3d). The animal's estimated current threshold for each test session is the mean of the four series' thresholds.

The response latency is defined as the average time in seconds that elapsed between the delivery of the electrical stimulus and the turning of the wheel manipulandum for all of the trials that led to a positive response. A quarter turn of the wheel manipulandum is considered one response.

Extra responses are recorded during the 2-s period after a positive response. Additional responses during this 2-s period have no consequences. Timeout responses are defined as responses that occur during the intertrial interval. The intertrial interval is initiated after the 7.5-s limited hold has elapsed or after the 2 s that followed a positive response.

Experimental questions typically focus on ICSS thresholds, often reporting latencies as an indication that motor performance was not disrupted by the experimental manipulation. Extra and timeout responses are not typically reported because they are often not affected by experimental manipulations. However, timeout responses reflect the ability to withhold inappropriate responses and may therefore be considered indicative of changes in impulsive behavior (24). Increases in extra responses reflect increased vigor in turning the wheel manipulandum because only a quarter turn of the wheel is required to record it as a response.

For analyses of the data:

1. Establish a baseline variable by averaging the thresholds of 3–5 sessions before the test day or the beginning of implementation of manipulations.

2. Obtain variable(s) for the test day(s).

3. Calculate the test day variable as a percentage of the baseline variable. Because of large variability among thresholds and response latencies in individual mice, threshold and response latency data are most frequently reported as percent changes from baseline. However, extra and timeout responses are often expressed as difference scores from baseline values or as absolute values for analysis. Calculating the percentage change from baseline is often not applicable to these two measures because the baseline values could potentially equal zero.

4. Perform appropriate analyses of variance (ANOVAs).

5.5. Rate-Frequency Curve-Shift Procedure

5.5.1. Training

Training and testing for the ICSS rate-frequency curve-shift training procedure in mice is explained here as previously described by the laboratory of Carlezon (16, 25), but also see the methodology described by the laboratory of Roy Wise (9, 26). Although the main procedures that have been used for testing between these two laboratories are similar, they have differences, such as different parameters of the electrical stimulation.

The following parameters are applied for the training sessions. The duration of the electrical stimulation train is set to 500 ms. The electrical current is set to 150 µA, and the pulse duration is set to 100 µs at a frequency of 141 Hz. Each stimulation train is followed by a 500 ms timeout period, during which additional responses have no consequences, and no additional stimulation can be received. The duration of a typical training session is 60–90 min. Each session begins with five priming stimulation trains, after which the mouse is allowed to operate the manipulandum to obtain further stimulation rewards on a fixed-ratio 1 schedule of reinforcement. The electrical current intensity is varied by the researcher during these training sessions, with the goal of determining the lowest value at which reliable responding can be maintained (~40 rewards per minute) for at least 3 consecutive days. The current intensity thus obtained is used throughout the remainder of the experiment.

5.5.2. Testing

Upon establishment of the minimal rewarding current intensity, mice are subsequently tested in the rate-frequency curve-shift procedure. Each trial has a 60-s duration and begins with the delivery of a 5-s priming period, during which five noncontingent electrical stimuli are administered to allow the mouse to experience the obtainable stimulation frequency for this trial. This priming period is followed by a 50-s testing phase, during which the number of positive responses is recorded. The testing period is then followed by a 5-s timeout interval before the next trial is initiated using a different stimulus frequency. A training session involves testing the mice at 15 descending stimulation frequencies, with frequencies lowered by 0.05 log unit steps (equals 10%) per trial. This 15-min procedure is repeated six consecutive times, such that a total daily training session lasts 90 min. The initial frequency is set so that the mice are responsive during the first six or seven stimulation trials.

5.5.3. Data Analysis

The threshold value of each series is defined as the frequency at which the stimulation is rewarding to the mouse. The threshold is established by plotting the response rate as a function of the log of the pulse frequency (Fig. 4). Frequencies at which responding is sustained at, for example, 20, 30, 40, 50, and 60% (16) or 20 and 80% (26) of the maximum response rate, are defined and plotted. When a curve is plotted through these frequencies, the value at the intersection of this line with the x-axis defines the T_0 threshold (Fig. 4). Another means of establishing the threshold is to calculate the M_{50}, which is the stimulation frequency that maintains the half-maximal rate of responding. For establishment of this threshold, a horizontal line is drawn from the half-maximum wheel turns on the y-axis of the graph until it intersects with the plotted curve. The x-axis value at this intersection is the M_{50} threshold (Fig. 4).

For analyses of the data:

1. Establish a baseline threshold by averaging the thresholds of 3–5 sessions before the test day.

2. Obtain threshold(s) for the test day(s).

3. Calculate the test day threshold as a percentage of the baseline threshold.

4. Perform appropriate ANOVAs.

5.6. Histology

Depending on the experimental question addressed, the electrode placements may need to be determined. For example, if the intent is to determine brain sites that support self-stimulation, then the exact site of stimulation must be ascertained. Alternatively, if a particular mouse strain did not perform well in the ICSS procedure, or if genetically engineered mice had different thresholds compared with their wildtype counterparts, then determining that very similar stimulation sites were targeted is important. By contrast, if the interest is in within-subjects manipulations, then performing histology is usually unnecessary because all mice that exhibited stable thresholds throughout the study are assumed to have accurate electrode placements. For histology, brains are fixed with paraformaldehyde, sliced, and stained with cresyl violet to confirm placement of the electrodes.

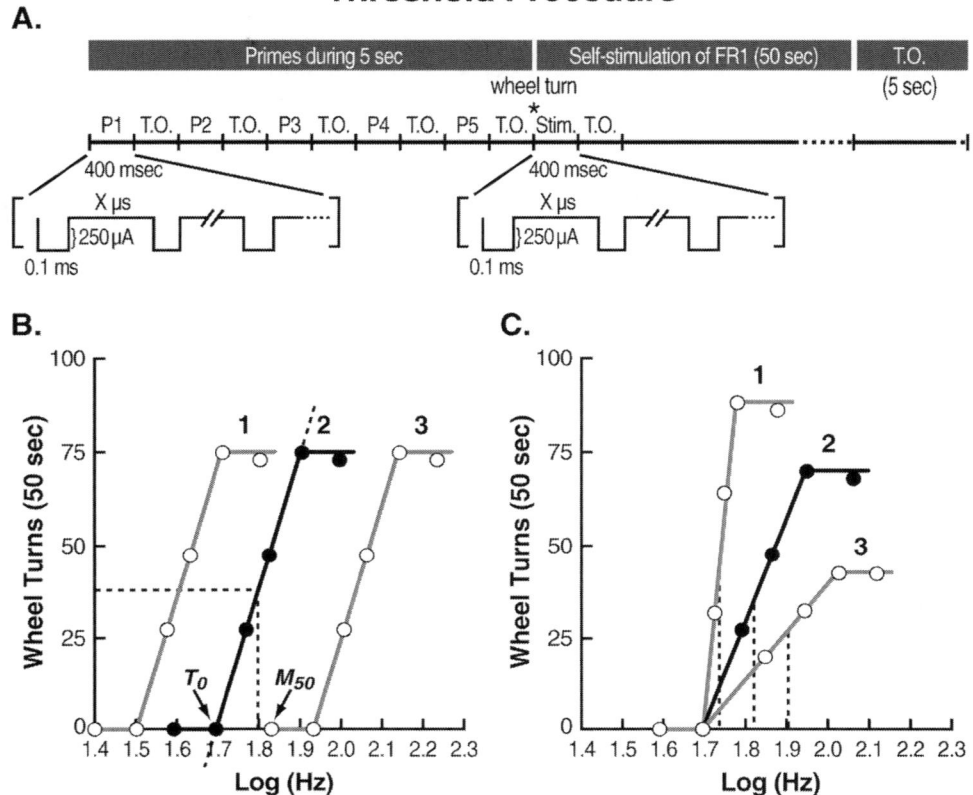

Fig. 4. The rate-frequency curve-shift procedure. (**a**) Schematic depiction of an ICSS trial of the rate-frequency curve-shift procedure: each trial has a 60-s duration and starts with a 5-s priming period. Each of the five noncontingent electrical stimuli (P1–P5) is followed by a timeout (TO) period. After the five primes, a 50-s testing phase begins, during which the number of positive responses is recorded. This testing period is followed by a 5-s timeout interval before the next trial is initiated. (**b**) Curve 2 illustrates the theoretical functions that relate the response rate (number of wheel turns) to the frequency of the electrical stimulation (Log Hz). Frequencies at which responding is sustained at 20, 30, 40, 50, and 60% of the maximum response rate are plotted. T_0 and M_{50} can both be used as brain reward thresholds. T_0 measures the frequency at which the electrical stimulation becomes rewarding and is defined as the value of the curve (Log Hz) at which the plotted curve intersects with the x-axis. M_{50} is defined as the stimulation frequency that maintains the half-maximum rate of responding. Curve 1 demonstrates a lowering in brain reward thresholds (i.e., a lower frequency is required to obtain a similar number of wheel turns compared with Curve 2). Conversely, Curve 3 illustrates an example of elevations in brain reward thresholds. (**c**) Motoric effects caused by experimental manipulations may result in increased responding (Curve 3) or decreased responding (Curve 1). In this scenario, M_{50} can be artificially affected because it is a direct function of maximal response rates. T_0, however, is an indirect function of the maximal response rates and is therefore not affected by motoric effects of experimental manipulations. Adapted with permission from Carlezon and Chartoff (25).

6. Anticipated Results

Thresholds should be stable (standard deviation over 3 consecutive days less than 10% of the mean thresholds during these 3 days) before any experimental manipulation is applied to allow for proper evaluation of the results. When ICSS thresholds are lowered, less electrical stimulation is required for the mouse to perceive the stimulation as rewarding. Lowered thresholds are therefore considered to reflect the increased value of the stimulation. This phenomenon is often referred to as reward enhancement, reward facilitation, or reward potentiation. By contrast, elevations in ICSS thresholds reflect a decrease in the value of the stimulation.

Differences in latency may reflect changes in locomotor performance. Increases in latency must be studied carefully because these can be caused by manipulation-induced motoric effects or sickness. Increases in extra responses often reflect increased force at turning the wheel manipulandum. Increases in timeout responses may reflect increased impulsive behavior and inability to withhold inappropriate responding (24).

The acute administration of drugs of abuse, such as amphetamine (6, 9, 14), cocaine (15–17), and morphine (9), lowers thresholds in laboratory mice and therefore enhances brain reward function in this species. Elevations of brain reward thresholds after the administration of drugs of abuse are indicative of a negative mood state and suggest that the dose used may be too high to induce the acute euphorigenic effects of relatively low doses of drugs of abuse. Similarly, elevations in brain reward thresholds may be induced by the administration of pharmacological compounds other than drugs of abuse, reflecting the aversive properties of these compounds, which occurs, for example, with the κ-opioid receptor agonist U50,488 (12). Discontinuation of prolonged administration of drugs of abuse, such as nicotine (18, 19) and cocaine (unpublished data), similarly elevates brain reward thresholds, indicating decreased brain reward function or anhedonia in mice.

When using an experimental design with mice that differ genetically (either different inbred strains of mice or genetically engineered mice compared with their wildtype littermates), the initial brain stimulation required to produce a rewarding effect may be different. The electrical current required to produce the M_{50} threshold did not differ between C57BL/6J and DBA/2J mice (26), but dopamine D_2 receptor knockout mice required a higher stimulation frequency than their wildtype littermates (9). Mice with disruption in *CLOCK* gene function required a lower stimulation frequency than control mice (13).

7. Experimental Variables and Troubleshooting

7.1. Mice Fail to Nosepoke or Lever-Press for Electrical Stimulation

Although head-dipping (14), place-learning (27), and lever-pressing (4) procedures have been applied in mouse ICSS studies, personal communications with other ICSS laboratories revealed that wheel manipulanda have been the most successful in maintaining sustained responding in mouse studies. The use of a wheel manipulandum is therefore recommended.

7.2. Mice Do Not Acquire the Fixed-Ratio 1 Schedule of Reinforcement

Because the rewarding effects of ICSS are very powerful, the stimulation intensity may be too high for some mice to immediately begin self-stimulation. Alternatively, the electrode may have been misplaced during surgery or displaced if the head mount was not steady, the electrode assembly may have come undone, or the mouse may have an infection after the surgery, all of which result in poor self-stimulation. When a mouse does not reach successful acquisition of the fixed-ratio 1 schedule, it is excluded from further experimentation.

7.3. Mice Do Not Train Well in the Discrete-Trial Current-Intensity Procedure

After having been trained on a fixed-ratio 1 schedule of reinforcement, mice often become distracted after they progress to the discrete-trial current-intensity procedure, in which not every wheel turn leads to reinforcement. Not giving the mice too many non-contingent electrical rewarding stimuli if they remain unresponsive is important. When the mice are no longer turning the wheel, the program should be discontinued and only restarted when the mouse begins to regain its focus on the wheel. Additionally, 24-h food restriction may be beneficial during training. Importantly, however, mice should be food restricted only when they have made at least 70% correct responses out of at least 100 trials in the first step of the discrete-trial current intensity procedure (3-s intertrial interval with a 1-s penalty timeout). If the mice do not learn the contingency of the procedure, then they will not benefit from food restriction.

7.4. Mice Show Motor Responses upon Brain Stimulation

Upon electrical stimulation, mice will occasionally show motor responses that may affect training. The stimulation may be too intense, or the electrode may not have been placed in the correct position. Motor responses may include shaking of the front paws or twitching of the head. Such "side-effects" of the stimulation may cause the mouse to turn away involuntarily from the manipulandum upon receiving brain stimulation. For the rate-frequency curve-shift procedure, the electrical current may simply be reduced. In the discrete-trial current-intensity procedure, however, reducing the current may affect the rate of training. Although mice that show mild motor artifacts upon electrical stimulation can still be successfully trained, such training may be more time-consuming. The stimulus intensity described in the training protocol above is often most successful when training mice, even with the occurrence of motor responses. However, while closely monitoring the mouse's behavior, lowering the current intensity or frequency of the stimulation may be effective. When the mouse has successfully learned the ICSS procedure, the stimulation intensity can be decreased during the threshold establishment phase, and motor artifacts will no longer occur.

7.5. Mice Suddenly Fail to Respond During the Testing Phase

From one session to the next, mice may suddenly stop responding during testing. First, check to ensure that the lead is properly placed and in good condition. The lead may be loosely attached, reversed, or damaged. Second, inspect the commutator, stimulator, and swivel and all of the connections to ensure that they are working properly. If all of the connections and equipment are working properly, then the problem is most likely with the mouse itself. The electrode assembly may have moved, thus mislocating the electrode, or an infection may be causing sickness behavior. In this case, the mouse should be excluded from further experimentation.

7.6. Mice Fail to Stabilize Their Reward Thresholds

If thresholds fail to stabilize (i.e., standard deviation over three consecutive sessions is more than 10% of the average threshold) over a prolonged period of testing, then the mice should be excluded from further experimentation because such mice will yield very noisy data.

7.7. Very Low Thresholds Result in Variable Data

In the discrete-trial current-intensity procedure, excessively low baseline thresholds may lead to highly variable data or failure of the manipulations to affect these low thresholds compared with the higher thresholds of other subjects. Changing the parameters of the electrical stimulus for a particular mouse will elevate the baseline threshold values (Fig. 1) and thus alleviate this problem. For the discrete-trial procedure, the frequency of the stimulation should be decreased when the mouse has a threshold below 70 μA during the first seven test sessions. The frequency could be set to half (i.e., 100 μs pulse duration, 20,000 μs interpulse interval, 25 pulses) of the original frequency (i.e., 100 μs duration, 10,000 μs interpulse interval, 50 pulses), which would be a decrease from 100 to 50 Hz. When the frequency of the stimulation is decreased within the first seven sessions, mice will acquire stable performance. When changes are made to the frequency of the stimulation after this time, however, thresholds are more likely to remain unstable at half-frequency reinforcement.

7.8. Coordinates for Electrode Placement in Different Mouse Strains

When using different mouse strains, different coordinates may need to be used for placement of the electrode. Especially when comparing different strains of mice in a single study, the electrode should be positioned in the same anatomical location as closely as possible, which may require different coordinates for each of the strains. Roy Wise's laboratory has demonstrated an elegant approach for such an experimental setup (26).

8. Conclusion

ICSS is a valuable and unique behavioral procedure that provides us with the opportunity to quantitatively measure reward function. Although this assay has historically focused on the laboratory rat as the subject, the mouse has become increasingly popular in recent years with the opportunities that this species offers for genetic engineering. The ICSS procedure has numerous benefits, including quantitative assessment of the involvement of neuroanatomy, neurochemistry, and genetics in reward and motivational processes. Concerns about the feasibility of this procedure in mice have been alleviated by the successful use of mouse ICSS procedures in numerous studies. In conclusion, the laboratory mouse serves as a valuable subject for the study of reward and motivational processes using the ICSS procedure.

Acknowledgments

This work was supported by National Institute on Drug Abuse grant R01DA232090 to AM. We wish to thank Dr. Nurith Amitai and Dr. Berend Olivier for their insightful comments and input during the preparation of this manuscript, Mr. Mike Arends for his editorial assistance, and Ms. Janet Hightower for her assistance with figure preparation.

References

1. Olds J (1956) Pleasure center in the brain. Sci Am 195: 105–116

2. Olds J, Milner P (1954) Positive reinforcement produced by electrical stimulation of septal area and other regions of rat brain. J Comp Physiol Psychol 47(6): 419–427

3. Yeomans JS (1990) Principles of Brain Stimulation. New York: Oxford University Press

4. Cazala P, Cardo B (1972) Etude préliminaire du comportement d'autostimulation chez la souris. Physiol Behav 9(2): 255–257

5. Cazala P, Cazals Y, Cardo B (1974) Hypothalamic self-stimulation in three inbred strains of mice. Brain Res 81(1): 159–167

6. Cazala P (1976) Effects of d- and l-Amphetamine on Dorsal and Ventral Hypothalamic Self-Stimulation in Three Inbred Strains of Mice. Pharmacol Biochem Behav 5(5): 505–510

7. Tran AH, Tamura R, Uwano T et al (2002) Altered accumbens neural response to prediction of reward associated with place in dopamine

D2 receptor knockout mice. Proc Natl Acad Sci USA 99(13): 8986–8991

8. Tran AH, Tamura R, Uwano T et al (2005) Dopamine D1 receptors involved in locomotor activity and accumbens neural responses to prediction of reward associated with place. Proc Natl Acad Sci USA 102(6): 2117–2122

9. Elmer GI, Pieper JO, Levy J et al (2005) Brain stimulation and morphine reward deficits in dopamine D2 receptor-deficient mice. Psychopharmacology (Berl) 182(1): 33–44

10. American Psychiatric Association (2000) Diagnostic and statistical manual of mental disorders. 4th edn, text rev. Washington, DC: American Psychiatric Press.

11. Bruijnzeel AW (2009) kappa-Opioid receptor signaling and brain reward function. Brain Res Rev 62(1): 127–146

12. DiNieri JA, Nemeth CL, Parsegian A et al (2009) Altered sensitivity to rewarding and aversive drugs in mice with inducible disruption

of cAMP response element-binding protein function within the nucleus accumbens. J Neurosci 29(6): 1855–1859

13. Roybal K, Theobold D, Graham A et al (2007) Mania-like behavior induced by disruption of CLOCK. Proc Natl Acad Sci USA 104(15): 6406–6411

14. Kokkinidis L, Zacharko RM (1980) Intracranial self-stimulation in mice using a modified hole-board task: effects of d-amphetamine. Psychopharmacology (Berl) 68(2): 169–171

15. Fish EW, Riday TT, McGuigan MM et al (2010) Alcohol, cocaine, and brain stimulation-reward in C57Bl6/J and DBA2/J mice. Alcohol Clin Exp Res 34(1): 81–89

16. Gilliss B, Malanga CJ, Pieper JO et al (2002) Cocaine and SKF-82958 potentiate brain stimulation reward in Swiss-Webster mice. Psychopharmacology (Berl) 163(2): 238–248

17. Straub CJ, WA Jr, Rudolph U (2010) Diazepam and cocaine potentiate brain stimulation reward in C57BL/6J mice. Behav Brain Res 206(1): 17–20

18. Johnson PM, Hollander JA, Kenny PJ (2008) Decreased brain reward function during nicotine withdrawal in C57BL6 mice: evidence from intracranial self-stimulation (ICSS) studies. Pharmacol Biochem Behav 90(3): 409–415

19. Stoker AK, Semenova S, Markou A (2008) Affective and somatic aspects of spontaneous and precipitated nicotine withdrawal in C57BL/6J and BALB/cByJ mice. Neuropharmacology 54(8): 1223–1232

20. Kornetsky C, Esposito RU (1979) Euphorigenic drugs: effects on the reward pathways of the brain. Fed Proc 38(11): 2473–2476

21. Campbell KA, Evans G, Gallistel CR (1985) A microcomputer-based method for physiologically interpretable measurement of the rewarding efficacy of brain stimulation. Physiol Behav 35(3): 395–403

22. Gill BM, Knapp CM, Kornetsky C (2004) The effects of cocaine on the rate independent brain stimulation reward threshold in the mouse. Pharmacol Biochem Behav 79(1): 165–170

23. Paxinos G, Franklin KBJ, The mouse brain in stereotaxic coordinates (second ed.). 2001, Academic Press: San Diego.

24. Amitai N, Semenova S, Markou A (2009) Clozapine attenuates disruptions in response inhibition and task efficiency induced by repeated phencyclidine administration in the intracranial self-stimulation procedure. Eur J Pharmacol 602(1): 78–84

25. Carlezon WA Jr, Chartoff EH (2007) Intracranial self-stimulation (ICSS) in rodents

to study the neurobiology of motivation. Nat Protoc 2(11): 2987–2995

26. Elmer GI, Pieper JO, Hamilton LR et al (2010) Qualitative differences between C57BL/6J and DBA/2J mice in morphine potentiation of brain stimulation reward and intravenous self-administration. Psychopharmacology (Berl) 208(2): 309–321

27. Ikeda K, Moss SJ, Fowler SC et al (2001) Comparison of two intracranial self-stimulation (ICSS) paradigms in C57BL/6 mice: head-dipping and place-learning. Behav Brain Res 126(1–2): 49–56

28. Markou A, Koob GF (1992) Construct validity of a self-stimulation threshold paradigm: effects of reward and performance manipulations. Physiol Behav 51(1): 111–119

29. Cazala P (1980) Effect of clonidine and phentolamine on self-stimulation behavior in the dorsal and ventral regions of the lateral hypothalamus in mice. Psychopharmacology (Berl) 68(2): 173–177

30. Cazala P, Guenet JL (1980) The recombinant inbred strains: a tool for the genetic analysis of differences observed in the self-stimulation behaviour of the mouse. Physiol Behav 24(6): 1057–1060

31. Garrigues AM, Cazala P (1983) Central catecholamine metabolism and hypothalamic self-stimulation behaviour in two inbred strains of mice. Brain Res 265(2): 265–271

32. Zacharko RM, Bowers WJ, Kokkinidis L et al (1983) Region-specific reductions of intracranial self-stimulation after uncontrollable stress: possible effects on reward processes. Behav Brain Res 9(2): 129–141

33. Bowers W, Hamilton M, Zacharko RM et al (1985) Differential effects of pimozide on response-rate and choice accuracy in a self-stimulation paradigm in mice. Pharmacol Biochem Behav 22(4): 521–526

34. Kokkinidis L, Zacharko RM, Anisman H (1986) Amphetamine withdrawal: a behavioral evaluation. Life Sci 38(17): 1617–1623

35. Zacharko RM, Lalonde GT, Kasian M et al (1987) Strain-specific effects of inescapable shock on intracranial self-stimulation from the nucleus accumbens. Brain Res 426(1): 164–168

36. Zacharko RM, Gilmore W, MacNeil G et al (1990) Stressor induced variations of intracranial self-stimulation from the mesocortex in several strains of mice. Brain Res 533(2): 353–357

37. Zacharko RM, Kasian M, MacNeil G et al (1990) Stressor-induced behavioral alterations in intracranial self-stimulation from the ventral

tegmental area: evidence for regional variations. Brain Res Bull 25(4): 617–621

38. Zacharko RM, Kasian M, Irwin J et al (1990) Behavioral characterization of intracranial self-stimulation from mesolimbic, mesocortical, nigrostriatal, hypothalamic and extra-hypothalamic sites in the non-inbred CD-1 mouse strain. Behav Brain Res 36(3): 251–281

39. Wolfe C, Zacharko RM (1991) Desmethylimipramine promotes recovery of self-stimulation from the prefrontal cortex following footshock. Brain Res Bull 27(5): 601–604

40. Hebb AL, Zacharko RM, Anisman H (1998) Self-stimulation from the mesencephalon following intraventricular interleukin-2 administration. Brain Res Bull 45(6): 549–556

41. Yavich L, Tiihonen J (2000) In vivo voltammetry with removable carbon fibre electrodes in freely-moving mice: dopamine release during intracranial self-stimulation. J Neurosci Methods 104(1): 55–63

42. Yavich L, Tiihonen J (2000) Patterns of dopamine overflow in mouse nucleus accumbens during intracranial self-stimulation. Neurosci Lett 293(1): 41–44

43. Hebb AL, Zacharko RM, Gauthier M et al (2003) Exposure of mice to a predator odor increases acoustic startle but does not disrupt the rewarding properties of VTA intracranial self-stimulation. Brain Res 982(2): 195–210

44. Hebb AL, Zacharko RM (2003) Central D-Ala2-Met5-enkephalinamide μ/δ-opioid receptor activation blocks behavioral sensitization to cholecystokinin in CD-1 mice. Brain Res 970(1–2): 20–34

45. Oksman M, Tanila H, Yavich L (2006) Brain reward in the absence of alpha-synuclein. Neuroreport 17(11): 1191–1194

46. Takahashi T, Zhu Y, Hata T et al (2009) Intracranial self-stimulation enhances neurogenesis in hippocampus of adult mice and rats. Neuroscience 158(2): 402–411

Chapter 20

The Female Urine Sniffing Test (FUST) of Reward-Seeking Behavior

Oz Malkesman

Abstract

Abnormal reward-seeking behavior is a key feature in several psychiatric and neuroscience diseases. Though there are numerous paradigms for measuring reward-seeking behavior in rodents, each has limitations that affect the ability of the researcher to make conclusions on the reward-seeking behavior per se. Here, we describe a novel approach for monitoring reward-seeking behavior in rodents: the female urine-sniffing test (FUST).

Lately, we found that sniffing of estrus female urine by male mice is a preferred activity of numerous mice strains and rats. In addition, this preferred activity was found to be accompanied by biological changes linked to hedonic and rewarding activities. The FUST was also been found to be sensitive to behavioral and genetic manipulation and to drug treatment related to depression and mania.

Key words: Hedonia, Reward-seeking behavior, Sniffing, Estrus female urine

1. Backgrounds and Historical Overview

One of the key features of many widespread psychiatric diseases is abnormal hedonic behavior. Anhedonia is one of the two core symptoms of major depressive disorder (1). By contrast, manic symptoms in bipolar patients commonly enhance hedonic drive and motivation (see (2) for review). In addition, drug addictions, including alcoholism, have highly rewarding components (for review see (3)). Therefore, reward/pleasure-seeking behavior and hedonia have been the subject of research of numerous animal and human studies (4).

Though these terms are related and sometimes are used interchangeably in the literature, they have a complex meaning and should be used carefully.

Todd D. Gould (ed.), *Mood and Anxiety Related Phenotypes in Mice: Characterization Using Behavioral Tests, Volume II*, Neuromethods, vol. 63, DOI 10.1007/978-1-61779-313-4_20, © Springer Science+Business Media, LLC 2011

Anhedonia can be defined as a "sharp, pervasive impairment of the capacity to experience pleasure or to respond affectively to anticipation of pleasure" ((5), p. 349). Only humans can describe their experiences and aversive/pleasurable feelings, and therefore the term hedonia/anhedonia is usually used only in the context of human studies.

Reward-seeking behavior can be defined as "the ability to elicit an approach behavior, similar to incentive motivation (an incentive is defined as any stimulus that activates an approach behavior)... Liking and positive affect are intrinsically rewarding and are measurable by approach, consummatory responses, or affective reactions" ((6), pp. 15–16). Thus, reward-seeking behavior is the term saved for animal studies.

In the literature, one can find three different approaches for studying reward-seeking behavior in laboratory animals: operant paradigms, non-operant paradigms, and brain stimulation reward techniques (7).

Widely used operant paradigm, for example, is the conditioned place preference (CPP). In this paradigm, a potentially rewarding stimulus is repeatedly paired with a set of distinct environmental cues, while a neutral control stimulus is repeatedly paired with a different set of environmental cues. Consequently, the animal will show a preference for the "rewarding" cues over the neutral when given a free choice between them (8–10).

An often-employed non-operant method of assessing reward sensitivity is the measurement of an animal's preference for sweetened fluids (sucrose/saccharin) vs. water (10–12). Mammals (including laboratory rodents) find the sweet taste of sugar has a potent motivational stimulus (13). For example, rats will not only consume sweet solution freely available but also conduct numerous tasks in an attempt to obtain this solution: press levers, run down alleys, etc. In general, as the sweet solution is more concentrated, the rodents will work more and harder in an attempt to obtain it (13).

Intracranial self-stimulation (ICSS) is an invasive, quantitative method for studying the mechanism of reward processing and regulation by brain stimulation. The animal is required to engage levels of activities to receive a pleasurable electrical current delivered through a probe, typically implanted in the nucleus accumbens (NAc) (14).

Though widely used and well known, each of these reward-seeking behavior paradigms has its own limitations, some of which are critical: Most of the operant paradigms are heavily dependent on animal ability of learning/memory and motor activity (7). Thus these operant paradigms may not reflect changes in hedonic behavior per se.

Forbes et al. (35) showed that in the non-operant paradigm for measuring an animal's preference for sweetened fluids, reduced sucrose consumption in stressed rats might result solely from diminished body weight, rather than exposure to stress, and concluded that sucrose consumption is not a valid index of responsiveness. In addition, differences in this paradigm may result from alterations in taste sensory and appetite by gene manipulation and/or pharmacological treatment, instead of changes in reward-seeking state.

At last, using brain stimulation reward techniques involves surgeries, and the results can be influenced by an animal's ability to recover without deficits in motor learning and coordination (15).

Therefore, the design of a novel approach for monitoring reward-seeking behavior is appropriate and particularly relevant. The female urine-sniffing test (FUST) is a non-operant complementary method for assessing reward-seeking behavior in rodents, which does not incorporate components of taste, activity, or memory (16).

There are numerous natural behaviors that have been found to be reinforcing approaching behavior in rodents, including feeding, drinking, sex, nest building, etc. (7). In addition to the male rodent approaching behavior to sexually receptive females, estrus female rodent urine was also found to extract behavioral displays analogous to reward incentive stimuli: increased number of approaches, longer time spent with the stimulus, and shortened approaching time to the stimulus (for review see (17)). Reward state is mediated by opioids both in male and female rodents, and the medial preoptic area of the anterior hypothalamus has been found to be a crucial site for sexual reward (for a review, see (18)). The hormonal factors and neural circuitry that control sexual behavior and reward are similar in rodents (17), thus we hypothesized that sexual incentive could be used to effectively evaluate and measure hedonic behavior combining two previous areas of investigation: the olfactory habituation/dishabituation test (19–23) and using estrus urine or bedding as a sexual incentive stimulus (24, 25).

The sensitivity and utility of the FUST paradigm was assessed by testing it on several rodent strains (C57BL/6J mice, 129S1/SVImJ mice, and Wistar-Kyoto rats) as well as in several preclinical experimental paradigms, including the learned helplessness (LH) paradigm (26); chronic treatment with citalopram (a selective serotonin reuptake inhibitor [SSRI]) for mice that underwent the LH paradigm; stress manipulation; and knockout mice strain (GluR6) which exhibits phenocopies of manic-like symptoms (27). In addition, dopamine levels in the NAc were recorded and analyzed by using a microdialysis while conducting the FUST paradigm.

Males from the tested strains spent significantly longer duration sniffing female urine than water. In addition, male mice showed significantly elevated NAc dopamine levels while sniffing urine in the FUST paradigm. LH males as well as stressed mice spent significantly less time sniffing female urine compared to mice that did not undergo the LH paradigm or the stress manipulation, and citalopram treatment alleviated this reduced levels in the "depressed-like" animals. At last, GluR6KO males spent significantly longer duration sniffing female urine, and showed enhanced saccharin preference compared to the wild-type control mice (16).

Therefore, it seems that the FUST paradigm is sensitive to behavioral and genetic manipulation and to relevant drug treatment.

2. Equipment, Materials, and Setup

2.1. Animals

The FUST has been found to be affective so far to several inbred and outbred mouse strains and knockouts, including C57BL/6J, 129S1/SVImJ mice and GluR6 knockout mice (males, 20–25 g), as well as to rats (males, 300–350 g): Wistar-Kyoto rats. In addition, this paradigm has been found to be sensitive to behavioral manipulation and drug treatment to these stains, including stress, learned-helplessness, and SSRI treatment – citalopram (16).

2.2. Test Environment

Animals are singly housed 1 week prior to the FUST in a conventional vivarium (temperature $22 \pm 1c$; 12 h light-dark cycle; food and water ad libitum). For the FUST procedure, rodents are transferred to a dimly lit room (~3 lux lighting).

2.3. Estrus Urine Determination

In the morning (8:00–9:00 am) of the experiment, the vaginal estrus condition of the female donors is assessed by inserting the tip of a plastic pipette filled with 10 µL of saline (NaCl 0.9%) into the mouse vagina and collecting vaginal secretions. Then the vaginal fluid is transferred to a glass slide and mounted under a coverslip with a trace of methylene blue. By examining the proportion and the morphology of leukocytes and epithelial cells present in the smear under 400× magnification of a light microscope, the four stages (proestrus, estrus, metaestrus, and diestrus) of the estrus cycle are determined (28).

2.4. Estrus Urine Collection

Female mice in estrus are being hold on by their skin on the nape and are placed on a flat surface covered with clean aluminum foil. Handling is usually sufficient to induce urination; otherwise, the ventral area should be rubbed in an anterior-to-posterior direction (29). The urine on the foil is collected immediately, and ~60 µL is pipetted directly from the aluminum surface onto a clean cotton swab (30). The experimenter gloves as well as the aluminum foil should be replaced after each urine collection.

Fig. 1. The female urine sniff test (FUST). Illustration of the female urine sniffing test (FUST). A male C57BL/6J mouse sniffs a cotton-tipped applicator dipped into urine from a female mouse in estrus to assess reward-seeking behavior in rodents. Only when the mouse/rat nose was adjacent to the cotton-tipped applicator (as seen in the figure) and the researcher could detect that the rodent was sniffing was the behavior counted as sniffing duration.

2.5. Setup

For the test, each mouse in his home-cage is transferred separately to a dimly lit room (the test must be conducted for one mouse at a time). The top of the cage and the bedding are removed (Fig. 1), and a sterile cotton tip is being taped to one of the home-cages wall in angle of 45° directed downward to the direction of the floor cage.

2.6. Dependent Measures

The FUST paradigm is using a volatile stimuli – odor, and due to the nature of this stimuli and the sensitivity of the rodent olfactory system, two different olfactory choices (water and female estrus urine) are presented sequentially rather than simultaneously. In that case, the dependent measure is the sniffing duration, however, since the same animal in presented with two stimulus, one after the other, a statistical test that take in consideration this factor should be used (e.g., paired t-tests, repeated measures ANOVA/MANOVA, etc.).

3. Procedure

One hour prior to the test, rodents are habituated to a sterile cotton-tipped applicator inserted into their home cage. After 1 h, the sterile cotton-tipped applicator is removed and each rodent, separately, is transferred to a dimly lit room (~3 lux lighting). The FUST has three phases: (1) *Water* – one exposure (3 min) to a cotton tip dipped in sterile water, during which sniffing duration is measured; (2) *Habituation* – the rodent cage is transferred from the experiment room to the habituation area for an interval of 45 min during which no cotton tip is presented; and (3) *Urine* – the animal is transferred back to the experiment room for one exposure (3 min) to a cotton-tipped applicator infused with fresh urine collected from females in estrus (of the same strain), during which sniffing duration is measured.

4. Data Analysis and Anticipated Results

Here we will present an example for the effect of the LH paradigm on reward-seeking behavior in male C57BL/6J mice in the FUST paradigm (16). In this example, the LH paradigm was conducted as previously described (31). The number of escape failures and latency to escape was recorded for each mouse. Mice were defined as helplessness when they showed at least 20 escape failures – so eventually we had two different groups: LH mice (at least 20 escape failures) and non-learned helplessness mice (NLH – less the 20 escape failures). Reward-seeking behavior in the FUST paradigm was compared between LH and NLH mice.

In the LH paradigm, after uncontrollable and inescapable foot-shocks, the majority of mice (50–80%) exhibit failure to escape – also known as LH. This failure behavior persists for weeks and can be treated by chronic (but not acute) antidepressants treatment (32). Numerous studies showed that the LH paradigm is a suitable model to study depression-like symptoms and their underlying mechanisms in rodents, including anhedonic behavior (33). Other studies as well have showed that different stressors elicit anhedonic behavior in rodents (34).

As mentioned above, since the same animal in presented with two stimuli (water and urine), one after the other, a statistical test that take in consideration this factor was used. In this example, two-way repeated measures ANOVA was used to assess the sniffing duration (of water and urine) in the LH paradigm.

In the specific example, both NLH and LH animals spent, overall, more time sniffing urine than water (urine vs. water: $F(1,38) = 35.55$; $p < 0.01$, Fig. 2). Overall, NLH mice sniffed more than LH mice ($F(1,38) = 50.75$; $p < 0.05$). Most importantly, the interaction showed that while there were no significant differences between NLH and LH mice in duration of water sniffing, LH males sniffed urine significantly less than NLH males ($F(1,38) = 6.67$; $p < 0.05$). According to these results, the FUST paradigm is sensitive to the well-validated model of depression – the LH paradigm (33).

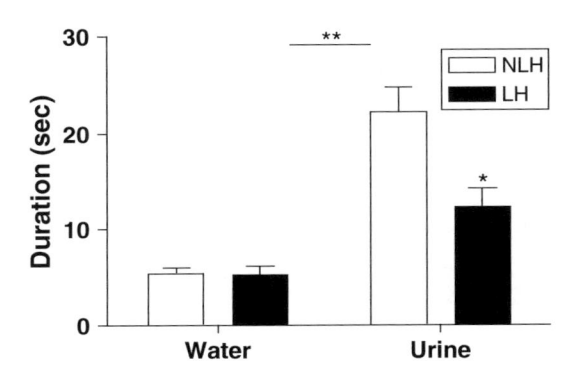

Fig. 2. Time spent sniffing for learned helplessness (LH) and non-learned helplessness C57BL/6J mice. After undergoing the LH paradigm, both LH and NLH male C57BL/6J mice underwent the FUST paradigm. Both the LH and NLH male C57BL/6J mice spent significantly more time sniffing estrus female urine than water ($**p < 0.01$). NLH C57BL/6J mice spent significantly longer time sniffing the estrus female urine than the LH mice ($*p < 0.05$; $n = 8$–12). No significant differences were found between these groups for time spent sniffing the water-dipped applicator.

Indeed, as expected, the LH animals in the current example spent less time sniffing estrus urine, but not water, than NLH mice (Fig. 2), exhibiting impaired reward-seeking behavior, which resembles to the human depressive symptom – anhedonia.

5. Experimental Variables and Troubleshooting

5.1. Large Number of the Mice Spent Very Short Time (2–5 s) Sniffing the Cotton-Tipped Applicator Infused with Fresh Urine Collected from Females in Estrus

If a large number of naïve mice spent very short time sniffing the cotton-tipped applicator infused with fresh urine collected from females in estrus, it can be due to a couple of reasons:

– The mice have previous experience with females – mice supposed to be "sexually naïve" and never met females/females scent.

– The mice are too young and not sexually matured – mice supposed to be sexually matured – at least 60 PND.

5.2. Large Number of Mice Exhibits Anxiety-Like Behavior in the FUST Paradigm

If a large number of naïve mice exhibit anxiety-like behavior – sitting in one corner of the cage without moving ("freezing behavior"), the researcher should eliminate the following causes:

– Illumination – lights might be too bright – a low dim light should be kept (~3 lux).

– Noise – the experiment environment should be very quiet, avoid any loud noise.

5.3. Large Number of the Mice Spent Very Long Time (25–30 s) Sniffing the Cotton-Tipped Applicator Dipped in Water

If a large number of naïve mice spent very long time (25–30 s) sniffing the cotton-tipped applicator dipped in water, the main reason might be the fact that the room is already saturated with female scent. The experimenter should pay attention to two factors that might contribute to this condition:

1. First conduct the water phase in the FUST paradigm with all the mice that participate in the experiment, only then conduct the urine phase with all the mice. Once a female scent is introduced to the room, the following mice can detect it.

2. Avoid overconducting the FUST paradigm in the same room – the experimenter should avoid conducting the urine phase with too many mice one after the other (usually no more than 12 animals per experiment). The room is getting saturated with female scent and while the first mouse was introduced to a room free of female scent, the last mouse in the experiment is being introduced to a room saturated with female scent, which might affect its behavior in a different way.

References

1. American Psychiatric Association (2000): Diagnostic and Statistical Manual of Mental Disorders, Fourth edition. Washington, D.C: American Psychiatric Association.

2. Hasler G, Drevets WC, Gould TD, Gottesman, II, Manji HK (2006): Toward constructing an endophenotype strategy for bipolar disorders. *Biol Psychiatry* 60:93–105.

3. Koob GF, Ahmed SH, Boutrel B, Chen SA, Kenny PJ, Markou A, O'Dell LE, Parsons LH, Sanna PP (2004). Neurobiological mechanisms in the transition from drug use to drug dependence. Neuroscience and Biobehavioral Reviews. 27:739–749.

4. Barbano MF, Cador M (2007): Opioids for hedonic experience and dopamine to get ready for it. *Psychopharmacology (Berl)* 191:497–506.

5. Klein DF (1974): Endogenomorphic depression. A conceptual and terminological revision. *Arch Gen Psychiatry* 31:447–454.

6. Paredes RG (2009): Evaluating the neurobiology of sexual reward. *Ilar J* 50:15–27.

7. Crawley JN (2007): What's Wrong With My Mouse: Behavioral Phenotyping of Transgenic and Knockout Mice. Wiley-Liss, New York.

8. Calcagnetti, D.J. & Schechter, M.D. (1992). Place conditioning reveals the rewarding aspect of social interaction in juvenile rats. *Physiol. Behav.* 51, 667–672.

9. Tzschentke, T.M. (1999). The medial prefrontal cortex as a part of the brain reward system. Amino Acids. 19, 211–219.

10. Willner P, Towell A, Sampson D, Sophokleous S, Muscat R. (1987). Reduction of sucrose preference by chronic unpredictable mild stress, and its restoration by a tricyclic antidepressant. Psychopharmacology 93(3):358–64.

11. Pucilowski, O., Overstreet, D.H., Rezvani, A.H., Janowsky, D.S. (1993). Chronic Mild Stress-Induced Anedonia: Greater Effect in a Genetic Rat Model of Depression. Physiology & Behavior. 54, 1215–1220.

12. Harris, R.B., Zhou, J., Youngblood, B.D., Smagin, G.N., & Ryan, D.H. (1998). Failure to change Exploration or Saccharin Preference In Rats Exposed to Chronic Mild Stress. Physiology & Behavior. 63(1), 91–100.

13. Sclafani A (2006): Sucrose motivation in sweet "sensitive" (C57BL/6J) and "subsensitive" (129P3/J) mice measured by progressive ratio licking. Physiol Behav 87:734–744.

14. Greenshaw, A. J. (1993). "Differential effects of ondansetron, haloperidol and clozapine on electrical self-stimulation of the ventral tegmental area." *Behav Pharmacol* 4(5): 479–485.

15. Sanchis-Segura C, Spanagel R (2006): Behavioural assessment of drug reinforcement and addictive features in rodents: an overview. *Addict Biol* 11:2–38.

16. Malkesman O, Scattoni ML, Paredes D, Tragon T, Pearson B, Shaltiel G, Chen G, Crawley JN, Manji HK (2010): The Female Urine Sniffing Test: A Novel Approach for Assessing

Reward-Seeking Behavior in Rodents. *Biol Psyc* 67:864–871.

17. Martinez-Garcia F, Martinez-Ricos J, Agustin-Pavon C, Martinez-Hernandez J, Novejarque A, Lanuza E (2009): Refining the dual olfactory hypothesis: pheromone reward and odour experience. *Behav Brain Res* 200:277–286.

18. Sipos ML, Kerchner M, Nyby JG (1992): An ephemeral sex pheromone in the urine of female house mice (Mus domesticus). *Behav Neural Biol* 58:138–143.

19. Gregg B, Thiessen DD (1981): A simple method of olfactory discrimination of urines for the Mongolian gerbil, Meriones unguiculatus. *Physiol Behav* 26:1133–1136.

20. Luo AH, Cannon EH, Wekesa KS, Lyman RF, Vandenbergh JG, Anholt RR (2002): Impaired olfactory behavior in mice deficient in the alpha subunit of G(o). *Brain Res* 941:62–71.

21. Chadman KK, Gong S, Scattoni ML, Botluck SE, Gandhy SU, Heintz N, et al. (2008): Minimal aberrant behavioral phenotypes of neuroligin-3 R451C knockin mice. *Austism Res* 1:147–158.

22. Crawley JN, Chen T, Puri A, Washburn R, Sullivan TL, Hill JM, et al. (2007): Social approach behaviors in oxytocin knockout mice: comparison of two independent lines tested in different laboratory environments. *Neuropeptides* 41:145–163.

23. Stack CM, Lim MA, Cuasay K, Stone MM, Seibert KM, Spivak-Pohis I, et al. (2008): Deficits in social behavior and reversal learning are more prevalent in male offspring of VIP deficient female mice. *Exp Neurol* 211:67–84.

24. Agmo A, Pfaff DW (1999): Research on the neurobiology of sexual behavior at the turn of the millennium. *Behav Brain Res* 105:1–4.

25. Hull EM, Dominguez JM (2006): Getting his act together: Roles of glutamate, nitric oxide, and dopamine in the medial preoptic area. *Brain Research* 1126:66–75.

26. Cryan JF, Mombereau C (2004): In search of a depressed mouse: utility of models for studying depression-related behavior in genetically modified mice. *Mol Psychiatry* 9:326–357.

27. Shaltiel G, Maeng S, Malkesman O, Pearson B, Schloesser RJ, Tragon T, et al. (2008): Evidence for the involvement of the kainate receptor subunit GluR6 (GRIK2) in mediating behavioral displays related to behavioral symptoms of mania. *Mol Psychiatry* 13:858–872.

28. Rugh R (1990): *The Mouse: Its Reproduction and Development.* Oxford, UK: Oxford University Press.

29. Nyby J, Wysocki CJ, Whitney G, Dizinno G, Schneider J (1979): Elicitation of male mouse (Mus musculus) ultrasonic vocalizations: I. Urinary cues. *J Comp Physiol Psychol* 93:957–975.

30. Hoffman F, Musolf K, Penn DJ (2009): Freezing urine reduces its efficacy for eliciting ultrasonic vocalizations from male mice. *Physiol Behav* 96:602–605.

31. Maeng S, Hunsberger JG, Pearson B, Yuan P, Wang Y, Wei Y, et al. (2008): BAG1 plays a critical role in regulating recovery from both manic-like and depression-like behavioral impairments. *Proc Natl Acad Sci USA* 105:8766–8771.

32. Gambarana C, Scheggi S, Tagliamonte A, Tolu P, De Montis MG (2001): Animal models for the study of antidepressant activity. *Brain Res Brain Res Protoc* 7:11–20.

33. Henn FA, Edwards E, Muneyyirci J (1993): Animal models of depression. *Clin Neurosci* 1:152–156.

34. Strekalova T, Gorenkova N, Schunk E, Dolgov O, Bartsch D (2006): Selective effects of citalopram in a mouse model of stress-induced anhedonia with a control for chronic stress. *Behav Pharmacol* 17:271–287.

35. Forbes NF, Stewart CA, Matthews K, Reid IC (1996): Chronic mild stress and sucrose consumption: validity as a model of depression. *Physiol Behav* 60:1481–1484.

Chapter 21

Measuring Impulsive Choice Behaviour in Mice

Claire L. Dent and Anthony R. Isles

Abstract

Impulsive behaviour is a fundamental component of numerous psychiatric illnesses including mood disorders. In order to measure "impulsivity" and understand the complex neurological underpinnings of this behavioural construct, it is beneficial to employ the use of mouse models. Neuropsychological tasks used to measure impulsivity in humans have been successfully translated into behavioural tests to characterise impulsivity in mice. This has lead to the development of the delayed reinforcement paradigm, which specifically measures "impulsive choice". This, combined with genetic and pharmacological manipulations, allows insight into possible biological markers associated with impulsive choice. This chapter provides a description of the equipment and procedures required to measure impulsive choice in a delayed reinforcement paradigm, as well as examples of results and troubleshooting advice to optimise the behavioural data.

Key words: Delay discounting, Reinforcer, Operant chamber, Inbred strains, Psychopharmacology

1. Background and Historical Significance

The term "impulsivity" is defined as action without adequate forethought and is a natural part of human behaviour (1). It is a complex behavioural construct and encompasses a variety of underlying behaviours, such as "actions that are poorly conceived, prematurely expressed, unduly risky, or inappropriate to the situation and that often result in undesirable outcomes" (2). It seems that in certain individuals the impulsive behaviour becomes pathological and maladaptive. Research has led to deficits in impulsivity being recognised as a clinical trait of many psychiatric disorders, including schizophrenia, attention deficit/hyperactivity disorder (AD/HD) and obsessive compulsive disorder (OCD). Impulsivity is also a key cognitive deficit of mood disorders such as Bipolar (3), with impulsive behaviours thought to increase during the manic phase (4).

Todd D. Gould (ed.), *Mood and Anxiety Related Phenotypes in Mice: Characterization Using Behavioral Tests, Volume II*, Neuromethods, vol. 63, DOI 10.1007/978-1-61779-313-4_21, © Springer Science+Business Media, LLC 2011

The neuroanatomical correlates of impulsivity have been well defined in humans (5), and pre-clinical research has assisted the identification of this circuitry in rodents. Primarily the prefrontal cortical, striatal and limbic brain regions have been found to play an important role in impulsivity in rodents. Specifically, the prefrontal cortex in conjunction with the nucleus accumbens plays a role in the structuring of impulsive behaviour (6–8).

Measuring impulsive behaviour in humans has been successfully established using response inhibition tasks. With the increased recognition of impulsive behaviour being a trait of psychiatric illness, these tasks have proved extremely useful in identifying and understanding disorders such as AD/HD in children. These "Go/No go" tasks specifically target individuals' ability to inhibit a pre-learnt response, for example making a motor-response to a certain stimuli ("go" trial) and refraining from making the motor response to a stimulus during a "no-go" trial. This basic concept has been further developed through the "Stop signal reaction time task" (SSRTT), which has been used widely in the clinical setting to measure impulsivity. For example, children with AD/HD are slower to inhibit their responses than normal children, as indicated by increases in their stop signal reaction time (e.g. (9)), and similarly fail to inhibit their "go" response on the "no-go" trials in go/no-go tasks (1).

The scientific quest to identify the neurological basis and possible genetic predispositions of psychiatric illness has led to the increasing use of genetically modified animals in measuring behaviour. Therefore, the neuropsychological tasks used in humans, such as the SSRTT, have been successfully translated to rat models (10). Recent experimental paradigms have led to the development of novel techniques that allow the characterisation and measurement of "impulsive" behaviour in mice. As a result of pre-clinical research, it appears the heterogeneous underlying traits of "impulsivity" are dissociable both behaviourally and neurologically; whereby, the different behaviours are underpinned by distinct biological mechanisms. Therefore, different behavioural tasks can target and measure these separate behaviours under the umbrella of "impulsivity". These can be broadly divided into two categories: those measuring impulsive choice or decision-making, and those measuring impulsive action or motoric impulsivity (11).

One of the most prominent features of impulsive choice is an intolerance of a delay of a reward or gratification. One method used to target the measurement of this behaviour is the "delayed-reinforcement task" (12), a variation of the "delay-discounting" paradigm used to measure impulsivity in humans (13) and rats (14). Delay discounting tasks measure impulsivity using a forced choice in which the animals have to choose between rewards that are relatively small but available immediately, and rewards that are larger but progressively delayed. The selection of smaller immediate reward in preference to larger delayed reward has been considered to reflect "impulsive" behaviour, whereas the opposite bias towards delayed gratification has been taken to indicate increasing "self-control".

Ongoing research has contributed to the successful translation of the delay-discounting task previously used in humans, to rodents and specifically transgenic mice. Furthermore, experimental research has provided the finding that certain genetic characteristics promote impulsivity in mice; whereby, "impulsive" mice have a propensity to choose a small immediate reward over a more beneficial, larger delayed reward (15). The use of the delayed reinforcement task in rodent models allows the identification and measurement of this specific aspect of impulsive behaviour. Moreover, by applying mice carrying specific gene deletions or transgenes there is the possibility of delineating the genetic contribution underlying the neurobiology and molecular substrates of impulsive behaviours (16–19). Importantly, in combination with psychopharmacological manipulation, this can lead to the innovation and development of drug therapies for the psychiatric disorders in which abnormal impulsivity is a feature.

2. Equipment, Materials and Set-Up

Rodent models of delay discounting normally utilise operant chambers (6, 7, 12), although this behaviour has been studied using a T-maze in rats (20). In rats, the task is normally run using lever-pressing. However, our experience is that mice are better suited to nose-poking. Consequently, we have developed a delayed reinforcement task using the 9-hole operant chambers (Campden Instruments Ltd.). These chambers consist of a curved panel, with nine holes in which stimulus lights can be illuminated. For the delayed reinforcement task configuration, holes 3, 5 and 7 are open (counted from the left, see Fig. 1). The mice are trained to respond to visual stimuli, recessed into the holes, with a nose-poke detected by an infrared beam crossing the entrance vertically (Fig. 2a). Mounted above the operant chambers are infrared sensitive cameras (Watac WM6, Tracksys Ltd., Nottingham, UK). Illumination of the chambers when dark is provided by four roof-mounted infrared LEDs. The control of stimuli and recording of responses is managed by an Acorn Archimedes computer running custom-written BASIC programmes with additional interfacing by ARACHNID (Cambridge Cognition Ltd.). However, PC-based control systems are now available (Campden Instruments Ltd.).

Fig. 1. The 9-hole operant box employed when assessing delay discounting behaviour in the mouse using a delayed reinforcement paradigm. For the delayed reinforcement task configuration, holes 3, 5 and 7 are open (counted from the *left*). The mice are trained to respond to visual stimuli, recessed into the holes, with a nose-poke detected by an infra-red beam crossing the entrance vertically.

3. Procedures

In the mouse task delays can be incremented onto the response leading to the delivery of the larger reinforcer within a session. Others have successfully used tasks where the delay associated with the larger reinforcer is varied *between* sessions (16, 17). However, our design encourages flexible "trial discrete" choices and mini-mises the possibility of behaviour being controlled by rigid, habit-like strategies (14). For this reason, we chose to use a variant of the blocked design delayed reinforcer task developed by Evenden for use in rats (14), where the delay increased between blocks but was constant within blocks of trials (Fig. 2b).

Prior to any experimental work all animals are handled daily for 2 weeks and their body weight monitored. After this time, the animals are placed on a 20-h water deprivation schedule for 4 days, and then on a 22-h water deprivation for a further 10 days until body weight had stabilised. The animals are habituated to the liq-uid reinforcer used in the operant procedures (in our case 10% solution of condensed milk, Nestle Ltd., UK; but a 8–10% sucrose solution can be used). Habituation to the liquid reinforcer is performed outside the operant chambers; briefly, the subjects are given four, separate 10-min sessions over 4 days with an excess of either water or the reinforcer presented in two small bowls (apart from the first session where two samples of water are presented).

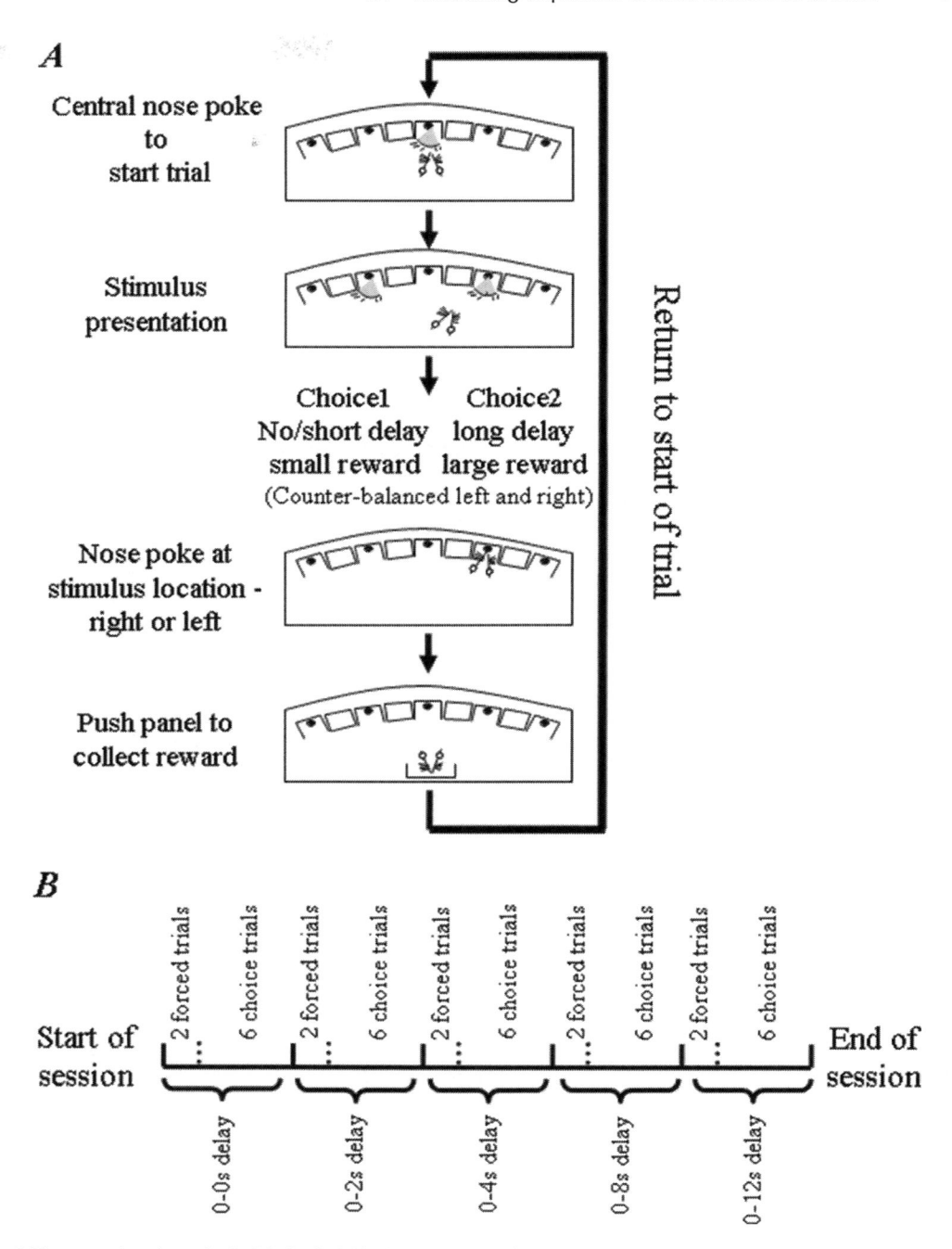

Fig. 2. The procedure for a single "choice" trial in the delayed reinforcement task (**a**). A typical session consists of 40 trials, separated into five blocks (**b**). Within each block the subject first experiences a number of "forced" information trials in which only one of the two response options is available. This is designed to provide the subjects with prior notice of the extent of any delay associated with the larger reward response. Then the subject is given a number of "choice" trials; in this example, there are 2 "forced" and 6 "choice" trials. Under baseline conditions, the delay associated with the response leading to the larger reinforcer increases as the session progresses.

Animals are maintained on this restricted water access regime throughout testing (apart from ad libitum access task manipulation, see below) in order to motivate behaviour.

3.1. Training to Baseline Performance

Mice require much more shaping than rats, and initial training is carried out over 20 sessions (one session per day). The first five sessions involve general habituation, panel pressing and learning that food is available in the magazine. Here food is simply delivered to the food hopper every 10 s. In sessions 6–12 the subject is required to make a contingent nose-poke in the centre hole in response to a 10-s light stimulus in order to initiate food delivery. In the final 8 sessions the subject has to respond sequentially to a 10-s light stimulus presented first in the central location and then (pseudo-randomly) either to the left or right of the centre before food is delivered. In this way, the centre nose-poke acts as a cue, signalling the beginning of a trial and focusing the behaviour of the mice towards the subsequent presentation of stimuli in the lateral locations. The animals then move on to the delayed reinforcement task proper where, critically, in the choice component of the task stimuli in the lateral holes are presented simultaneously.

The task comprises sequential blocks of trials in which (following the initial centre nose-poke) a nose-poke response in one direction (left or right, counterbalanced across animals) results in the delivery of 50 μL reinforcer, with a response in the other direction resulting in the delivery of 25 μL reinforcer (Fig. 2a). The number of blocks of trials, and also the number of trials within a block may vary. We normally use 5 blocks of 8 trials; however, we have also successfully used 3 blocks of 10 trials.

In block 1 a response in either direction leads to the immediate delivery of reinforcer. In subsequent blocks increasing delays are introduced between the response and the delivery of the larger reinforcer. Initially, the mice are trained on delay pairs of (small reward response first) 0 vs. 0 s; 0 vs. 0.5 s; 0 vs. 1 s; 0 vs. 2 s; 0 vs. 4 s for 10 days. They then move on to 0 vs. 0 s; 0 vs. 1 s; 0 vs. 2 s; 0 vs. 4 s; 0 vs. 8 s for 10 days, and finally the full delay range 0 vs. 0 s; 0 vs. 2 s; 0 vs. 4 s; 0 vs. 8 s; 0 vs. 12 s for a further 10 days to baseline performance. The longest delay that may be tolerated is determined by the cohort of mice being used; as we discuss below inbred strains of mice vary in their impulsiveness and this may determine the final range of delays. Trials 1 and 2 in any block are "forced" information trials, whereby after the central nose-poke only one of the two response options is available. This was designed to provide the subjects with prior notice of the extent of any delay associated with the larger reward response. In the next 6 trials, the animals have a choice of either right or left as shown in Fig. 2b. Both stimuli are presented for 10 s each, and are extinguished once an animal has made a nose-poke. No central nose-poke at the initiation of a trial is recorded as "non-started trials"; commission of the central nose poke but a failure to choose a left or right nose-poke response is recorded as an "omission". Regardless of the outcome of a trial, a new trial is started 45 s after the presentation of the previous central stimulus. The food-hopper light is illuminated when the reinforcer is delivered; however, there is no light or programmed signal during the delayed period between the choice nose-poke and delivery of the food.

3.2. Task Manipulations

Once stable baseline performance has been achieved, a series of manipulations of the basic task parameters may be performed to assess the sensitivity of the mice to the contingency between reward and delay, and the extent to which behaviour is controlled in a trial discrete manner independently of satiation effects. These manipulations consist of probe conditions where zero (or equal) delay persisted on the large reward response throughout a session ("zero delay condition"), and reversal of delay order (12, 8, 4, 2, 0 s – "reversed delay condition"). Others may include alteration of motivation via pre-feeding or allowing animals ad libitum access to water for 24 h prior to testing.

4. Data Handling

In the delayed reinforcement task, the primary measure of interest is the bias to the large reinforcer, defined as the number of times the large reinforcer was chosen as a proportion of the total number of choices made (*not* the possible number of choices) within a given delay block. Additional measures include non-started trials (no centre nose-poke, so a trial was not started), omissions (centre nose-poke but no choice made), initiation latency (time taken to centre nose-poke once the centre light was turned on) and response latency (time taken to make a choice nose-poke after the centre nose-poke). If the operant boxes are fitted with infrared beams across the chamber, then a measure of activity can also be obtained by analysing the number of beam breaks. The results from the delayed reinforcement task are generally analysed by ANOVA with a within-subject factor DELAY (delay associated with large reinforcer); other factors could include condition (e.g. baseline vs. zero delay), DOSE (*d*-amphetamine, 0.4, 0.6, 0.8 and 1 mg/kg i.p.) or STRAIN (inbred strain of mouse).

5. Results Examples

5.1. Baseline Performance and Zero Delay Manipulation

Typical choice behaviour at baseline performance is shown in Fig. 3a. It can be seen that, under the schedule of increasing delay on the large reward response, the mice exhibited a systematic change in response bias from a maximum of ca. 89% in favour of

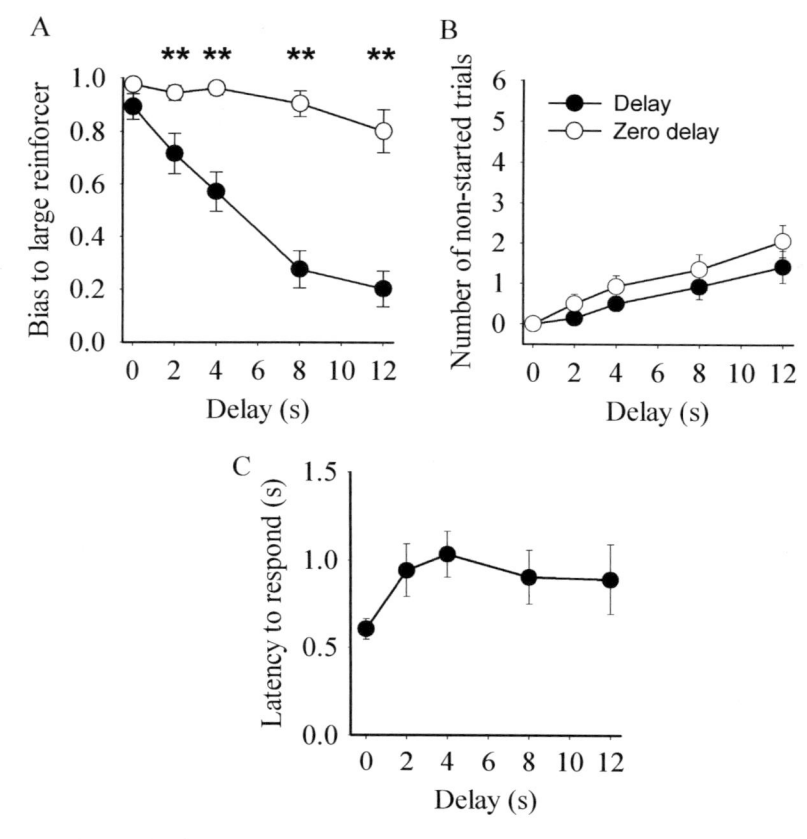

Fig. 3. Baseline performance in the delayed reinforcement task. When there is an increasing delay associated with the larger reinforcer choice of this large reinforcer reduces as the session progresses (**a**). However, when the delays are equivalent (0 s) then choice of the large reinforcer is high throughout the session. There are no differences in non-started trials in the delay and zero-delay conditions (**b**). Stimulus control is high, as indicated by the latency to respond (**c**).

the large reward response at 0 delay to ca. 20% in favour of a delay at 12 s (main effect of DELAY, $F_{4,65} = 15.402$, $p < 0.001$). This progressive change in response bias away from the larger but delayed reinforcer to the immediate, smaller reinforcer was associated with a small but significant increase (Fig. 3b) in the number of non-started trials (main effect of DELAY, $F_{4,65} = 4.915$, $p < 0.01$). In our experience, the mice make very few omissions (an average of 0.07 ± 0.06 SEM out of a possible 30 choice trials per session). Generally, the mice show a high degree of stimulus control in performing the task as demonstrated by the response latencies (Fig. 3c).

The effects of maintaining a zero delay on the large reward response (as well as the small reward response) is shown in Fig. 3a. Although the delay associated with the larger reinforcer is 0 s throughout the session, the task is still run and presented in five separate blocks in order to compare with baseline, delay conditions. Under zero delay conditions, the response bias alters such that there is a systematic change towards the large reward response that was present across the entire session (effect of DELAY, $F_{4,65} = 2.039$, NS). Importantly, other behavioural measures such as non-started trials (Fig. 3b), omissions and latencies are similar to those observed in the delay baseline condition. This indicates the sensitivity of the mice to the contingencies between reward and delay, the trial-discrete nature of the task, and the lack of confounds arising from satiety effects.

5.2. Psychopharmacology: d-Amphetamine Sulphate

The continuous performance nature of the delayed reinforcement task (and indeed most operant paradigms) makes it amenable to within-subject testing of pharmacological challenges. In this example, the sensitivity of the response bias function to dopaminergic manipulations was determined by systemic administration with low doses of the indirect dopamine agonist d-amphetamine. As shown in Fig. 4a, d-amphetamine gives complex, dose-dependent effects, whereby at the two lowest doses (0.4/0.6 mg/kg) the response bias shifted towards the delayed reward response (consistent with decreased impulsivity) but at the 0.8 mg/kg dose the response bias shifted towards the immediate reward response (consistent with increased impulsivity). The statistical analysis of these data indicated significant main effects of DOSE ($F_{4,325} = 11.31$, $p < 0.001$). At these doses the drug effects on response bias were completely independent of effects on non-started trials (Fig. 4b). However, in our hands higher doses (1.0 mg/kg) also increase non-responding (as measured by omissions and non-started trials) (12) and thus confound the choice bias data.

5.3. Genetic Differences: Variation in Inbred Strains

Variation in impulsive choice behaviour in the delayed reinforcement task may also be affected by genetic background. We have clearly demonstrated this by examining behaviour in four different

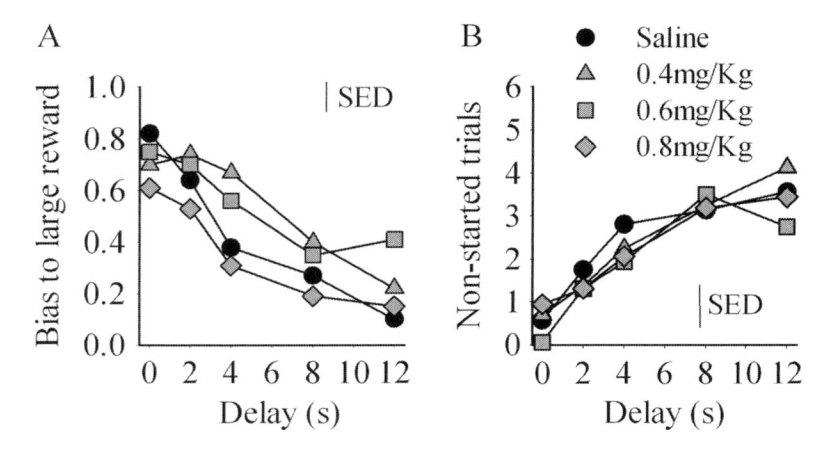

Fig. 4. The effect of d-amphetamine administration on behaviour in the delayed reinforcement task. Dosing with d-amphetamine had a complex set of effects on choice (**a**), with lower doses increasing choice of the larger, delayed reinforcer (i.e. making animals less impulsive). This effect disappears at higher doses. However, there is no differential effect of d-amphetamine on non-started trials (**b**).

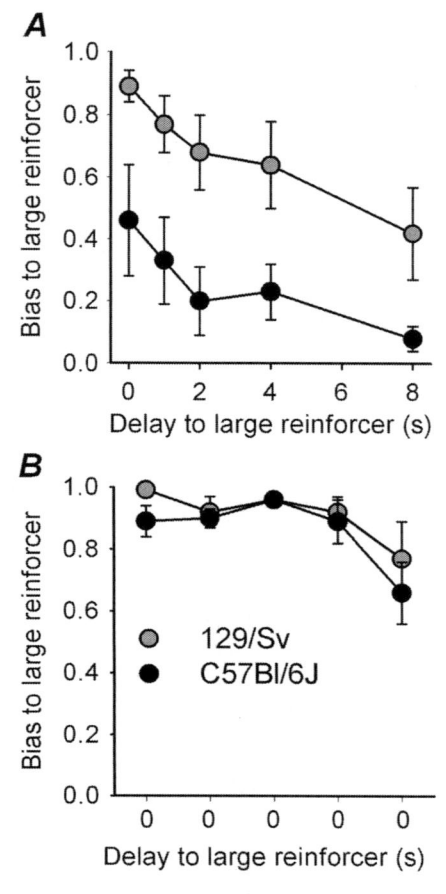

Fig. 5. Behaviour in the delayed reinforcement task of two inbred strain of mice (C57Bl/6J and 129/Sv) commonly used in generating gene knock-outs and transgenics. When increasing delay is associated with the response leading to the larger reinforcer, there is clear difference between the two strains in terms of choice bias, with C57Bl/6J showing more impulsive choice behaviour than 129/Sv (a). However, when the delay associated with the response leading to the larger reinforcer is equal to that of the smaller reinforcer (0 s), then there is no difference between the two strains (b).

inbred strains of mice (C57Bl/6J, 129/Sv, CBA/Ca and BALB/c) (15). Choice behaviour in the delayed reinforcement task at stable baseline levels of performance is shown for the two inbred strains most commonly used in gene targeting and transgenic studies, C57Bl/6J and 129/Sv, in Fig. 5a. Both C57Bl/6J and 129/Sv exhibited a systematic change in choice bias away from the response leading to the delayed (larger) reinforcer towards the immediate (smaller) reinforcer with increasing delay (main effect of DELAY, $F_{4,44} = 6.75$, $P < 0.001$). However, the choice bias functions were different across the groups (main effect of STRAIN, $F_{1,11} = 12.08$, $P < 0.001$) with C57Bl/6J mice exhibiting a pattern consistent with a relatively greater, and 129/Sv strains exhibiting a pattern consistent with a relatively lesser degree, of impulsive choice.

The strain differences in choice bias were dependent on the increasing delay imposed on the large reward response as evidenced by the effects of the "zero delay" probe manipulation (where both small and large reward responses were reinforced immediately). As illustrated in Fig. 5b, both C57Bl/6J and 129/Sv strains of mice were equally responsive to this manipulation, maintaining a preference for the large reward response across the entire session. Under zero delay conditions, there were no strain differences in either stable performance levels (STRAIN, $F_{1,11} = 1.38$, NS). Again this dissociation points to the fact that the difference in behaviour between the strains is not in the way they have acquired the task, but is dependent on the contingency between the delay and the reward.

6. Troubleshooting

Several practical recommendations, summarised here, may help the researchers to obtain more reliable and reproducible behavioural data.

1. As with all operant tasks, early shaping is crucial. Mice are particularly susceptible to being "spooked" and this can lead to non-responding throughout a session. Unfortunately, mice that show this behaviour need to be excluded from the study; however, this is usually detected early on in training.

2. Mice are also prone to developing "superstitious" behaviours. One key problem is the development of a side bias when moved from the training paradigms to the main task. This may develop as a consequence of the subject learning the task in a slightly different manner. These problems may be eliminated by taking the subject back to a previous training stage in order to re-acquire the task. However, in some subjects the side bias is intractable and these animals will need to be eliminated from the study. In our experience numbers are quite low (5–10%).

3. The value of the reinforcer is obviously important in delayed reinforcement tasks. This may be adjusted to suit the investigator's needs and can be altered for any given session as a task manipulation. However, it is important that floor and ceiling effects are not induced by the reinforcers associated with the large and small delays being too similar or too different, respectively. In this respect, it is strongly recommended that the nature of the reinforcer associated with the large and small delays be the same, and that amount or concentration is varied. For instance, it is inadvisable to use water to reinforce the choice leading to a small delay and a sucrose/milk solution to reinforce the choice associated with a larger delay.

4. As described above, there are clear differences between inbred strains of mice in impulsive choice behaviour as measured on the delayed reinforcement paradigm. This will have important consequences when using transgenic mice on this task and care needs to be taken when selecting a delay range. Typically, in the analogous rat delayed reinforcement task, this period will extend over ca. 60 s (14). With mice, this period is much shorter; a range of delays between 0 and 12 s seems to be appropriate for most mouse strains. However, even a delay range 0–12 s may result in "floor" effects in high impulsive strains such as C57Bl/6J. By contrast, for knock-out mice on the less impulsive 129 Sv background we have used a range of delays between 0 and 24 s.

7. Conclusion

It is increasingly apparent that there are behaviourally and neurally dissociable aspects to what is generally termed "impulsive" behaviour (11). Here we have described the equipment and procedure for examining what is termed "choice" impulsivity in mice. We utilise a delayed reinforcement paradigm in which delays can be incremented onto the response leading to the delivery of the larger reinforcer within a session. This design encourages flexible "trial discrete" choices and minimises the possibility of behaviour being controlled by rigid, habit-like strategies (14). Moreover, as we demonstrate above, the delayed reinforcement task is also ideal for examining the effects of pharmacological manipulation; the continuous nature of the operant task allows for within-subject comparisons, and changes can be measured across delays within a session (12).

The main reason for using mouse models is their genetic tractability. Illustrating this is the fact that different inbred strains of mice also show variability in choice impulsivity (15) and a number of gene knockout models also show altered impulsive behaviour as measured on versions of delay discounting tasks (16, 17). We underline this here by showing that behaviour in the delayed reinforcement task is very different for two commonly used inbred strains of mice.

References

1. C. A. Winstanley, D. M. Eagle, T. W. Robbins, Behavioral models of impulsivity in relation to ADHD: translation between clinical and preclinical studies. *Clin Psychol Rev* **26**, 379 (Aug, 2006).

2. J. L. Evenden, Varieties of impulsivity. *Psychopharmacology* **146**, 348 (1999).

3. A. C. Swann, M. Lijffijt, S. D. Lane, J. L. Steinberg, F. G. Moeller, Increased trait-like impulsivity and course of illness in bipolar disorder. *Bipolar disorders* **11**, 280 (May, 2009).

4. A. C. Swann, Impulsivity in mania. *Curr Psychiatry Rep* **11**, 481 (Dec, 2009).

5. S. Kim, D. Lee, Prefrontal cortex and impulsive decision making. *Biological Psychiatry* **69**, 1140 (2011).

6. R. N. Cardinal, D. R. Pennicott, C. L. Sugathapala, T. W. Robbins, B. J. Everitt, Impulsive choice induced in rats by lesions of the nucleus accumbens core. *Science* **292**, 2499 (2001).

7. S. Mobini *et al.*, Effects of lesions of the orbitofrontal cortex on sensitivity to delayed and probabilistic reinforcement. *Psychopharmacology (Berl)* **160**, 290 (2002).

8. C. A. Winstanley, D. E. Theobald, R. N. Cardinal, T. W. Robbins, Contrasting roles of basolateral amygdala and orbitofrontal cortex in impulsive choice. *J Neurosci* **24**, 4718 (May 19, 2004).

9. K. L. Purvis, R. Tannock, Phonological processing, not inhibitory control, differentiates ADHD and reading disability. *Journal of the American Academy of Child and Adolescent Psychiatry* **39**, 485 (Apr, 2000).

10. D. M. Eagle, T. W. Robbins, Inhibitory control in rats performing a stop-signal reaction-time task: effects of lesions of the medial striatum and d-amphetamine. *Behav Neurosci* **117**, 1302 (Dec, 2003).

11. C. A. Winstanley, J. W. Dalley, D. E. Theobald, T. W. Robbins, Fractionating impulsivity: contrasting effects of central 5-HT depletion on different measures of impulsive behavior. *Neuropsychopharmacology* **29**, 1331 (Jul, 2004).

12. A. R. Isles, T. Humby, L. S. Wilkinson, Measuring impulsivity in mice using a novel operant delayed reinforcement task: effects of behavioural manipulations and d-amphetamine. *Psychopharmacology (Berl)* **170**, 376 (Dec, 2003).

13. J. Crean, J. B. Richards, H. de Wit, Effect of tryptophan depletion on impulsive behavior in men with or without a family history of alcoholism. *Behav Brain Res* **136**, 349 (Nov 15, 2002).

14. J. L. Evenden, C. N. Ryan, The pharmacology of impulsive behaviour in rats: the effects of drugs on response choice with varying delays of reinforcement. *Psychopharmacology (Berl)* **128**, 161 (1996).

15. A. R. Isles, T. Humby, E. Walters, L. S. Wilkinson, Common genetic effects on variation in impulsivity and activity in mice. *J Neurosci* **24**, 6733 (Jul 28, 2004).

16. L. Bevilacqua *et al.*, A population-specific HTR2B stop codon predisposes to severe impulsivity. *Nature* **468**, 1061 (Dec 23, 2010).

17. D. Brunner, R. Hen, Insights into the neurobiology of impulsive behavior from serotonin receptor knockout mice. *Annals of the New York Academy of Sciences* **836**, 81 (1997).

18. C. M. Helms, N. R. Gubner, C. J. Wilhelm, S. H. Mitchell, D. K. Grandy, D4 receptor deficiency in mice has limited effects on impulsivity and novelty seeking. *Pharmacology, biochemistry, and behavior* **90**, 387 (Sep, 2008).

19. A. R. Isles *et al.*, An mTph2 SNP gives rise to alterations in extracellular 5-HT levels, but not in performance on a delayed-reinforcement task. *Eur J Neurosci* **22**, 997 (Aug, 2005).

20. J. C. Bizot, M. H. Thiebot, C. Le Bihan, P. Soubrie, P. Simon, Effects of imipramine-like drugs and serotonin uptake blockers on delay of reward in rats. Possible implication in the behavioral mechanism of action of antidepressants. *J Pharmacol Exp Ther* **246**, 1144 (Sep, 1988).

Chapter 22

Assessment of Male Sexual Behavior in Mice

Jin Ho Park

Abstract

Sexual dysfunction is prevalent among patients with poor emotional health and has been strongly associated with antidepressant medications. This chapter provides an overview of the general experimental approaches for assessing male sexual behavior (MSB) in male mice. One of the most common procedures to assess MSB is the pairing of a male mouse with an ovariectomized female mouse that has been hormonally primed to be receptive in either the male's home cage or in an enclosed arena. The more common outcomes of interest include measures of male sexual performance behaviors (also referred to as consummatory behaviors), which can include latencies and frequencies of mounts, intromissions, and ejaculation. A brief historical background and a recommended experimental protocol are presented as well as caveats that can be considered when conducting these experiments.

Key words: Male sexual behavior, Copulation, Mounting, Intromission, Ejaculation, Post-ejaculatory interval, Testosterone, Estradiol

1. Background and Historical Overview

1.1. Relevance of Male Sexual Behaviors to Mood and Anxiety Disorders

Sexual dysfunction in humans has long been associated with numerous mood and anxiety disorders, and it may be caused by both psychological and physiological factors. Sexual dysfunction, which can be defined as a disruption of the normal physiological and psychological processes involved in sexual functioning, may include a range of distinct conditions, such as inhibited sexual desire or excitement, inhibited orgasm, premature ejaculation in men, pain during intercourse, or hypersexuality (1).

Sexual dysfunction is prevalent among patients with poor emotional health and has been strongly associated with antidepressant medications. For example, depression has been associated with sexual problems, and comparative studies have consistently reported higher levels of sexual dysfunction in patients with depression than

Todd D. Gould (ed.), *Mood and Anxiety Related Phenotypes in Mice: Characterization Using Behavioral Tests, Volume II*,
Neuromethods, vol. 63, DOI 10.1007/978-1-61779-313-4_22, © Springer Science+Business Media, LLC 2011

in normal controls (reviewed in (2)). Loss of sexual interest, characterized by loss of libido or decrease of sexual desire or potency, has been reported by a large proportion of patients diagnosed with depression (3). Although accurate assessment of antidepressant-induced sexual dysfunction is subject to many difficulties, impaired sexual behaviors have been reported to be a side effect of many antidepressant drugs, such as tricyclic antidepressants, selective serotonin reuptake inhibitors, and monoamine oxidase inhibitors (4–8). In the case of anxiety disorders, the prevalence of sexual dysfunction is a poorly studied area. Several studies report that anxiety may have adverse effects on sexual function, although other studies have suggested anxiety may actually enhance sexual arousal in some patient populations (9, 10). Another mental disorder, mania, has been well documented to be associated with an increase in male sexual behavior (MSB) (11, 12). In summary, the protocols outlined in this chapter may be beneficial to researchers interested in determining the potential side effects associated with sexual dysfunction of pharmacological intervention for many mood and anxiety disorders.

1.2. Description of Male Sexual Behavior in Male Mice

Although extensive studies have been conducted investigating MSB in rats (see (13) for review), there is a relative paucity of knowledge concerning sexual behaviors in male mice. More importantly, it is dangerous to assume that what is true for the rat is also true for the mouse, as numerous species differences have been documented (reviewed in (14)).

Beginning in the 1960s, Thomas E. McGill conducted pioneering studies of mouse MSB. His body of work currently persists as the best collective body of data on strain differences in mating behavior in male mice. An excellent review of the design of his studies and the types of data he collected, which were similar for over nearly 2 decades (15–18), is summarized in (19) and is briefly discussed below. Table 1 lists many of the components of MSB measured by McGill along with several other investigators documenting strain differences in MSB. It should be noted that there are discrepancies in protocol among numerous studies of MSB in mice since the studies conducted by McGill (20–25). Variables known to influence expression of MSB are significantly different among several of these studies, making it difficult to determine when strain or procedural differences are responsible for disparity in recorded behavior.

MSB includes both motivational (also referred to as appetitive) and performance (also referred to as consummatory) behaviors. Motivational behaviors can include ultrasonic vocalizations, urine marking, latency to approach a sexually receptive female, and preferences for females or their odors. In this chapter, motivational behaviors will not be extensively discussed, and the focus will be on the consummatory behaviors. In most of the species studied in the laboratory, there are three distinguishable consummatory behavioral

Table 1
Strain differences in male mouse copulatory behavior

Strain	ML (s)	IL (s)	EL (s)	Mounts	Intro missions	References
BALB/c	62–110	12	1,302–3,645	18	36	(28, 87, 88)
C3H/HeJ	274	17	6,943	12	142	(89)
AKR/J	73–112	8–9	2,160–3,541	8–55	32–47	(89, 90)
C57BL/6J	29–59	15	888–1,258	10–18	17–23	(87, 88, 91)
DBA/2J	56–130	17–20	594–1,946	12–16	5–13	(4, 90, 91)
C57BL/6J×DBA/2J	42	19	1,091	7	18	(91)
AKR/J×DBA/2J	75	9	1,700	20	39	(89, 90)

Modified from (19)
ML mount latency from the introduction of the female; *IL* mount to intromission interval; *Mounts* number of mounts without intromission; *EL* latency from the first intromission to ejaculation; *Intromissions* number of intromissions preceding ejaculation; Values are ranges of median scores

components to the act of male copulation: mounting, intromission, and ejaculation (Fig. 1).

When a male mouse is placed in an arena with a receptive female, the male initiates contact with the female and engages in chemoinvestigation and rooting, which consists of the male putting his head under the female and raising the head to sniff. As a consequence, the female is often lifted or pushed. Mounting then occurs when the male assumes a copulatory position, with his front paws on the back of the females, but does not necessarily include insertion of the penis into the female's vagina. Intromissions occur when the penis enters the vagina during a mount; this component has also been described as deep thrusts. An intromission is defined as an ejaculation when it includes the expulsion of semen from the distal urethra. The temporal arrangements of the components of a mating bout ending in ejaculation have species-typical features. Male mice tend to perform more bouts of mounting and intromissions before a final bout that includes an ejaculation. In mice, the male grasps the female and remains completely still for ~10–30 s after the ejaculation. Another aspect of MSB is the post-ejaculatory interval (PEI) in which the male enters a period of sexual quiescence. This interval can be as long as hours to days in other species including some commonly used mouse strains (26).

Some consummatory behaviors that are commonly tabulated and reported include the numbers of animals that display mounts, mounts with thrusts, intromissions during mounts, and ejaculation. Latencies to and frequencies of the individual behaviors, as well as intervals between behaviors, are also commonly reported.

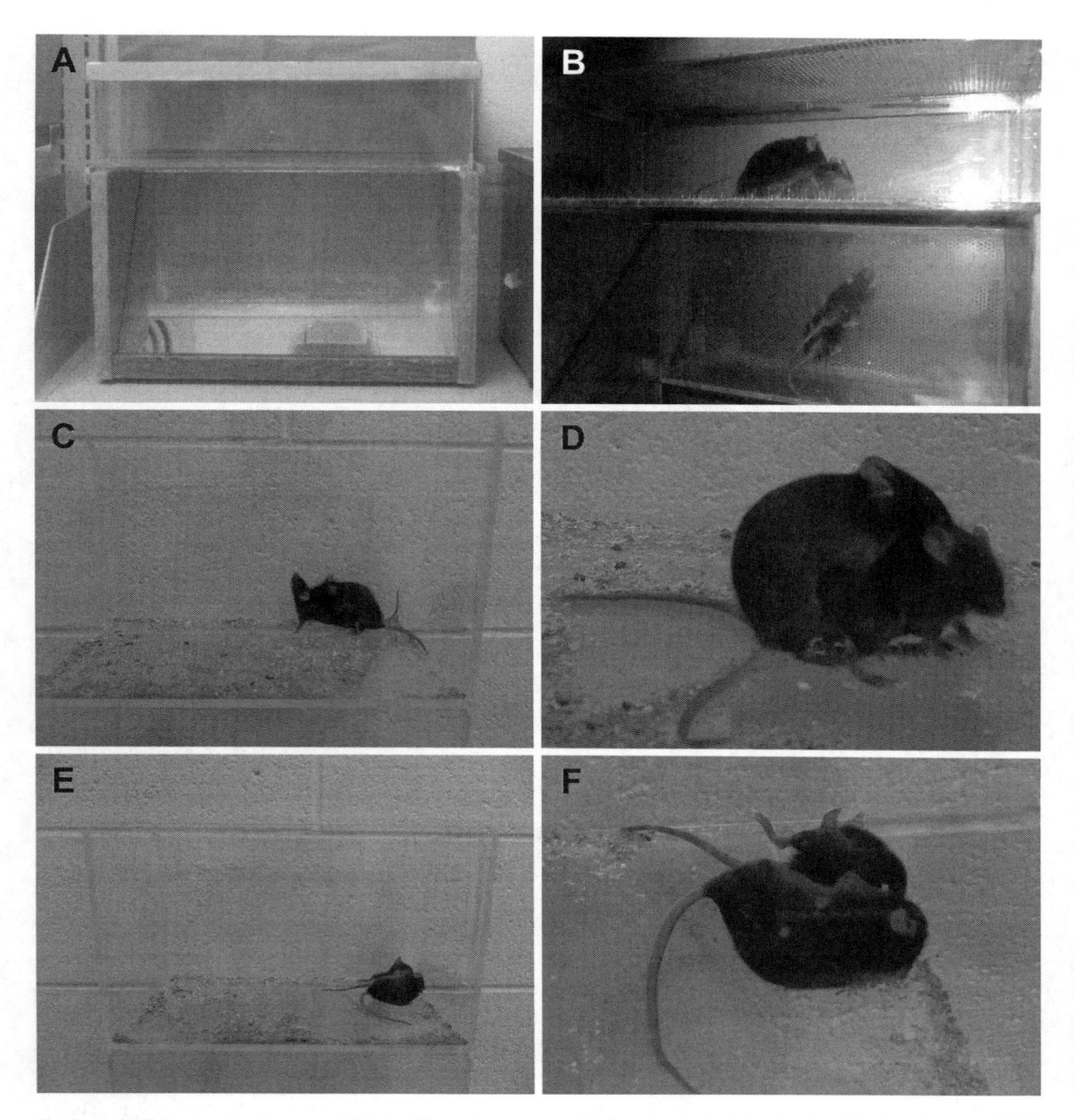

Fig. 1. (**a**, **b**) Behavioral testing arena fitted with a mirror mounted below for ventral viewing to aid in the viewing of MSB components, particularly intromissions, which can be difficult to ascertain when viewing from the side. (**c**, **d**) Male mouse mounting and intromitting a receptive female. One distinguishing feature of an intromission from a mount is the deep thrusting that occurs during intromissions. (**e**, **f**) A male mouse that has ejaculated. Typically, as depicted here, the male grasps the female and falls to the side as he remains completely still for ~10–30 s after the ejaculation.

MSB within a test arena can be monitored either by a trained observer, or by recording the behavior using a video camera and then scoring the behavior at another time. Video recording is convenient when running multiple behavior tests simultaneously. If recording behavior in the dark during the dark phase of the animals' light-dark cycle (which is recommended), the video camera should have the capability to record in the dark under dim red lights. Software analysis programs, such as Noldus (Ethovision; Wageningen, The Netherlands), can be utilized to record the specific times and numbers of the different components of MSB.

1.3. The Role of Gonadal Steroids and Their Receptors in Male Sexual Behavior

It is well established that gonadal steroids are essential in modulating a myriad of male social behaviors that contribute to successful reproduction and survival across a wide range of mammalian species, including man (reviewed in (13, 27–29)). Space constraints force me to focus on the activational impact of hormones on male adult sexual behaviors; however, the reader should be made aware of the vast number of developmental factors, particularly hormones, that can sculpt the neural circuitry underlying MSB in adulthood.

In males the major steroid secreted by the testes is testosterone (T). Testosterone can be converted by aromatase enzyme to estradiol or to the nonaromatizable androgen, dihydrotestosterone (DHT), by 5α-reductase enzymes. It is important to note that in general both androgen and estrogen receptor (ER) activation may be required for complete MSB (reviewed in (13)), the experimenter should be aware of species specificity that exists among rats, mice, and hamsters along with specificity among inbred strains of mice.

In the vast majority of mammalian species, castration diminishes expression of sexual behaviors, both motivational and performance aspects. This decline in sexual behavior is reversible with T treatment, which typically restores sex behavior within 5–10 days in rats (30–33). In mice, the restorative effects of T on MSB are strain dependent (Table 2). The elimination of MSB after castration has been documented in all rodent models of MSB studied in the laboratory with the exception of one hybrid strain of mice (B6D2F1 hybrid male mouse (18, 34)) and the Siberian hamster (35). In these two animal models, retention of the ejaculatory reflex long after castration has been reported for as long as 26 weeks after castration in approximately 30% of castrated males. Thus, the persistence of MSB after long-term castration in the B6D2F1 hybrid male mouse allows for the investigation of gonadal steroid-independent MSB in a rodent model in which the complete genome has been mapped out (36). Investigation of this model has important clinical implications for treatment of male sexual dysfunction, as striking interindividual variation in the response to castration has been documented in numerous species, including humans (37).

In rats and mice, exogenously administered estradiol is sufficient to reinstate most copulatory behavior in castrated males. However, there are strain differences (Table 2); for example,

Table 2
Activational effects of steroid hormones on the consummatory components of male sexual behavior

	Copulatory behaviors					
	Mount		**Intromission**		**Ejaculation**	
Testosterone						
Swiss–Webster	*	(10, 30, 38, 92)	*	(10, 30, 38, 92)	×	(10, 38)
DBA/2J	*	(65)	*	(64, 65)	*	(65)
	×	(64)				
CD-1	*	(31, 38)	*	(38)	×	(38)
	×	(31)				
C57BL6J	–		*	(64)	×	(64)
C57BL/6J×ARK/J	–		–		–	
BALB/c AnN	*	(28)	*	(28)	*	(28)
Dihydro-testosterone						
Swiss–Webster	*	(10, 30, 38)	*	(10, 30)	×	(10, 38)
	×	(92)	×	(38, 92)		
DBA/2J	–		–		–	
CD–1	×	(38, 93)	×	(38, 93)	×	(38)
C57BL/6J×ARK/J	–		–		–	
BALB/c AnN	×	(28)	×	(28)	×	(28)
Estrogen						
Swiss–Webster	*	(29, 30)	×	(29, 30)	×	(29)
DBA/2J	–		–		–	
CD-1	*	(10, 31)	*	(10, 31)	–	
C57BL/6J×ARK/J	–		–		–	
BALB/c AnN	×	(28)	×	(28)	×	(28)

Modified from (19)
* restored behavior; × did not restore behavior; – no data. Numbers refer to references

castrated BALB/c AnN male mice failed to restore mounting behavior when provided with exogenous estradiol benzoate (38) whereas mounting behavior is restored in CD-1 and Swiss–Webster mice (22, 39–41). In a number of species, DHT alone is sufficient to maintain or restore male copulatory behavior (rabbits (42); guinea pigs (43, 44); deer mice (45); monkeys (46, 47)). In mice, strain differences exist as DHT alone was sufficient to restore mounting behavior in Swiss–Webster mice (22, 40, 48) whereas in castrated BALB/c AnN male mice, DHT alone was not sufficient to restore or maintain copulation (49–51). These species- and strain-specific differences underscore the fact that model organisms may not yield all the answers.

Androgen receptor (AR) knockout mice (ARKO) show less masculine sexual behavior than their wild-type control counterparts, suggesting a role for ARs in MSB (52). However these mice have female-like external genitalia which certainly influence their behavior. Male mice with a testicular feminization mutation (Tfm), caused by a spontaneous mutation in the AR gene, display impaired MSB similar to ARKO mice (53), but the impairment in mounts and thrusts is reversed after estradiol replacement in adulthood (54), suggesting that sufficient ER stimulation may overcome a lack of AR activation. Tfm males also act like wild-type females in showing no partner preferences for females or their odors (54). Unlike Tfm mice, partner preference is masculinized in Tfm male rats (55–59), again raising awareness of the dangers of assuming generality of MSB across species.

Of the two subtypes of ERs, ERα plays a more substantial role in MSB in mice than ERβ. Male mice lacking the ERα (ERα "knock out" mice, or ERαKO) mount occasionally but perform few mounts with intromissions and rarely display ejaculations (60–63). Castrated ERαKO males treated with either T (64) or DHT (61) do not display ejaculations. Male mice with the aromatase enzyme knockout (ArKO) do not display much sexual behavior in adulthood (52). However, estradiol treatment either during the early neonatal period or prior to testing in adulthood can restore much of their MSB (65, 66). The role of ERβ in MSB is less substantial than ERα, as there were no deficits in adult mating behavior of ERβ knockout mice (67). However, these males have delayed behavioral puberty compared with wild-type littermates, as indicated by attainment of their first ejaculation at an older age (68).

In addition to the influence of gonadal hormones on MSB, the experimenter should also be aware of numerous neurotransmitters and neuropeptides that work along with steroid hormones to regulate MSB. For example, systemic administration with both T and apomorphine, a dopamine agonist, reinstates MSB in ERαKO mice (64); when ERαKO mice are treated with apomorphine alone via intracerebroventricular infusion, other components of MSB (but not ejaculations) are restored (19). Lastly, various transgenic mice have been developed, and investigation of their MSB has given important insight into potential novel genes that may play a role in MSB (see (69, 70) for review).

2. Equipment, Materials, and Setup

2.1. Room

It is preferable for the behavioral testing room to be isolated from sound and unintentional interruptions. It is also recommended that behavioral testing occur during the dark phase of the light-dark cycle under dim red lighting.

2.2. Arena

The arena (Fig. 1) can be constructed using any number of materials, including plastic, glass, metal, or wood, with at least one side made from a clear material since many behavioral tests are viewed or video-taped from one side. The general dimensions of the testing arena should be sufficient for the male and female mouse to move around freely (the dimensions of the arenas that I have used have been $18 \times 30 \times 14$ cm). The arenas are usually covered with a top that prevents the rodents from jumping out of the arena and escaping.

Rectangular, circular, or hemicircular arenas can be used. Circular and hemicircular arenas reduce the option for stimulus female mice to spend time in the corners which may at times make it more difficult for males to mount.

Arenas can also be fitted with mirrors mounted below for ventral viewing to aid in the viewing of MSB components, particularly intromissions, which can be difficult to ascertain when viewing from the side.

2.3. Method to Acquiring Data

This can be done manually or the behavioral tests can be video recorded.

2.4. New Cage for Mice After Test

If the mice are group housed, it is preferable to place the rodent that has just been tested in a new cage rather than back with his cage mates. Reintroduction of the rodent and the odors from the behavioral testing with a stimulus female may modify behavior of rodents not yet tested.

2.5. Material for Clean Up

Generally includes a trash bin and paper towels to clean up urine and to wipe down the arenas.

2.6. Disinfectant

A spray bottle with a disinfectant such as 10–70% ethanol can be used for a general wipe down of the arena. However, note that ethanol should be used with caution since it can cause cracks in many plastics and dissolves glues used to hold plastics and other materials together.

3. Procedure

3.1. Preparing Stimulus Females for MSB Testing

There are various methods to prepare stimulus ovariectomized female mice so that they are hormonally primed to be receptive for planned paired behavioral testing for MSB. Two of the more common methods is to (1) inject female mice that are ovariectomized sc with 0.5 μg estradiol benzoate (dissolved in 0.05 mL of sesame oil) 48 h prior to testing. Then, 3–5 h prior to testing, stimulus females are injected sc with 500 mg progesterone (dissolved in 0.05 mL of sesame oil). Another method is to implant a Silastic capsule (1.96 mm i.d. × 3.18 mm o.d.) filled with estradiol (50 μg dissolved in 30 μL of sesame oil) into an ovariectomized female and then inject the stimulus female with progesterone (500 mg) 3–5 h prior to testing. This method allows for one set of injections rather than two in preparing stimulus females.

Female BALB/c female mice tend to show greater sexual receptivity relative to other strains, and can be considered if the strain of the stimulus female will not be considered an experimental variable.

3.2. Habituation

If you plan on testing for MSB outside of the home cage, habituation to the arena prior to the introduction of the stimulus female is highly recommended, preferably for 30–60 min. In addition, soiled bedding from the home cage can also be added to the test arena to help habituate; however, this is not recommended if the arenas are fitted with mirrors.

If you are planning to test mice in their home cage, it is critical to not change the bedding for at least up to 24 h prior to testing to reduce novelty or stress which will modify MSB. If testing weekly or bi-weekly, it is recommended to not change the cages for several days prior to testing.

3.3. The Following Is a General Procedure in Testing MSB in Mice

1. Transport mice to the testing room at least 1 h prior to testing, and place the test male in the behavioral testing arena to acclimate to the experimental room and the arena. If utilizing soiled bedding from the home cage of the male mouse, place the bedding in the arena prior to placing the male mouse in the arena.

2. Prior to the introduction of the stimulus female, turn on video camera and start recording.

3. Introduce the stimulus female by placing her in the arena and start a timer. If 10 min pass without the male mounting the female, replace the first female with a second stimulus female; if this change fails to elicit copulation, a third female can be introduced. Failure to mount the third female may indicate deficits in MSB and the test may be terminated.

4. The tester should leave the room if the tests are being videotaped or remain still and quiet throughout each trial if recording behavior manually.

5. At the completion of the trial, stop the video recording, remove the stimulus female from the arena and then the male mouse from the arena, and place in a new cage.

6. Empty the arena of soiled bedding. Clean the arena with soap and water and then with disinfectant.

7. Wait until the disinfectant has fully dried prior to placing the next mouse in the arena.

8. Latencies to mount, intromit and ejaculate and PEI, number of mounts, and intromissions are parameters that can be reliably

measured. If the tests were digitally recorded, multiple tests can be conducted simultaneously if the resolution of the recordings is sufficiently high enough so that when panned back, mounts, intromissions, and ejaculations are easily detectable.

9. The primary outcomes of the experiments are generally considered as continuous variables and analyzed with a t-test or analysis of variance (ANOVA) depending on the number of groups or experimental conditions compared.

4. Anticipated Results

An assessment of several of the performance aspects of MSB of various mouse strains is presented in Table 1. In general, mice will generally investigate the stimulus female almost immediately after she is introduced into the arena. The male will then proceed to display mounts, intromissions, and ejaculations, in that order. Because the measurements of MSB can vary due to any number of factors, including both individual and strain differences, it is recommended to allow for sufficient time for all the major components of MSB to be observed. The range of time needed to display all of the major components of MSB, from the time of the introduction of the female to ejaculation, in gonadally intact male mice can take as long as a few minutes to over an hour (Table 1).

5. Experimental Variables

5.1. Genetic Background

Many strains of mice have been shown to vary in their MSB (Table 1; see (19) for review). When working with any mouse, especially mice with genetic manipulations, it will be important to consider the background strain the mouse line is on, when designing and conducting experiments investigating MSB (71).

In a study by McGill (15), three inbred strains (C57BL, BALB/c, and DBA/2J) were compared on 16 different MSBs. Strain differences were noted for the majority (12 of 16) of the measures. The two strains that had the most pronounced behavioral differences were C57BL/6J and DBA/2J males. C57BL/6J males had shorter latencies to mount and mount with thrust and shorter intervals between intromissions. Furthermore, these males displayed more intromissions and thrusts than DBA/2J males. McGill concluded that C57BL/6J males are faster to begin mating than the other two strains. DBA/2J males took longer to begin to mate, but ejaculated after fewer thrusts and intromissions than the other two strains and were the only strain that displayed aggression toward the females after the ejaculation (see Table 1 for an overview). McGill also commented that ejaculation typically terminated the males' interest in sexual activity for 24 h. However, if a novel female was presented to a satiated male, sexual behavior could resume after 2–3 h.

Mosig and Dewsbury (72) assessed AKR/J and C3H/HeJ male copulatory behavior. The behavior observed by the AKR/J males during the first ejaculatory bout was very similar to that reported by McGill and Ransom (73). In addition, Mosig and Dewsbury allowed pairs to continue to mate after the first ejaculation. In these two strains, PEIs were shorter than those of strains studied by McGill. Male C3H/HeJ mice had a 60-min PEI, while in AKR/J males the range was 17–42 min. In the second bout, ejaculation occurred more quickly and after fewer mounts, thrusts, and intromissions than in the initial bout. In addition, the substitution of the original female with a novel second female decreased the time between the two ejaculations. In AKR/J, but not in C3H/HeJ, the males ejaculated with the novel female during the first insertion of the penis.

The experimenter should also be aware that different strains of male mice may have different activational requirements for steroid hormones. For example, DBA/2J and C57BL/6J males seem to be more responsive to smaller doses of T than the other strains tested (64, 74, 75).

5.2. Age

Male mammals typically experience a progressive decline in their sexual activity during middle or advanced age (76), and this decline in MSB is generally correlated with decreased levels of circulating gonadal steroids that occurs with advanced age. Studies have documented declined sexual arousal (21) and sexual performance in aged male mice (77). Interestingly, age-related declines in copulatory behavior in mice were not associated with altered patterns of hormone secretion, and have been suggested that deficits in sexual behavior and episodic LH release in old males result from neural senescence rather than diminished testicular support of reproduction (78).

5.3. Housing Conditions

Another variable to consider is the housing conditions of the mice prior to assessment of MSB. Typically, mice are housed several animals to a cage; this is due mainly to ethological concerns as mice are social animals and show signs of stress when assessed in behavioral tests after being single-housed. The effect of isolation on MSB in rats is unclear. Several studies have found isolation to lead to decreased MSB (79, 80), whereas other studies have found that isolation has little to no effect on MSB (81, 82). Enriched housing seems to have a beneficial effect on reproductive performance in C57BL/6T Tac mice (83), but no effect on Swiss–Weber mice (84). It seems that the effect of social and physical enrichment on MSB in male mice may be strain dependent.

5.4. Behavioral History

Social experience is a critical variable to consider when testing for MSB. Prior sexual experience facilitates sexual behavior in rats (85, 86) and in mice (18). Prior sexual experience also facilitates retention of MSB after castration in rats relative to sexually naïve rats that have been castrated (85, 86). In fact, one study reported that significantly more sexually experienced male rats maintained male copulatory behavior 1 month longer after castration as compared to nonexperienced males (87). Notably, precastration sexual experience was not necessary for expression of MSB after castration in B6D2F1 hybrid mice (34), nor was weekly behavioral testing after castration necessary for the retention of ejaculatory reflexes (88).

Exposure to male and female mice for short periods of time prior to the experiment (social experience paradigm) is also known to expedite the initiation of sexual behavior in sexually naïve male subjects.

5.5. Lighting and Time of Testing

There is a circadian rhythm in the display of MSB in male rats and Syrian hamsters. Male rats display more mounting behavior and show shorter latencies to first ejaculation as the dark period progresses (89–91). Male hamsters have been found to be quicker to initiate copulation and to ejaculate in the dark phase than in the light phase of a daily light-dark cycle (92). Although it remains to be elucidated whether a circadian rhythm of MSB exists in male mice, it is recommended that behavioral testing for MSB occur during the dark phase of the daily light-dark cycle and recording of the behavior occur in the dark with dim red light for illumination since mice are nocturnal animals.

5.6. Arena (Without Soiled Bedding) vs. Home Cage

The location of the behavioral testing can also be an important variable to consider. For example, when tested in their home cage, 75% of castrated B6D2F1 hybrid males retained the ability to ejaculate, but there was a significant decrease in the percentage of castrated males displaying male copulatory behavior when tested in a test arena without soiled bedding, indicating that environmental factors impact this behavioral phenotype in some of the castrated B6D2F1 hybrid mice (88).

6. Trouble Shooting

6.1. High Latency to Engage in MSB

Below are some common problems one may encounter when assessing MSB.

1. Confirm that the mice have not experienced stress, particularly a recent change of bedding.

2. Because mice are nocturnal animals, the time of testing may have an effect on MSB. It is recommended that MSB be

assessed during the dark phase of the light-dark cycle under dim red lights.

3. Ambient environment. Confirm that there are no environmental stressors such as changes in humidity or temperature.

4. If the hormonally primed stimulus female seems unresponsive or is aggressive toward the male and MSB has not begun after 10 min, replace the stimulus female. Repeat this procedure at least 3 times, and if there is still no MSB, then the lack of receptivity from the female mouse is probably not the cause of the lack of MSB.

5. Modify the shape of the arena. If the stimulus females are backed into a corner of the arena, making it difficult for the male to mount, then a circular or hemicircular arena is recommended.

6. Consider providing social experience prior to MSB testing. A social experience paradigm, in which sexually naïve male subjects are exposed to male and female mice for short periods of time prior to the experiment, is known to expedite the initiation of sexual behavior.

7. Lengthen the time of testing. When a short (e.g., 10–20 min) pair test is used, later components of sexual behavior, such as intromission and ejaculation, may be uncommon and therefore not measured. Pair tests that permitted longer male–female interaction (e.g., 30–60 min, or even longer) are more likely to include measures of the entire sexual behavior sequence. For measuring MSB in B6D2F1 hybrid males prior to castration, tests were run for 1 h and for tests after castration, tests were run for 2 h to ensure sufficient time to record most of the components of MSB (36, 93).

8. Modify the time of habituation. Generally, increasing the time of habituation to a new arena will decrease the latency of the male to begin investigating the newly introduced stimulus female.

6.2. Lack of Sexual Behavior

1. There are well-characterized strain differences of MSB in mice (Table 1). Background strain is a crucial variable when evaluating MSB in transgenic mice. In addition, if testing the activational effects of steroid hormones, there is differential sensitivity to different steroids among strains. For example, a single dose of estradiol (1 μg) in CD-1 castrates displayed full restoration of the proportion of males showing the entire copulatory sequence, including ejaculation (41), whereas treatment with estradiol benzoate in castrated BALB/c AnN male mice (38) failed to restore several components of MSB, including intromissions and ejaculations. Although estradiol benzoate failed to restore MSB, administration of testosterone propionate to BALB/c AnN castrated male mice restored full MSB (38).

2. If exogenous steroids are administered, adjust dosage. For example, castrated C57BL/6J mice that received 12.5 µg T daily for 14 days showed a restoration of intromission behavior but not ejaculation (74), whereas a 4-week treatment with implants that yielded a physiological level of T in plasma instated the full copulatory sequence in sexually naïve castrated C57BL/6J mice (62, 64).

3. Adjust the method of administration of exogenous steroids (if applicable). Tests of chronic vs. acute administration of T (T in powered form packed in Silastic capsules vs. T dissolved in either oil or ethanol, respectively) have revealed that constant levels and short spikes of hormone may have different effects on the expression of sexual behavior (41).

4. Confirm that all the mice have the same testing history and are within a few weeks of age of each other. Prior sexual experience greatly influences MSB in mice (34).

References

1. Nathan, S.G., *The epidemiology of the DSM-III psychosexual dysfunctions.* J Sex Marital Ther, 1986. **12**(4): p. 267–81.

2. Montgomery, S.A., D.S. Baldwin, and A. Riley, *Antidepressant medications: a review of the evidence for drug-induced sexual dysfunction.* J Affect Disord, 2002. **69**(1–3): p. 119–40.

3. Casper, R.C., et al., *Somatic symptoms in primary affective disorder. Presence and relationship to the classification of depression.* Arch Gen Psychiatry, 1985. **42**(11): p. 1098–104.

4. Baldwin, D.S. and J. Birtwistle, *The side effect burden associated with drug treatment of panic disorder.* J Clin Psychiatry, 1998. **59 Suppl 8**: p. 39–44; discussion 45–6.

5. Goldstein, B.J. and P.J. Goodnick, *Selective serotonin reuptake inhibitors in the treatment of affective disorders--III. Tolerability, safety and pharmacoeconomics.* J Psychopharmacol, 1998. **12**(3 Suppl B): p. S55–87.

6. Lane, R.M., *A critical review of selective serotonin reuptake inhibitor-related sexual dysfunction; incidence, possible aetiology and implications for management.* J Psychopharmacol, 1997. **11**(1): p. 72–82.

7. Margolese, H.C. and P. Assalian, *Sexual side effects of antidepressants: a review.* J Sex Marital Ther, 1996. **22**(3): p. 209–17.

8. Segraves, R.T., *Antidepressant-induced sexual dysfunction.* J Clin Psychiatry, 1998. **59 Suppl 4**: p. 48–54.

9. Kafka, M.P., *The monoamine hypothesis for the pathophysiology of paraphilic disorders: an update.* Ann N Y Acad Sci, 2003. **989**: p. 86–94; discussion 144–53.

10. Palace, E.M. and B.B. Gorzalka, *The enhancing effects of anxiety on arousal in sexually dysfunctional and functional women.* J Abnorm Psychol, 1990. **99**(4): p. 403–11.

11. Adelson, S., *Psychodynamics of hypersexuality in children and adolescents with bipolar disorder.* J Am Acad Psychoanal Dyn Psychiatry, 2010. **38**(1): p. 27–45.

12. Spalt, L., *Sexual behavior and affective disorders.* Dis Nerv Syst, 1975. **36**(12): p. 644–7.

13. Hull, E.M., R.I. Wood, and K.E. McKenna, *Neurobiology of male sexual behavior*, in *Knobil and Neill's Physiology of Reproduction, Third Edition*, J.D. Neill, et al., Editor. 2006: Academic Press, New York. p. 1729–1824.

14. Bonthuis, P.J., et al., *Of mice and rats: key species variations in the sexual differentiation of brain and behavior.* Front Neuroendocrinol, 2010. **31**(3): p. 341–58.

15. McGill, T.E., *Sexual behavior in three inbred strains of mice.* Behaviour, 1962. **19**: p. 341–350.

16. McGill, T.E., *A double replication of a "small sample" study of the sexual behavior of DBA-2J male mice.* Anat Rec, 1967. **157**(2): p. 151–3.

17. McGill, T.E. and W.C. Blight, *The sexual behaviour of hybrid male mice compared with the sexual behaviour of males of the inbred parent strains.* Anim Behav, 1963. **11**: p. 480–483.

18. McGill, T.E. and A. Manning, *Genotype and retention of the ejaculatory reflex in castrated*

male mice. Animal Behavior, 1976. **24**(3): p. 507–18.

19. Burns-Cusato, M., E.M. Scordalakes, and E.F. Rissman, *Of mice and missing data: what we know (and need to learn) about male sexual behavior.* Physiol Behav, 2004. **83**(2): p. 217–32.

20. Bronson, F.H. and C. Desjardins, *Endocrine responses to sexual arousal in male mice.* Endocrinology, 1982. **111**(4): p. 1286–91.

21. Craigen, W. and F.H. Bronson, *Deterioration of the capacity for sexual arousal in aged male mice.* Biol Reprod, 1982. **26**(5): p. 869–74.

22. Hall, N.R. and W.G. Luttge, *Maintenance of sexual behavior in castrate male SW mice using the anti-androgen, cyproterone acetate.* Pharmacol Biochem Behav, 1975. **3**(4): p. 551–5.

23. Levine, L., G.E. Barsel, and C.A. Diakow, *Mating behaviour of two inbred strains of mice.* Anim Behav, 1966. **14**(1): p. 1–6.

24. Robertson, K.M., et al., *Characterization of the fertility of male aromatase knockout mice.* J Androl, 2001. **22**(5): p. 825–30.

25. Sprott, R.L., *Behavioral characteristics of C57BL/6J. DBA/2J, and B6D2F1 mice which are potentially useful for gerontological research.* Exp Aging Res, 1975. **1**(2): p. 313–23.

26. Dewsbury, D.A., *Patterns of copulatory behavior in male mammals.* Q Rev Biol, 1972. **47**(1): p. 1–33.

27. Davidson, J.M., C.A. Camargo, and E.R. Smith, *Effects of androgen on sexual behavior in hypogonadal men.* The Journal of clinical endocrinology and metabolism, 1979. **48**: p. 955–958

28. Salmimies, P., et al., *Effects of testosterone replacement on sexual behavior in hypogonadal men.* Archives of sexual behavior, 1982. **11**(4): p. 345–53.

29. Young, W., *The hormones and mating behavior,* in Sex and Internal Secretions, W. Young, Editor. 1961, Williams and Wilkins: Baltimore. p. 1173–1239

30. Davidson, J.M., *Characteristics of sex behaviour in male rats following castration.* Anim Behav, 1966. **14**(2): p. 266–72.

31. Beach, F.A. and A.M. Holz-Tucker, *Effects of different concentrations of androgen upon sexual behavior in castrated male rats.* Journal of comparative and physiological psychology, 1949. **42**(6): p. 433–53.

32. McGinnis, M.Y. and R.M. Dreifuss, *Evidence for a role of testosterone-androgen receptor interactions in mediating masculine sexual behavior in male rats.* Endocrinology, 1989. **124**(2): p. 618–26.

33. Putnam, S.K., et al., *Testosterone restoration of copulatory behavior correlates with medial preoptic dopamine release in castrated male rats.* Hormones and Behavior, 2001. **39**(3): p. 216–24.

34. Manning, A. and M.L. Thompson, *Postcastration retention of sexual behaviour in the male BDF1 mouse: the role of experience.* Animal Behavior, 1976. **24**(3): p. 523–33.

35. Park, J.H., et al., *Long-term persistence of male copulatory behavior in castrated and photo-inhibited Siberian hamsters.* Hormones and Behavior, 2004. **45**(3): p. 214–21.

36. Park, J.H., et al., *Amyloid beta precursor protein regulates male sexual behavior.* J Neurosci, 2010. **30**(30): p. 9967–72.

37. Zverina, J., et al., *Hormonal status and sexual behaviour of 16 men after surgical castration.* Urological, nephrological, and andrological sciences, 1990. **62**(1): p. 55–8.

38. Arteaga-Silva, M., et al., *Effects of castration and hormone replacement on male sexual behavior and pattern of expression in the brain of sex-steroid receptors in BALB/c AnN mice.* Comp Biochem Physiol A Mol Integr Physiol, 2007. **147**(3): p. 607–15.

39. Edwards, D.A. and K.G. Burge, *Estrogenic arousal of aggressive behavior and masculine sexual behavior in male and female mice.* Horm. Behav., 1971. **2**: p. 239–245.

40. Nunez, A.A. and D.T. Tan, *Courtship ultrasonic vocalizations in male Swiss-Webster mice: effects of hormones and sexual experience.* Physiol Behav, 1984. **32**(5): p. 717–21.

41. Dalterio, S., A. Bartke, and K. Butler, *A single injection of 17 beta-estradiol facilitates sexual behavior in castrated male mice.* Horm Behav, 1979. **13**(3): p. 314–27.

42. Agmo, A. and P. Sodersten, *Sexual behaviour in castrated rabbits treated with testosterone, oestradiol, dihydrotestosterone or oestradiol in combination with dihydrotestosterone.* Journal of Endocrinology, 1975. **67**(3): p. 327–32.

43. Alsum, P. and R.W. Goy, *Actions of esters of testosterone, dihydrotestosterone, or estradiol on sexual behavior in castrated male guinea pigs.* Hormones and Behavior, 1974. **5**(3): p. 207–17.

44. Butera, P.C. and J.A. Czaja, *Effects of intracranial implants of dihydrotestosterone on the reproductive physiology and behavior of male guinea pigs.* Hormones and Behavior, 1989. **23**(3): p. 424–31.

45. Clemens, L.G. and S.M. Pomerantz, *Testosterone acts as a prohormone to stimulate male copulatory behavior in male deer mice (peromyscus maniculatus bairdi).* Journal of comparative and physiological psychology, 1982. **96**(1): p. 114–22.

46. Michael, R.P., D. Zumpe, and R.W. Bonsall, *Comparison of the effects of testosterone and*

dihydrotestosterone on the behavior of male cynomolgus monkeys (Macaca fascicularis). Physiology & behavior, 1986. **36**(2): p. 349–55.

47. Phoenix, C.H., Effects of dihydrotestosterone on sexual behavior of castrated male rhesus monkeys. Physiology & behavior, 1974. **12**(6): p. 1045–55.

48. Luttge, W.G. and N.R. Hall, Differential effectiveness of testosterone and its metabolites in the induction of male sexual behavior in two strains of albino mice. Horm. Behav., 1973. **4**: p. 31–43.

49. McDonald, P., et al., Failure of 5alpha-dihydrotestosterone to initiate sexual behaviour in the castrated male rat. Nature, 1970. **227**(5261): p. 964–5.

50. Whalen, R.E. and W.G. Luttge, Testosterone, androstenedione and dihydrotestosterone: effects on mating behavior of male rats. Hormones and Behavior, 1971. **2**: p. 117–125.

51. Beyer, C., et al., Androgen structure and male sexual behavior in teh castrated rat. Hormones and Behavior, 1973. 7: p. 99–108.

52. Sato, T., et al., Brain masculinization requires androgen receptor function. Proc Natl Acad Sci USA, 2004. **101**(6): p. 1673–8.

53. Ono, S., L.N. Geller, and E.V. Lai, TfM mutation and masculinization versus feminization of the mouse central nervous system. Cell, 1974. **3**(3): p. 235–42.

54. Bodo, C. and E.F. Rissman, Androgen receptor is essential for sexual differentiation of responses to olfactory cues in mice. The European journal of neuroscience, 2007. **25**(7): p. 2182–90.

55. Hamson, D.K., et al. Mating behavior and CNS morphology in rats carrying the testicular feminization mutation. in Soc. Neurosci. 2005;320.19.

56. Bardin, C.W., et al., Androgen metabolism and mechanism of action in male pseudohermaphroditism: a study of testicular feminization. Recent Prog Horm Res, 1973. **29**: p. 65–109.

57. Naess, O., et al., Androgen receptors in the anterior pituitary and central nervous system of the androgen "insensitive" (Tfm) rat: correlation between receptor binding and effects of androgens on gonadotropin secretion. Endocrinology, 1976. **99**(5): p. 1295–303.

58. Sherins, R.J. and C.W. Bardin, Preputial gland growth and protein synthesis in the androgen-insensitive male pseudohermaphroditic rat. Endocrinology, 1971. **89**(3): p. 835–41.

59. Charest, N.J., et al., A frameshift mutation destabilizes androgen receptor messenger RNA in the Tfm mouse. Mol Endocrinol, 1991. **5**(4): p. 573–81.

60. Ogawa, S., et al., Behavioral effects of estrogen receptor gene disruption in male mice. Proc Natl Acad Sci USA 1997. **94**(4): p. 1476–81.

61. Ogawa, S., et al., Modifications of testosterone-dependent behaviors by estrogen receptor-alpha gene disruption in male mice. Endocrinology, 1998. **139**(12): p. 5058–69.

62. Wersinger, S.R., et al., Masculine sexual behavior is disrupted in male and female mice lacking a functional estrogen receptor alpha gene. Horm Behav, 1997. **32**(3): p. 176–83.

63. Dominguez-Salazar, E., H.L. Bateman, and E.F. Rissman, Background matters: the effects of estrogen receptor alpha gene disruption on male sexual behavior are modified by background strain. Horm Behav, 2004. **46**(4): p. 482–90.

64. Wersinger, S.R. and E.F. Rissman, Dopamine activates masculine sexual behavior independent of the estrogen receptor alpha. The Journal of Neuroscience, 2000. **20**(11): p. 4248–54.

65. Bakker, J., et al., Restoration of male sexual behavior by adult exogenous estrogens in male aromatase knockout mice. Horm Behav, 2004. **46**(1): p. 1–10.

66. Toda, K., et al., Oestrogen at the neonatal stage is critical for the reproductive ability of male mice as revealed by supplementation with 17beta-oestradiol to aromatase gene (Cyp19) knockout mice. J Endocrinol, 2001. **168**(3): p. 455–63.

67. Ogawa, S., et al., Survival of reproductive behaviors in estrogen receptor beta gene-deficient (betaERKO) male and female mice. Proc Natl Acad Sci USA 1999. **96**(22): p. 12887–92.

68. Temple, J.L., et al., Lack of functional estrogen receptor beta gene disrupts pubertal male sexual behavior. Hormones and Behavior, 2003. **44**(5): p. 427–34.

69. Park, J.H. and E.R. Rissman, Behavioral Neuroendocrinology of Reproduction in Mammals. Chapter in Hormones and Reproduction in Vertebrates, ed. D.O. Norris and K.H. Lopez, 2010. **5**: p. 139–174.

70. Nelson, R.J., The use of genetic "knockout" mice in behavioral endocrinology research. Horm Behav, 1997. **31**(3): p. 188–96.

71. Crawley, J.N., et al., Behavioral phenotypes of inbred mouse strains: implications and recommendations for molecular studies. Psychopharmacology (Berl), 1997. **132**(2): p. 107–24.

72. Mosig, D.W. and D.A. Dewsbury, Studies of the copulatory behavior of house mice (Mus musculus). Behav. Biol., 1976. **16**: p. 463–473.

73. McGill, T.E. and T.W. Ransom, Genotypic change affecting conclusions regarding the mode of inheritance of elements of behaviour. Anim. Behav., 1968. **16** p. 88–91.

74. Ogawa, S., et al., *Effects of testosterone and 7 alpha-methyl-19-nortestosterone (MENT) on sexual and aggressive behaviors in two inbred strains of male mice.* Horm Behav, 1996. **30**(1): p. 74–84.

75. Shrenker, P., S.C. Maxson, and B.E. Ginsburg, *The role of postnatal testosterone in the development of sexually dimorphic behaviors in DBA/1Bg mice.* Physiol Behav, 1985. **35**(5): p. 757–62.

76. Hafez, E.S.E., *Aging and Reproductive Physiology.* 1976, Ann Arbor, MI: Ann Arbor Science Press.

77. Bronson, F.H. and C. Desjardins, *Reproductive failure in aged CBF1 male mice: interrelationships between pituitary gonadotropic hormones, testicular function, and mating success.* Endocrinology, 1977. **101**(3): p. 939–45.

78. Coquelin, A. and C. Desjardins, *Luteinizing hormone and testosterone secretion in young and old male mice.* Am J Physiol, 1982. **243**(3): p. E257–63.

79. Duffy, J.A. and S.E. Hendricks, *Influences of social isolation during development on sexual behavior of the rat.* Anim Learn Behav, 1973. **1**(3): p. 223–7.

80. Chambers, K.C., et al., *Disruption of sexual behavior in socially isolated adult male rats.* Behav. Neural Biol., 1982. **34**: p. 215–220.

81. Beach, F.A., *Normal sexual behavior in male rats isolated at fourteen days of age.* J Comp Physiol Psychol, 1958. **51**(1): p. 37–8.

82. Swanson, H.H. and N.E. van de Poll, *Effects of an isolated or enriched environment after handling on sexual maturation and behaviour in male and female rats.* J Reprod Fertil, 1983. **69**(1): p. 165–71.

83. Whitaker, J., et al., *Effects of cage size and enrichment on reproductive performance and behavior in C57BL/6Tac mice.* Lab Anim (NY), 2009. **38**(1): p. 24–34.

84. Carvalho, A.F., et al., *Development and reproductive performance of Swiss mice in an enriched environment.* Braz J Biol, 2009. **69**(1): p. 153–60.

85. Bloch, G.J., et al., *Cholecystokinin stimulates and inhibits lordosis behavior in female rats.* Physiology and Behavior, 1987. **39**(2): p. 217–24.

86. Rabedeau, R.G. and R.E. Whalen, *Effects of copulatory experience on mating behavior in the male rat.* Journal of comparative and physiological psychology, 1959. **52**: p. 482–4.

87. Pfaus, J.G., T.E. Kippin, and S. Centeno, *Conditioning and sexual behavior: a review.* Hormones and Behavior, 2001. **40**(2): p. 291–321.

88. Wee, B.E. and L.G. Clemens, *Environmental influences on masculine sexual behavior in mice.* Physiology & behavior, 1989. **46**(5): p. 867–72.

89. Harlan, R.E., et al., *Sexual performance as a function of time of day in male and female rats.* Biol Reprod, 1980. **23**(1): p. 64–71.

90. Sodersten, P. and P. Eneroth, *Neonatal treatment with antioestrogen increases the diurnal rhythmicity in the sexual behaviour of adult male rats.* J Endocrinol, 1980. **85**(2): p. 331–9.

91. Stefanick, M.L., *The circadian patterns of spontaneous seminal emission, sexual activity and penile reflexes in the rat.* Physiol Behav, 1983. **31**(6): p. 737–43.

92. Eskes, G.A., *Neural control of the daily rhythm of sexual behavior in the male golden hamster.* Brain Res, 1984. **293**(1): p. 127–41.

93. Park, J.H., et al., *Androgen- and estrogen-independent regulation of copulatory behavior following castration in male B6D2F1 mice.* Horm Behav, 2009. **56**(2): p. 254–63.

INDEX

Todd D. Gould (ed.), *Mood and Anxiety Related Phenotypes in Mice: Characterization Using Behavioral Tests, Volume II*, Neuromethods, vol. 63, DOI 10.1007/978-1-61779-313-4, © Springer Science+Business Media, LLC 2011